PALEOMAGNETISM

This is Volume 73 in the
INTERNATIONAL GEOPHYSICS SERIES
A series of monographs and textbooks
Edited by RENATA DMOWSKA, JAMES R. HOLTON, and H. THOMAS ROSSBY

A complete list of books in this series appears at the end of this volume.

PALEOMAGNETISM

Continents and Oceans

MICHAEL W. McELHINNY

Gondwana Consultants
Hat Head, New South Wales, 2440 Australia

PHILLIP L. McFADDEN

Australian Geological Survey Organisation
Canberra, 2601 Australia

ACADEMIC PRESS

A Harcourt Science and Technology Company

San Diego   San Francisco   New York   Boston   London   Sydney   Tokyo

Academic Press
A Harcourt Science and Technology Company
525 B Street, Suite 1900, San Diego, California 92101-4495, U.S.A.
http://www.apnet.com

Academic Press
24-28 Oval Road, London NW1 7DX, UK
http://www.hbuk.co.uk/ap/

Harcourt/Academic Press
A Harcourt Science and Technology Company
200 Wheeler Road, Burlington, Massachusetts 01803
http://www.harcourt-ap.com

Library of Congress Catalog Card Number: 99-65104

International Standard Book Number: 0-12-483355-1

International Standard Serial Number: 0074-6142

Printed and bound in the United Kingdom
Transferred to Digital Printing, 2011

Contents

Chapter 3 **Methods and Techniques**

Chapter 6 **Continental Paleomagnetism**

Chapter 7 **Paleomagnetism and Plate Tectonics**

Preface

This book is the sequel to *Palaeomagnetism and Plate Tectonics* written by Michael W. McElhinny, first published in 1973. The aim of that book was to explain the intricacies of paleomagnetism and of plate tectonics and then to demonstrate that paleomagnetism confirmed the validity of the new paradigm. Today it is no longer necessary to explain plate tectonics, but paleomagnetism has progressed rapidly over the past 25 years. Furthermore, magnetic anomaly data over most of the oceans have been analyzed in the context of sea-floor spreading and reversals of the Earth's magnetic field. Oceanic data can also be used to determine paleomagnetic poles by combining disparate types of data, from deep-sea cores, seamounts, and magnetic anomalies. Our aim here is to explain paleomagnetism and its contribution in both the continental and the oceanic environment, following the general outline of the initial book. We demonstrate the use of paleomagnetism in determining the evolution of the Earth's crust.

Our intention has been to write a text that can be understood by Earth-science undergraduates at about second-year level. To make the text as accessible as possible, we have kept the mathematics to a minimum. The book can be considered a companion volume to *The Magnetic Field of the Earth* by Ronald T. Merrill, Michael W. McElhinny, and Phillip L. McFadden, which was published in the same series in 1996. There is inevitably some overlap between the books, occurring mostly in Chapter 4. However, the emphasis is different, with this text concentrating more on the geological aspects.

Chapter 1 introduces geomagnetism and explains the basis of paleomagnetism in that context. It follows the original book quite closely. Chapter 2 is about rock magnetism and the magnetic minerals that are important in paleomagnetism. The theory of rock magnetism is an essential part of understanding how and why paleomagnetism works. Chapter 3 deals with field and laboratory methods and techniques. The chapter concludes with a summary of some methods for identifying magnetic minerals. Chapter 4 describes the evidence for magnetic field reversals and their paleomagnetic applications. The development of the geomagnetic polarity time scale and its application to magnetostratigraphy are highlighted, together with the analysis of reversal sequences.

Oceanic paleomagnetism, including the modeling and interpretation of marine magnetic anomalies, is discussed in Chapter 5. Methods for determining pole positions using oceanic paleomagnetic data are also covered. Global maps in color show the age of the ocean floor and of the evolution of the Pacific Ocean. Chapter 6 summarizes the results from continental paleomagnetism and includes methods of data selection and combination to produce apparent polar wander paths. Reference apparent polar wander paths are then compiled and presented for each of the Earth's major crustal blocks.

Chapter 7 puts it all together and relates the results to global tectonics. Here we emphasize only the major features of global tectonic history that can be deduced from paleomagnetism. Van der Voo (1993) gives an excellent detailed account of the application of paleomagnetism to tectonics, and it is not our intention, in a single chapter, to provide readers with that level of detail and analysis. Color paleogeographic maps illustrate continental evolution since the Late Permian. A new and exciting development in global tectonics is the hypothesis of a Neoproterozoic supercontinent named Rodinia. Paleomagnetism is playing and will continue to play an important role in determining its configuration and evolution. With this in mind we discuss Earth history from 1000 Ma to the present through a combination of geology with paleomagnetism.

In writing the book we have had discussions with many colleagues. We thank Jean Besse, Dave Engebretson, Dennis Kent, Zheng-Xiang Li, Roger Larson, Dietmar Müller, Andrew Newell, Neil Opdyke, Chris Powell, Phil Schmidt, Chris Scotese, Jean-Pierre Valet, and Rob Van der Voo for their assistance in providing us with materials. Our special thanks go to Charlie Barton, Steve Cande, Jo Lock, Helen McFadden, Ron Merrill, and Sergei Pisarevsky, who read parts of the book and made valuable comments. Mike McElhinny thanks Vincent Courtillot and the Institute de Physique du Globe de Paris for providing financial assistance for a visit to that institute in 1997, during which time he commenced writing the book. Phil McFadden thanks Helen Hunt and Christine Hitchman for their assistance in preparing the manuscript, and Neil Williams and Trevor Powell for their continued support.

Hat Head and Canberra   **Michael W. McElhinny**
April 1999   **Phillip L. McFadden**

Chapter One

Geomagnetism and Paleomagnetism

1.1 Geomagnetism

1.1.1 Historical

The properties of lodestone (now known to be magnetite) were known to the Chinese in ancient times. The earliest known form of magnetic compass was invented by the Chinese probably as early as the 2nd century B.C., and consisted of a lodestone spoon rotating on a smooth board (Needham, 1962; see also Merrill et al., 1996). It was not until the 12th century A.D. that the compass arrived in Europe, where the first reference to it is made in 1190 by an English monk, Alexander Neckham. During the 13th century, it was noted that the compass needle pointed toward the pole star. Unlike other stars, the pole star appeared to be fixed in the sky, so it was concluded that the lodestone with which the needle was rubbed must obtain its "virtue" from this star. In the same century it was suggested that, in some way, the magnetic needle was affected by masses of lodestone on the Earth itself. This produced the idea of polar lodestone mountains, which had the merit at least of bringing magnetic directivity down to the Earth from the heavens for the first time (Smith, 1968).

Roger Bacon in 1216 first questioned the universality of the north-south directivity of the compass needle. A few years later Petrus Peregrinus questioned the idea of polar lodestone deposits, pointing out that lodestone deposits exist in many parts of the world, so why should the polar ones have preference? Petrus Peregrinus reported, in his *Epistola de Magnete* in 1269, a remarkable series of experiments with spherical pieces of lodestone (Smith, 1970a). He defined the

1

concept of polarity for the first time in Europe, discovered magnetic meridians, and showed several ways of determining the positions of the poles of a lodestone sphere, each method illustrating an important magnetic property. He thus discovered the dipolar nature of the magnet, that the magnetic force is both strongest and vertical at the poles, and became the first person to formulate the law that like poles repel and unlike poles attract. The *Epistola* bears a remarkable resemblance to a modern scientific paper. Peregrinus used his experimental data as a source for new conclusions, unlike his contemporaries who sought to reconcile observations with pre-existing speculation. Although written in 1269 and widely circulated during the succeeding centuries, the *Epistola* was not published in printed form under Peregrinus' name until 1558.

Magnetic declination was known to the Chinese from about 720 A.D. (Needham, 1962; Smith and Needham, 1967), but knowledge of this did not travel to Europe with the compass. It was not rediscovered until the latter part of the 15th century. By the end of that century, following the voyages of Columbus, the great age of exploration by sea had begun and the compass was well established as an aid to navigation. Magnetic inclination (or dip) was discovered by Georg Hartmann in 1544, but this discovery was not publicized. In 1576 it was independently discovered by Robert Norman. Mercator, in a letter in 1546, first realized from observations of magnetic declination that the point which the needle seeks could not lie in the heavens, leading him to fix the magnetic pole firmly on the Earth. Norman and Borough subsequently consolidated the view that magnetic directivity was associated with the Earth and began to realize that the cause was not the polar region but lay closer to the center of the Earth.

In 1600, William Gilbert published the results of his experimental studies in magnetism in what is usually regarded as the first scientific treatise ever written, entitled *De Magnete*. However, credit for writing the first scientific treatise should probably be given to Petrus Peregrinus for his *Epistola de Magnete*; Gilbert, whose work strongly influenced the course of magnetic study, must certainly have leaned heavily on this previous work (Smith, 1970a). He investigated the variation in inclination over the surface of a piece of lodestone cut into the shape of a sphere and summed up his conclusions in his statement *"magnus magnes ipse est globus terrestris"* (the Earth globe itself is a great magnet). Gilbert's work, confirming that the geomagnetic field is primarily dipolar, thus represented the culmination of many centuries of thought and experimentation on the subject. His conclusions put a stop to the wild speculations that were then common concerning magnetism and the magnetic needle. Apart from the roundness of the Earth, magnetism was the first property to be attributed to the body of the Earth as a whole. Newton's theory of gravitation came 87 years later with the publication of his *Principia*.

1.1.2 Main Features of the Geomagnetic Field

If a magnetic compass needle is weighted so as to swing horizontally, it takes up a definite direction at each place and its deviation from geographical or true north is called the *declination* (or *magnetic variation*), D. In geomagnetic studies D is reckoned positive or negative according as the deviation is east or west of true north. In paleomagnetic studies D is always measured clockwise (eastwards) from the present geographic north and consequently takes on any angle between 0° and 360°. The direction to which the needle points is called *magnetic north* and the vertical plane through this direction is called the *magnetic meridian*. A needle perfectly balanced about a horizontal axis (before being magnetized), so placed that it can swing freely in the plane of the magnetic meridian, is called a dip needle. After magnetization it takes up a position inclined to the horizontal by an angle called the *inclination* (or *dip*), I. The inclination is reckoned positive when the north-seeking end of the needle points downwards (as in the northern hemisphere) or negative when it points upwards (as in the southern hemisphere).

The main elements of the geomagnetic field are illustrated in Fig. 1.1. The total intensity F, declination D, and inclination I, completely define the field at any point. The horizontal and vertical components of F are denoted by H and Z. Z is reckoned positive downwards as for I. The horizontal component can be

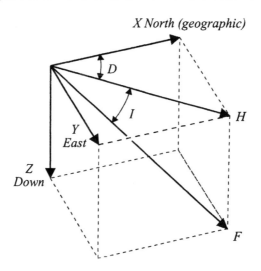

Fig. 1.1. The main elements of the geomagnetic field. The deviation, D, of a compass needle from true north is referred to as the *declination* (reckoned positive eastwards). The compass needle lies in the magnetic meridian containing the total field F, which is at an angle I, termed the *inclination* (or dip), to the horizontal. The inclination is reckoned positive downwards (as in the northern hemisphere) and negative upwards (as in the southern hemisphere). The horizontal (H) and vertical (Z) components of F are related as given by (1.1.1) to (1.1.3). From Merrill *et al.* (1996).

resolved into two components, X (northwards) and Y (eastwards). The various components are related by the equations:

$$H = F \cos I, \quad Z = F \sin I, \quad \tan I = Z / H ; \qquad (1.1.1)$$

$$X = H \cos D, \quad Y = H \sin D, \quad \tan D = Y / X ; \qquad (1.1.2)$$

$$F^2 = H^2 + Z^2 = X^2 + Y^2 + Z^2. \qquad (1.1.3)$$

Variations in the geomagnetic field over the Earth's surface are illustrated by isomagnetic maps. An example is shown in Fig. 1.2, which gives the variation of inclination over the surface of the Earth for the year 1995. A complete set of isomagnetic maps for this epoch is given in Merrill *et al.* (1996). The path along which the inclination is zero is called the *magnetic equator*, and the *magnetic poles* (or *dip poles*) are the principal points where the inclination is vertical, i.e. ±90°. The *north magnetic pole* is situated where $I = +90°$, and the *south magnetic pole* where $I = -90°$. The strength, or intensity, of the Earth's magnetic field is commonly expressed in Tesla (T) in the SI system of units (see §2.1.1 for discussion of magnetic fields). The maximum value of the Earth's magnetic field at the surface is currently about 70 μT in the region of the south magnetic pole. Small variations are measured in nanotesla (1 nT = 10^{-9} T).

Gilbert's observation that the Earth is a great magnet, producing a magnetic field similar to a uniformly magnetized sphere, was first put to mathematical analysis by Gauss (1839) (see §1.1.3). The best-fit geocentric dipole to the Earth's magnetic field is inclined at 10½° to the Earth's axis of rotation. If the axis of this geocentric dipole is extended, it intersects the Earth's surface at two points that in 1995 were situated at 79.3°N, 71.4°W (in northwest Greenland)

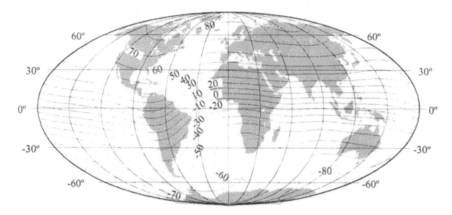

Fig. 1.2. Isoclinic (lines of constant inclination) chart for 1995 showing the variation of inclination in degrees over the Earth's surface.

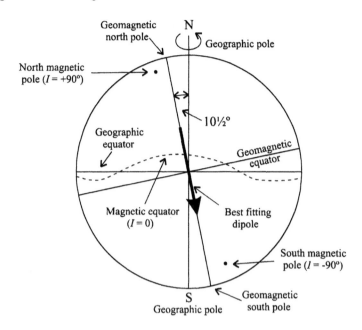

Fig. 1.3. Illustrating the distinction between the magnetic, geomagnetic, and geographic poles and equators. From McElhinny (1973a).

and 79.3°S, 108.6°E (in Antarctica). These points are called the *geomagnetic poles* (boreal and austral, or north and south respectively) and must be carefully distinguished from the magnetic poles (see preceding paragraph). The great circle on the Earth's surface coaxial with the dipole axis and midway between the geomagnetic poles is called the *geomagnetic equator* and is different from the magnetic equator (which is not in any case a circle). Figure 1.3 distinguishes between the magnetic elements (which are those actually observed at each point) and the geomagnetic elements (which are those related to the best fitting geocentric dipole).

In 1634, Gellibrand discovered that the magnetic inclination at any place changed with time. He noted that whereas Borough in 1580 had measured a value of 11.3°E for the declination at London, his own measurements in 1634 gave only 4.1°E. The difference was far greater than possible experimental error. The gradual change in magnetic field with time is called the *secular variation* and is observed in all the magnetic elements. The secular variation of the direction of the geomagnetic field at London and Hobart since about 1580 is shown in Fig. 1.4. At London the changes in declination have been quite large, from 11½°E in 1576 to 24°W in 1823, before turning eastward again. For a similar time interval the declination changes in Hobart have been less extreme.

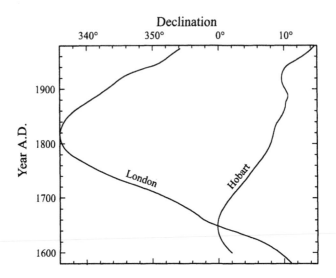

Fig. 1.4. Variation in declination at London, England (51.5°N) and at Hobart, Tasmania (42.9°S) from observatory measurements. The earliest measurement in the Tasmanian region was made by Abel Tasman at sea in 1642 in the vicinity of the present location of Hobart. Pre-observatory data have been derived also by interpolation from isogonic charts. From Merrill and McElhinny (1983).

The distribution of the secular variation over the Earth's surface can be represented by maps on which lines called *isopors* are drawn, joining points that show the same annual change in a magnetic element. These isoporic maps show that there are several regions on the Earth's surface in which the isoporic lines form closed loops centered around foci where the secular changes are the most rapid. For example, there are several foci on the Earth's surface where the total intensity of the geomagnetic field is currently changing rapidly, with changes of up to about 120 nT yr^{-1} (from -117 nT yr^{-1} at 48.0°S, 1.8°E to +56 nT yr^{-1} at 22.5°S, 70.8°E). Isoporic foci are not permanent but move on the Earth's surface and grow and decay, with lifetimes on the order of 100 years. The movements are not altogether random but have shown a westward drifting component in historic times. Because declination is the most important element for navigation, records of it have been kept by navigators since the early part of the 16th century. These records show that the point of zero declination on the equator, now situated in northeast Brazil, was in Africa four centuries ago.

Spherical harmonic analysis of the geomagnetic field (§1.1.3), first undertaken by Gauss in 1839, has been repeated several times since for succeeding and earlier epochs. When the field of the best fitting geocentric dipole (the main dipole) is subtracted from that observed over the surface of the Earth, the residual is termed the *nondipole field*, the vertical component of which is illustrated in Fig 1.5 for epoch 1995. The magnetic moment of the main dipole

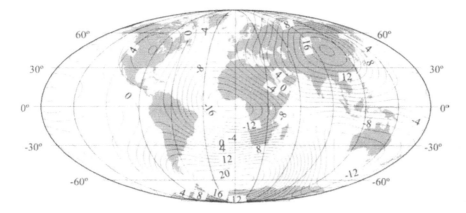

Fig. 1.5. The vertical component of the nondipole field for 1995. Contours are labeled in units of 1000 nT.

has decreased at the rate of about 6.5% per century since the time of Gauss' first analysis (§1.1.4). However, the largest (percentage) changes in the geomagnetic field are associated with the nondipole part of the field. Bullard *et al.* (1950) analyzed geomagnetic data between 1907 and 1945 and determined the average velocity of the nondipole field to be 0.18° per year westward, the so-called *westward drift of the nondipole field.* Bloxham and Gubbins (1985, 1986) used the records of ancient mariners to extend the spherical harmonic analyses back to 1715. The general view is that the westward drift is really only a recent phenomenon and has been decreasing up to the present time. The dominant feature of secular variation is, in fact, growth and decay.

1.1.3 Origin of the Main Field

In the absence of an appropriate analysis, it was not known whether the magnetic field observed at the surface of the Earth was produced by sources inside the Earth, by sources outside the Earth or by electric currents crossing the surface. Gauss (1839) was the first to express the problem in mathematical form, and to determine the general location of the source. In the absence of currents crossing the surface of the Earth, the field there can be derived from a potential function V that satisfies Laplace's equation (i.e., $\nabla^2 V = 0$) and can be expanded as a series of surface spherical harmonics. If the field is of internal origin (which means that the field should decrease as a function of increasing distance r from the center of the Earth) and the Earth is assumed to be a sphere of radius a, then the potential V (in units of ampère) at colatitude (i.e., 90° minus the latitude) θ and longitude ϕ can be represented as a series of spherical harmonics in the form:

$$V = \frac{a}{\mu_0} \sum_{l=1}^{\infty} \sum_{m=0}^{l} \left(\frac{a}{r}\right)^{l+1} P_l^m(\cos\theta)(g_l^m \cos m\phi + h_l^m \sin m\phi), \quad (1.1.4)$$

where P_l^m is the Schmidt quasi-normalized form of the associated Legendre function $P_{l,m}$ of degree l and order m. See Merrill *et al.* (1996) for more detail. The coefficients g_l^m and h_l^m are called the Gauss coefficients (Chapman and Bartels, 1940, 1962). In order to have the same numerical value for these coefficients as they had in the cgs emu system of units, they are now generally quoted in nanotesla (units of magnetic induction, see Table 1.1). Therefore, in (1.1.4) the factor μ_0 is included to correct the dimensions on the right-hand side for g, h in nT. It is apparent from (1.1.4) that the surface harmonic for a given r is simply a Fourier function around a line of latitude (colatitude) multiplied by an associated Legendre function along a line of longitude. In his analysis, Gauss (1839) included terms for sources outside the Earth, whose variation with distance from the center would take the form $(r/a)^l$ instead of $(a/r)^{l+1}$ as in (1.1.4). He showed there were no electric currents crossing the Earth's surface and, importantly, that any coefficients relating to a field of external origin were all zero. He concluded, therefore, that the magnetic field was solely of internal origin. In practice the external field is not totally absent but a small contribution from electric currents in the ionosphere is present, amounting to about 30 nT.

The International Association of Geomagnetism and Aeronomy (IAGA) publishes estimates of the values of the coefficients g_l^m and h_l^m at five-yearly intervals that are referred to as the International Geomagnetic Reference Field (IGRF), together with estimates of the secular variation to be expected in these

TABLE 1.1

IGRF 1995 Epoch Model Coefficients up to Degree 4

l	m	Main field (nT)		Secular change (nT yr^{-1})	
		g	h	$\dot{g}$	$\dot{h}$
1	0	-29,682.0	0.0	17.60	0.00
1	1	-1,789.0	5,318.0	13.00	-18.30
2	0	-2,197.0	0.0	-13.20	0.00
2	1	3,074.0	-2,356.0	3.70	-15.00
2	2	1,685.0	-425.0	-0.80	-8.80
3	0	1,329.0	0.0	1.50	0.00
3	1	-2,268.0	-263.0	-6.40	4.10
3	2	1,249.0	302.0	-0.20	2.20
3	3	769.0	-406.0	-8.10	-12.10
4	0	941.0	0.0	0.80	0.00
4	1	782.0	262.0	0.90	1.80
4	2	291.0	-232.0	-6.90	1.20
4	3	-421.0	98.0	0.50	2.70
4	4	116.0	-301.0	-4.60	-1.00

coefficients over the next 5 years. The 1995 epoch IGRF has Gauss coefficients truncated at degree 10 (corresponding to 120 coefficients) and degree 8 for the secular variation; this is regarded as a practical compromise to produce a well-determined main field model. The IGRF 1995 epoch model coefficients up to degree 4 are listed in Table 1.1; for degrees greater than 4 the magnitude of the coefficients falls off quite rapidly with increasing degree. Harmonics of order zero are referred to as *zonal harmonics*, with coefficients g_1^0, g_2^0, g_3^0, etc which are the coefficients for the *geocentric axial dipole, geocentric axial quadrupole, geocentric axial octupole*, and so on, respectively. All the other terms are the *nonzonal harmonics*. For convenience, the coefficients are typically referred to as if they were the harmonic; thus, g_l^m is used to refer to the harmonic of degree l and order m. The main field is dominated by the geocentric axial dipole term (g_1^0), then the *equatorial dipole* (g_1^1 and h_1^1). The latter causes the main dipole to be inclined to the axis of rotation by about 10½°. As a simplistic separation, the Gauss coefficients less than degree 14 are generally attributed to sources in the Earth's liquid core and those greater than degree 14 to sources in the Earth's crust. See Merrill *et al.* (1996) for more details on the Gauss coefficients and their analysis.

The dynamo theory of the Earth's magnetic field originates from a suggestion of Larmor (1919) that the magnetic field of the Sun might be maintained by a mechanism analogous to that of a self-exciting dynamo. Elsasser (1946) and Bullard (1949) followed up this suggestion proposing that the electrically conducting iron core of the Earth acts like a self-exciting dynamo and produces electric currents necessary to maintain the geomagnetic field. The action of such a dynamo is simplistically illustrated by the disc dynamo in Fig. 1.6. If a conducting disc is rotated in a small axial magnetic field, a radial electromotive

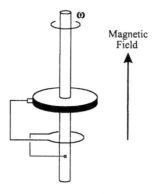

Fig. 1.6. The disc dynamo. A torque is applied to rotate a conducting disc at angular speed ω in a magnetic field aligned along the axis of the disc. An electric current, induced in the rotating disc, flows outward to the edge of the disc where it is tapped by a brush attached to a wire. The wire is wound back around the axis of the disc in such a way as to reinforce the initial field.

force is generated between the axis and the edge of the disc. A coil in the external circuit is placed coaxial with the disc so as to produce positive feedback so that the magnetic field it produces reinforces the initial axial field. This causes a larger current to flow because of the increased emf and the axial field is increased further, being limited ultimately by Lenz's law, the electrical resistance of the circuit, and the available mechanical power. The main point is that starting from a very small field, perhaps a stray one, it is possible to generate a much larger field.

In the simple disc dynamo of Fig. 1.6, the geometry (and therefore the current path) is highly constrained and all the parts are solid. That makes solution of the relevant equations, and understanding of the process, relatively simple. In the

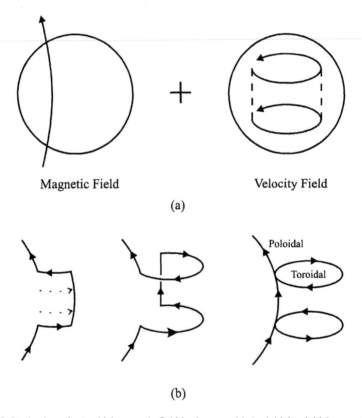

Magnetic Field   Velocity Field

(a)

(b)

Fig. 1.7. Production of a toroidal magnetic field in the core. (a) An initial poloidal magnetic field passing through the Earth's core is shown on the left, and an initial cylindrical shear motion of the fluid (i.e., with no radial component) is shown on the right. (b) The interaction between the fluid motion and the magnetic field in (a) is shown at three successive times moving from left to right. The fluid motion is only shown on the left by dotted lines. After one complete circuit two new toroidal magnetic field loops of opposite sign have been produced. After Parker (1955).

Earth there is a homogeneous, highly electrically conductive, rapidly rotating, convecting fluid that forms the dynamo. This highly unconstrained situation, together with the need to include equations such as the equation of state of the fluid and the Navier–Stokes equation, means that the geodynamo problem is exceptionally difficult to solve. Despite this, major advances have been made in recent years. Although the details are necessarily complex, several of the major concepts are reasonably accessible.

If a magnetic field exists in a perfectly conducting medium, then when the medium moves, it carries the magnetic field lines along with it according to the *frozen-in-field theorem* of Alfvén (1942, 1950). Although the core fluid is not a perfect conductor, there is still a strong tendency (certainly over short time scales) for the fluid to drag magnetic field lines along with it. This is central to dynamo theory because differential motions of the fluid stretch the magnetic field lines and thereby add energy to the magnetic field. Because the fluid is not a perfect conductor the magnetic field will diffuse away with time, and so it is necessary for there to be dynamo action to add energy back into the magnetic field to overcome this diffusion. Another central concept is that of *poloidal* and *toroidal* fields. Toroidal fields have no radial component and so it is not possible to observe at the Earth's surface a toroidal field in the Earth's core. Conversely, a poloidal field does have a radial component and the geomagnetic field at the Earth's surface is poloidal. The magnetic field can be written as the sum of a poloidal field and a toroidal field, and many of the concepts of dynamo theory revolve around the question of how to generate a toroidal field from a poloidal field and, conversely, how to generate a poloidal field from a toroidal field.

Figure 1.7 illustrates how a toroidal magnetic field can be generated from an initial poloidal magnetic field using a process referred to as the ω–*effect*. If the core fluid motion has a toroidal component (relative to the overall rotation of the Earth), then the highly conducting fluid drags the magnetic field lines along with it in its toroidal motion as shown in Fig. 1.7b. This stretches the magnetic field lines, thereby adding energy to the magnetic field, and draws the poloidal field lines out into toroidal loops. However, the ω–effect cannot generate a poloidal field from an initial toroidal field. Another process, known for historical reasons as an α–*effect*, is required for this.

The simplest picture of how the α–effect can occur is provided by convection in the core together with Alfvén's frozen flux theorem and *helicity*, as is illustrated in Fig. 1.8. The toroidal field will be affected by an upwelling of fluid. As the field line moves with the fluid the upwelling will produce a bulge, which stretches the field line. The field line is in tension so, just like an elastic band, energy is required to stretch the field line. By this process energy is added to the magnetic field. The Coriolis force will act to produce a rotation (known as helicity) in the fluid as it rises, counterclockwise in the northern hemisphere. The field line will be twisted with this rotation and a poloidal magnetic loop will be

Fig. 1.8. Production of poloidal magnetic field in the northern hemisphere. A region of fluid upwelling, illustrated by dotted lines on the left interacts with toroidal magnetic field (solid line). Because of the Coriolis effect the fluid exhibits helicity, rotating as it moves upward (thin lines center). The magnetic field line is carried with the conducting liquid and is twisted to produce a poloidal loop as on the right. After Parker (1955).

produced after 90° of rotation. Because the field gradients are large at the base of the loop, it can detach from the original field line to produce a closed flux loop. The process is inherently statistical, but eventually poloidal loops of this sort merge to produce a large poloidal loop. The above turbulent process provides a simple visualization of the generation of poloidal field from toroidal field. This particular turbulent process may not be the only contributor to the α–effect in the Earth's core (e.g., Roberts, 1992).

The combined action of the processes illustrated in Figs. 1.7 and 1.8 is referred to as an $\alpha\omega$–dynamo. It is worth noting that the α–effect can also generate poloidal field from an initial toroidal field. Thus it is possible to have α^2– and $\alpha^2\omega$–dynamos. Readers are referred to Merrill *et al.* (1996) for more details.

Roberts (1971) and Roberts and Stix (1972) pointed out that if the large-scale velocity shear that causes the ω–effect is symmetric with respect to the equator and if the α–effect is antisymmetric with respect to the equator (as might be expected since the Coriolis force changes sign across the equator), then the dynamo can be separated into two noninteracting systems made up of specific families of spherical harmonics. Gubbins and Zhang (1993) refer to these as the *antisymmetric* and *symmetric families*. Spherical harmonics whose degree and order sum to an odd number belong to the antisymmetric family and those whose degree and order sum to an even number belong to the symmetric family. The situation shown in Fig. 1.7 is the simplest one in which the initial poloidal field is antisymmetric with respect to the equator.

1.1.4 Variations of the Dipole Field with Time

The intensity of the dipole field has decreased at the rate of about 5% per century since the time of Gauss' first spherical harmonic analysis (Leaton and Malin, 1967; McDonald and Gunst, 1968; Langel, 1987; Fraser-Smith, 1987) (Fig. 1.9a). Indeed, Leaton and Malin (1967) and McDonald and Gunst (1968)

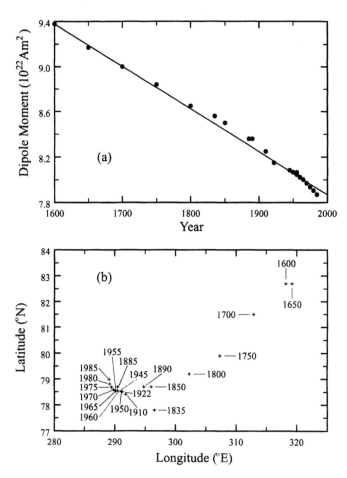

Fig. 1.9. Variations of the dipole field with time since A.D. 1600. (a) Variation of the dipole moment from successive spherical harmonic analyses. (b) Variation of the dipole axis as represented by the change in position of North Geomagnetic Pole. After Fraser-Smith (1987).

speculated on the demise of the main dipole around A.D. 3700 to 4000 if the present trends continue, but this change is probably just part of the natural variation as the dipole recovers from abnormally high values at about 2000 years ago (see Fig. 1.14). In contrast, the dipole axis, as represented by the position of the North Geomagnetic Pole (Fig. 1.9b), has hardly changed its position since the analysis of Gauss (Bullard *et al.*, 1950). Over the past 150 years there appears to have been a slow westward change of near 0.05° to 0.1° per year in azimuth angle but no progressive motion in polar angle (McDonald and Gunst, 1968; Fraser-Smith, 1987; Barton, 1989).

When there are only declination values available, it is still possible to obtain relative values of the Gauss coefficients. The most recent methods are those described by Bloxham and Gubbins (1985). To obtain estimates of the intensity Barraclough (1974) determined values of g_1^0 by extrapolating back in time from values determined since the time of Gauss. He fitted a straight line to values of g_1^0 from 170 spherical harmonic models of the field between 1829 and 1970 to derive the relation

$$g_1^0(t) = -31110.3 + 15.46(t - 1914) \ , \qquad (1.1.5)$$

where t is the epoch in years A.D. Barraclough (1974) then produced analyses of the geomagnetic field for epochs since 1600. His estimates of the positions of the North Geomagnetic Pole since 1600 as summarized by Fraser-Smith (1987) are plotted in Fig. 1.9b. They show that the dipole field has drifted westwards at about 0.08° per year since 1600 and has changed latitude at the much slower angular rate of 0.01° per year.

1.2 Paleomagnetism

1.2.1 Early Work in Paleomagnetism

The fact that some rocks possessed extremely strong remanent magnetization was noted as early as the late 18th century from their effect on the compass needle. Von Humboldt in 1797 attributed these effects to lightning strikes. Further investigations during the 19th century of these intense magnetizations occasionally found in rock exposures were also generally explained in this way. These were the first paleomagnetic phenomena to attract attention. The first studies of the direction of magnetization in rocks were made by Delesse in 1849 and Melloni in 1853. They both found that certain recent lavas were magnetized parallel to the Earth's magnetic field. The work of Folgerhaiter (1899) both extended and confirmed these earlier investigations. Chevallier (1925), from studies of the historical flows of Mt Etna, was able to trace the secular variation of the geomagnetic field over the past 2000 years.

David (1904) and Brunhes (1906) first investigated the material baked by lava flows, comparing the directions of magnetization of the flows with those of the underlying baked clay. They reported the first discovery of directions of magnetization roughly opposed to that of the present field. Confirmation that the baked clays were also reversely magnetized led to the first speculation that the Earth's magnetic field had reversed its polarity in the past. Mercanton (1926) then argued that if the Earth's magnetic field had reversed its polarity in the past, reversals should be found in rocks from all parts of the world. In studies of rocks of various ages from Spitsbergen, Greenland, Iceland, the Faroes, Mull, Jan Mayen Land and Australia, he found that some were magnetized in the same sense as the present field and some in the opposite sense. Concurrently Matuyama (1929) observed similar effects in Quaternary lavas from Japan and Manchuria, but noted that the reversely magnetized lavas were always older than those directed in the same sense as the present field (normally magnetized lavas). He concluded that during the early Quaternary the Earth's magnetic field was directed in a sense opposite to that of the present and gradually changed to its present sense later in the Quaternary.

Hospers (1955) appears to have been the first to suggest the use of reversals as a means of stratigraphic correlation but Khramov (1955; 1957) was the first to apply the concept. In his book, Khramov (1958; English translation 1960) suggested that it might be possible to determine a strict worldwide correlation of volcanic and sedimentary rocks and from that to create a single geochronological paleomagnetic time scale valid for the whole Earth. Khramov's seminal ideas clearly influenced early work on the development of polarity time scales (Glen, 1982).

By the mid-1920s, several important aspects of paleomagnetism in rocks had been established, culminating in the suggestion by Mercanton (1926) that, because of the approximate correlation of the present geomagnetic and rotational axes, it might be possible to test the hypothesis of polar wandering and continental drift. This inspiration was not put into practice until the 1950s, leading to the important papers by Irving (1956) and Runcorn (1956) showing that the apparent polar wander curve of Europe lay consistently eastwards of that of North America, clearly suggesting that continental drift had occurred. The first important reviews of paleomagnetic data that took the renewed activity in paleomagnetism into account were by Khramov (1958), Irving (1959), Blackett *et al.* (1960), and Cox and Doell (1960). It is interesting to note that the authors of the first three of these papers all took the view that the data supported the hypothesis of continental drift, whereas Cox and Doell (1960) took a more conservative view that the data could be interpreted in several ways, including changing magnetic fields. However, it was the application of paleomagnetism to the oceans in the analysis of marine magnetic anomalies (Vine and Matthews, 1963) that gave paleomagnetism and continental drift credibility, leading to the theory of plate tectonics.

Improved techniques and the undertaking of extensive paleomagnetic investigations in many parts of the world have dramatically increased the amount of paleomagnetic information. The magnetic anomalies across the world's oceans have now been extensively surveyed and analyzed. On land the number of independent investigations listed by Cox and Doell (1960) for the period up to the end of 1959 was about 200. By the end of 1963 this had increased to about 550 as listed by Irving (1964), and by the end of 1970 the figure had risen to about 1500 (McElhinny, 1973a). In 1987 the paleomagnetic section of IAGA decided to set up a computerized relational database of all paleomagnetic data. The first version of this *Global Paleomagnetic Database* was released in 1991 (McElhinny and Lock, 1990a,b; see also Lock and McElhinny, 1991, and McElhinny and Lock, 1996). The database is currently updated on roughly a 2-year basis. At the present time there are about 8500 paleomagnetic results derived from 3200 references listed in the database.

1.2.2 Magnetism in Rocks

The study of the history of the Earth's magnetic field prior to a few centuries ago relies on the record of the field preserved as fossil magnetization in rocks. Although most rock-forming minerals are essentially nonmagnetic, all rocks exhibit some magnetic properties due to the presence, as accessory minerals making up only a few percent of the rock, of (mainly) various iron oxides. The magnetization of the accessory minerals is termed the fossil magnetism, which, if acquired at the time the rock was formed, may act as a fossil compass and be used to estimate both the direction and the intensity of the geomagnetic field in the past. The study of fossil magnetism in rocks is termed paleomagnetism and is a means of investigating the history of the geomagnetic field over geological time. The study of pottery and baked hearths from archeological sites has been successful in tracing secular variation in historic times. This type of investigation is usually distinguished from paleomagnetism and is referred to as *archeomagnetism* (see §1.2.4).

Some of the common types of remanent magnetizations in rocks are listed in Table 1.2. The fossil magnetism initially measured in rocks (after preparation into suitably sized specimens) is termed the *natural remanent magnetization* or simply NRM. The mechanism by which the NRM was acquired depends on the mode of formation and subsequent history of rocks as well as the characteristics of the magnetic minerals. Magnetization acquired by cooling from high temperatures through the Curie point(s) of the magnetic mineral(s) is called *thermoremanent magnetization* (TRM) (see §2.3.5). If the magnetization is acquired by phase change or chemical action during the formation of magnetic oxides at low temperatures, it is termed *crystallization* (or *chemical*) *remanent magnetization* (CRM) (see §2.3.6). The alignment of detrital magnetic particles by the ambient magnetic field that might occur in a sediment during deposition gives rise to *detrital* (or *depositional*) *remanent magnetization* (DRM) or *post-depositional remanent magnetization* (PDRM) if the alignment takes place after deposition but before final compaction (see §2.3.7).

TABLE 1.2
Some Common Types of Remanent Magnetization (RM) in Rocks

Acronym	Type of Magnetization	Description
NRM	Natural Remanent Magnetization	The RM acquired by a sample under natural conditions
TRM	Thermoremanent Magnetization	The RM acquired by a sample during cooling from a temperature above the Curie temperature in an external field (usually in a weak field such as that of the Earth)
CRM	Crystallization (or Chemical) Remanent Magnetization	The RM acquired by a sample during a phase change or formation of a new magnetic mineral in an external field
DRM	Detrital (or Depositional) Remanent Magnetization	The RM acquired by sediments when grains settle in water in the presence of an external field
PDRM	Post-Depositional Remanent Magnetization	The RM acquired by the magnetic alignment of sedimentary grains after deposition but before final compaction
IRM	Isothermal Remanent Magnetization	The RM acquired in a short time at one temperature (usually room temperature) in a external field (usually strong)
VRM	Viscous Remanent Magnetization	The RM acquired over a long time in a weak external field
ARM	Anhysteretic Remanent Magnetization	The RM acquired when an alternating magnetic field is decreased from some large value to zero in the presence of a weak steady field

In nature *isothermal remanent magnetization* (IRM) usually refers to that magnetization acquired by rocks from lightning strikes, although it generally refers to that acquired in laboratory experiments aimed at determining the magnetic properties of samples (see §2.1.4 and §3.5.2). *Viscous remanent magnetization* (VRM) refers to the remanence acquired by rocks after exposure to a weak external magnetic field for a long time. Examples include that acquired by a sample after collection and before measurement or that acquired from deep burial and uplift (see §2.3.8). *Anhysteretic remanent magnetization* (ARM) is that produced by gradually reducing a strong alternating magnetic field in the presence of a weak steady magnetic field. To avoid heating samples (see §3.5.3), it is often used in laboratory experiments as an analog of TRM.

The component of NRM acquired when the rock was formed is termed the *primary magnetization*; this may represent all, part, or none of the total NRM. Subsequent to formation the primary magnetization may decay either partly or wholly and additional components may be added by several processes. These subsequent magnetizations are referred to as *secondary magnetization*. A major task in all paleomagnetic investigations is to identify and separate all the components be they primary or secondary.

1.2.3 Geocentric Axial Dipole Hypothesis

On the geological time scale the study of the geomagnetic field requires some model for use in analyzing paleomagnetic results, so that measurements from different parts of the world can be compared. The model should reflect the long-term behavior of the field rather than its more detailed short-term behavior. The model used is termed the *geocentric axial dipole* (GAD) *field* and its use in paleomagnetism is essentially an application of the principle of uniformitarianism. It is known from paleomagnetic measurements (see §6.3) that *when averaged over a sufficient time interval* the Earth's magnetic field for the past few million years has conformed with this model. However, there are second-order effects that cause departures from the model of no more than about 5%. Such an averaged field is referred to as the *time-averaged paleomagnetic field*. A basic problem that arises is to decide how much time is needed for the averaging process. In the early days of paleomagnetism it was generally thought that times of several thousands of years were sufficient, but it is now thought that much longer times may be required, possibly on the scale of hundreds of thousands of years (see discussion in Merrill *et al.*, 1996).

The GAD model is a simple one (Fig. 1.10) in which the geomagnetic and geographic axes coincide, as do the geomagnetic and geographic equators. For any point on the Earth's surface, the geomagnetic latitude λ equals the

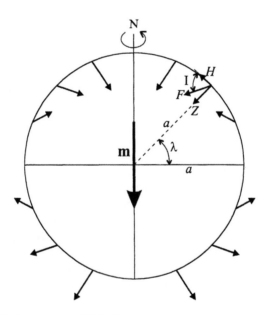

Fig. 1.10. The field of a geocentric axial dipole.

geographic latitude. If m is the magnetic moment of the dipole and a is the radius of the Earth, the horizontal (H) and vertical (Z) components of the field at latitude λ can be derived from the g_1^0 spherical harmonic term as

$$H = \frac{\mu_0 m \cos \lambda}{4\pi a^3} , \quad Z = \frac{2\mu_0 m \sin \lambda}{4\pi a^3} , \tag{1.2.1}$$

and the total field F is given by

$$F = (H^2 + Z^2)^{\frac{1}{2}} = \frac{\mu_0 m}{4\pi a^3}(1 + 3\sin^2 \lambda)^{\frac{1}{2}} . \tag{1.2.2}$$

Since the tangent of the magnetic inclination I is Z/H, then

$$\tan I = 2 \tan \lambda , \tag{1.2.3}$$

and, by definition,

$$D = 0° . \tag{1.2.4}$$

The colatitude p (90° minus the latitude) can be obtained from

$$\tan I = 2 \cot p \quad (0° \le p \le 180°) . \tag{1.2.5}$$

The relationship of (1.2.3) is central in paleomagnetism. It indicates that the GAD model, when applied to results from different geological periods, enables the paleomagnetic latitude to be derived simply from the mean inclination. The relationship between latitude and inclination is shown in Fig. 1.11.

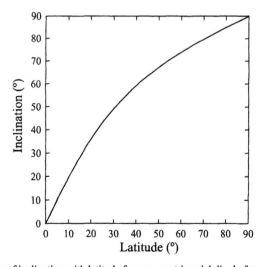

Fig. 1.11. Variation of inclination with latitude for a geocentric axial dipole from (1.2.3).

In order to compare paleomagnetic results from widely separated localities, it is convenient to calculate some parameter which, on the basis of the GAD model, should have the same value at each observing locality. The parameter used is called the *paleomagnetic pole* and represents the position where the time-averaged dipole axis cuts the surface of the Earth. The position of the paleomagnetic pole is referred to the present latitude–longitude grid. Thus, if the paleomagnetic *mean* direction (D_m, I_m) is known at some sampling locality S, with latitude and longitude (λ_s, ϕ_s), the co-ordinates of the paleomagnetic pole P (λ_p, ϕ_p) can be calculated from the following equations by reference to Fig. 1.12:

$$\sin \lambda_p = \sin \lambda_s \cos p + \cos \lambda_s \sin p \cos D_m \quad (-90^\circ \le \lambda_p \le +90^\circ) \quad (1.2.6)$$

$$\phi_p = \phi_s + \beta \qquad \text{when } \cos p \ge \sin \lambda_s \sin \lambda_p$$

or

$$\phi_p = \phi_s + 180 - \beta \quad \text{when } \cos p < \sin \lambda_s \sin \lambda_p$$

$$(1.2.7)$$

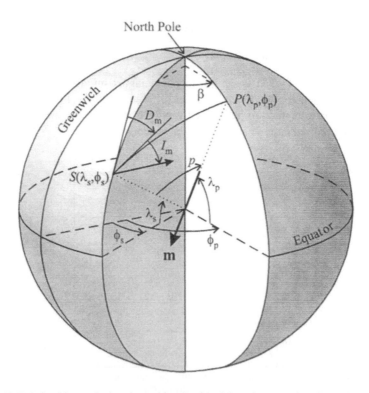

Fig. 1.12. Relationship to calculate the position (λ_p, ϕ_p) of the paleomagnetic pole P relative to the sampling site S at (λ_s, ϕ_s) with mean magnetic direction (D_m, I_m). After Merrill *et al.* (1996).

where $$\sin \beta = \sin p \sin D_m / \cos \lambda_p \quad (-90° \leq \beta \leq 90°) . \quad (1.2.8)$$

The paleocolatitude p is determined from (1.2.5). The paleomagnetic pole (λ_p, ϕ_p) calculated in this way implies that "sufficient" time averaging has been carried out. Alternatively, any instantaneous paleofield direction may be converted to a pole position using (1.2.6) – (1.2.8), in which case the pole is termed a *virtual geomagnetic pole* (VGP). The VGP can be regarded as a paleomagnetic analog of the geomagnetic poles of the present field. The paleomagnetic pole may then be calculated by finding the average of many VGPs, corresponding to many paleodirections. Table 1.3 gives a summary of the various types of poles used in geomagnetism and paleomagnetism.

Conversely, of course, given a paleomagnetic pole position with co-ordinates (λ_p, ϕ_p) the corresponding expected mean direction of magnetization (D_m, I_m) may be calculated for any site location (λ_s, ϕ_s) (Fig. 1.12). The paleocolatitude p is given by

$$\cos p = \sin \lambda_s \sin \lambda_p + \cos \lambda_s \cos \lambda_p \cos(\phi_p - \phi_s) , \quad (1.2.9)$$

and the inclination I_m may then be calculated from (1.2.5). The corresponding declination D_m is given by

$$\cos D_m = \frac{\sin \lambda_p - \sin \lambda_s \cos p}{\cos \lambda_s \sin p} , \quad (1.2.10)$$

where $$0° \leq D_m \leq 180° \quad \text{for} \quad 0° \leq (\phi_p - \phi_s) \leq 180°$$

and $$180° < D_m < 360° \quad \text{for} \quad 180° < (\phi_p - \phi_s) < 360° .$$

The declination is indeterminate (so any value may be chosen) if the site and the pole position coincide. If $\lambda_s = \pm 90°$ then D_m is defined as being equal to ϕ_p, the longitude of the paleomagnetic pole.

Tests for the validity of the GAD model over geological time can, in principle, be made in several ways. The simplest test for a dipole field (axial or not) is that the field, when viewed from regions of continental extent, should be consistent with that of a geocentric dipole. This requires that the paleomagnetic poles obtained from different rock units belonging to the same geological epoch be in close agreement, at least as good as that observed over the past few million years. Studies of the paleointensity of the paleomagnetic field as a function of paleolatitude should conform with (1.2.2) (see §1.2.5). Models of the latitude variation of paleosecular variation should also conform with paleomagnetic data (see §1.2.6). Testing for a geocentric *axial* dipole appeals to paleoclimatic evidence, which is independent of the GAD hypothesis. The Earth's climate is

TABLE 1.3
Summary of Poles Used in Geomagnetism and Paleomagnetism

North (south) magnetic pole	Point on the Earth's surface where the magnetic inclination is observed to be +90° (-90°). The poles are not exactly opposite one another and for epoch 1995 lie at 78.9°N, 254.9°E and 64.7°S, 138.7°E.
Geomagnetic north (south) pole	Point where the axis of the calculated best fitting geocentric dipole cuts the surface of the Earth in the northern (southern) hemisphere. The poles lie antipodal to one another and for epoch 1995 are calculated to lie at 79.3°N, 288.6°E and 79.3°S, 108.6°E.
Virtual geomagnetic pole (VGP)	The position of the equivalent geomagnetic pole calculated from a spot reading of the paleomagnetic field direction. It represents only an instant in time, just as the present geomagnetic poles are only an instantaneous observation.
Paleomagnetic pole	The pole of the paleomagnetic field averaged over periods sufficiently long so as to give an estimate of the geographic pole. Averages over times of 10^5 years or longer may be required. The pole may be calculated from the average paleomagnetic field direction or from the average of the corresponding VGPs.

controlled by the rotational axis and has an equator to pole distribution. It is warmer at the equator than at the poles. The paleolatitude spectra of various paleoclimatic indicators should all be appropriately latitude dependent to be consistent with the complete GAD hypothesis. Results from this type of investigation are discussed more fully in §6.3.4.

1.2.4 Archeomagnetism

Pottery and (more usefully) bricks from pottery kilns and ancient fireplaces, whose last dates of firing can be estimated from ^{14}C contents of ashes, have a TRM dating from their last cooling. Samples used in such studies, despite often having awkward shapes, can be measured by the usual techniques of paleomagnetism. Pioneer work in this field was undertaken by Folgerhaiter (1899) and Thellier (1937). The techniques commonly in use are those developed by Thellier and have been reviewed by Thellier (1966). Archeomagnetic investigations from different parts of the world are summarized in Creer *et al.* (1983) and are discussed in Merrill *et al.* (1996).

Cox and Doell (1960) observed that the average of VGPs calculated from observatory data around the world is close to the present geomagnetic pole. Unfortunately, archeomagnetic data are not evenly spaced around the world but are concentrated in the European region. Barbetti (1977) suggested that, to estimate the position of Recent geomagnetic poles, the effects of nondipole field variations could be averaged out if VGPs were averaged over 100-yr intervals for a limited number of regions of the Earth's surface.

The suggestion of Barbetti (1977) has been used by Champion (1980), Merrill and McElhinny (1983), and most recently by Ohno and Hamano (1992), who calculated the position of the North Geomagnetic Pole for successive times at 100-yr intervals for the past 10,000 yr. The results of the analysis of Ohno and Hamano (1992) are illustrated in Fig. 1.13 for each successive 2000-yr interval as well as for the entire 10,000 yr interval. Interestingly, the successive positions of the poles for 1600 to 1900 A.D. lie close to and have the same trend as the positions of the geomagnetic pole calculated from historical observations (Barraclough, 1974; Fraser-Smith, 1987; see §1.1.4 and Fig. 1.9b). Therefore, it seems reasonable to assume that 100-yr VGP means are indeed representative of positions of the North Geomagnetic Pole. Figure 1.13 shows that the mean VGP for each 2000-yr interval does not always average to the geographic pole, whereas the mean over 10,000 yr appears to do so. Thus, it appears that an interval of at least 10,000 yr is required for the dipole axis to average to the axis of rotation. Some caution is needed, however, because it is not at all clear that the motion of the dipole axis over the past 10,000 yr, as depicted in Fig. 1.13a,

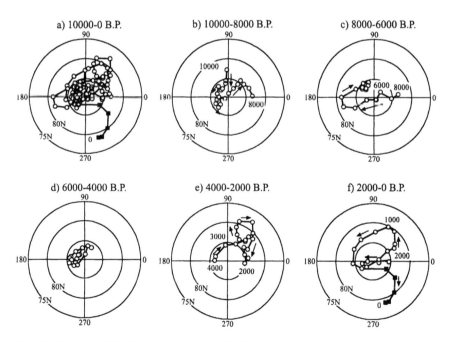

Fig. 1.13. Locations of the North Geomagnetic Pole (dipole axis) over the past 10,000 years at 100-year intervals estimated from archeomagnetic measurements. Locations are given for each 2000-yr interval as well as for the entire 10,000 yr interval. After Ohno and Hamano (1992).

can be regarded as a recurring feature, or that the average over the preceding 10,000 yr would also coincide with the geographic pole.

Studies of the intensity of the geomagnetic field over archeological times can be undertaken with a much wider variety of materials, such as pottery fragments, because the orientation of the samples need not be known. The method used is that developed by Thellier (see review by Thellier and Thellier, 1959a,b). Under the GAD hypothesis, measurements of ancient geomagnetic intensity are a function only of latitude and the magnitude of the Earth's dipole moment (1.2.2). Thus, paleointensity measurements from all over the world can be normalized by calculating an equivalent dipole moment, referred to as the *virtual dipole moment* (VDM). This is the intensity analog of the VGP. For a more detailed discussion see Merrill *et al.* (1996).

To smooth rapid variations of the nondipole field at any one locality, the virtual dipole moments must be averaged not only from different parts of the world but also in class intervals of a few hundred years, as has been done in the analysis of archeomagnetic VGPs. McElhinny and Senanayake (1982) calculated the variation in the Earth's dipole moment over the past 10,000 yr by averaging over 500-yr intervals back to 4000 yr B.P. and over 1000-yr intervals prior to that. The results are shown in Fig. 1.14 together with 95% confidence bars.

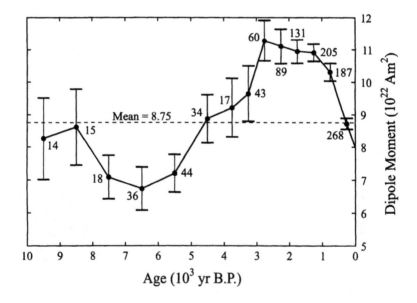

Fig. 1.14. Global dipole moment versus time estimates obtained from 500-year period averages from 0-4000 yr B.P. and then 1000-year period averages . The number of measurements averaged is shown at each point together with 95% confidence bars. After McElhinny and Senanayake (1982).

The mean dipole moment for the past ten 1000-yr intervals is 8.75×10^{22} Am^2 with an estimated standard deviation of 18.0%, which may be attributed to dipole intensity fluctuations. There is a maximum around 2500 yr B.P. and a minimum around 6500 yr B.P. Cox (1968) had previously thought that the data summary as illustrated in Fig. 1.14 was indicative of variations in the dipole moment with a simple periodicity of between 8000 and 9000 yr with maxima and minima respectively about 1.5 and 0.5 times the present dipole moment. However, the data available for times before 10000 yr B.P. show clearly that this is not the case and data for the interval 0–5 Ma are also inconsistent with the expectations of a periodic variation (Kono, 1972; McFadden and McElhinny, 1982; Merrill *et al.*, 1996).

1.2.5 Paleointensity over Geological Times

The problems of determining the paleointensity of the geomagnetic field are much more complex than those associated with paleodirectional measurements and become increasingly difficult the older the rocks studied. The presence of secondary components and the decay of the original magnetization all serve to complicate the problem. Kono and Tanaka (1995), Tanaka *et al.* (1995), and Perrin and Shcherbakov (1997) analyzed all the available measurements in terms of VDMs. In Fig. 1.15, the best estimate of the variation of the Earth's dipole moment over the whole of geological time (Kono and Tanaka, 1995) is summarized for the past 400 Myr averaged at 20-Myr intervals (Fig. 1.15a) and prior to that at 100-Myr intervals (Fig. 1.15b).

Prévot *et al.* (1990) first suggested that there was an extended period during the Mesozoic when the Earth's dipole moment was low, at about one-third of its Cenozoic value. Further measurements for the Jurassic (e.g. Perrin *et al.*, 1991; Kosterov *et al.*, 1997) supported low values. During the Cenozoic the dipole moment was similar to its present value. For the period prior to 400 Ma, the number of paleointensity measurements is much fewer. Of particular interest is the oldest paleointensity measurement, which is for the 3500 Ma Komati Formation Lavas in South Africa (Hale, 1987). The average VDM of $2.1 \pm 0.4 \times 10^{22}$ Am^2 is about 27% of the present dipole moment. This result clearly demonstrates the existence of the Earth's magnetic field at 3.5 Ga. The range of variation of the Earth's dipole moment is about $2 - 12 \times 10^{22}$ Am^2 and is approximately the same for Phanerozoic and Precambrian times (Prévot and Perrin, 1992). With the present data set, no very-long-term change in dipole moment is apparent. Kono and Tanaka (1995) point out that it is remarkable that the dipole intensity seems to have been within a factor of 3 of its present value for most of geological time.

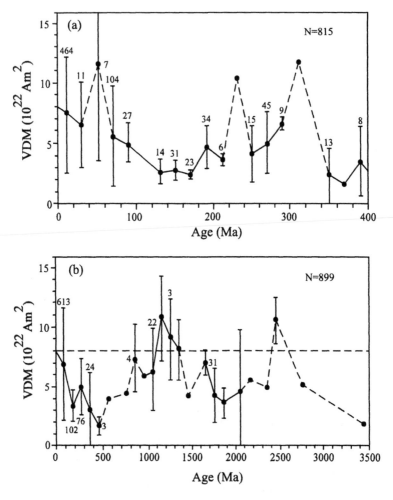

Fig. 1.15. Variation of VDM with geological time. (a) The past 400 Myr averaged over 20-Myr intervals. (b) The past 3500 Myr averaged over 100-Myr intervals. Error bars give the standard error of the mean for each interval and the number of measurements in each interval is indicated. The horizontal dashed line in (b) indicates the present dipole moment (8×10^{22} Am²). After Kono and Tanaka (1995).

1.2.6 Paleosecular Variation

Secular variation of the geomagnetic field in pre-archeological times has been investigated through paleomagnetic studies of Recent lake sediments. Long-period declination oscillations in cores taken from the postglacial organic sediments deposited at the bottom of Lake Windermere in England were first discovered by Mackereth (1971). Since that time, many such studies have been made throughout Europe, North America, Australia, Argentina, and New Zealand. Such studies are generally referred to as studies of *paleosecular variation* (PSV). Extensive investigations of lakes in England and Scotland have enabled a master curve of changes in declination and inclination in Great Britain over the past 10,000 years to be determined (Turner and Thompson, 1981, 1982) as illustrated in Fig. 1.16. Further details on such studies and their interpretation are summarized in Creer *et al.* (1983) and discussed by Merrill *et al.* (1996).

The GAD model takes no account of secular variation, although its effect must be averaged out before paleomagnetic measurements are said to conform with the model. The secular variation in paleomagnetic studies is expressed by the statistical scatter in paleomagnetic results after the effects of experimental errors have been removed. To estimate this scatter it is necessary to be sure that each measurement is a separate instantaneous record of the ancient geomagnetic field. Sediments cannot readily be used for this purpose because even small samples may have already averaged the field over the thickness of sediment covered by

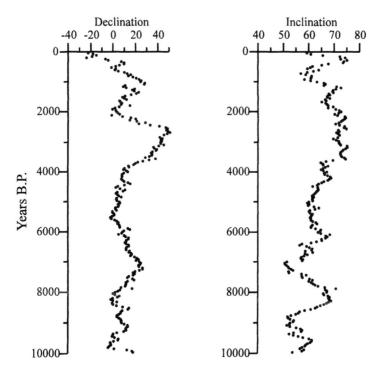

Fig. 1.16. Master curves for declination and inclination for Great Britain. After Turner and Thompson (1981, 1982).

the sample. Therefore study of the paleosecular variation through scatter in paleomagnetic results is restricted to investigations of lava flows and is referred to as *paleosecular variation from lavas* (PSVL).

The scatter in paleomagnetic results from lavas is measured by the angular dispersion either of paleomagnetic directions or, more commonly, of the corresponding VGPs. Several models for the latitude variation of this angular dispersion have been suggested and are summarized in detail by Merrill *et al.* (1996). McFadden *et al.* (1988a) have shown that the concept of separation of the dynamo into two approximately independent families (symmetric and antisymmetric – see §1.1.3) can be useful in modeling PSVL. In this model, the total angular dispersion (S) of VGPs is given by

$$S^2 = S_A^2 + S_S^2 \ , \tag{1.2.11}$$

where S_A is the angular dispersion due to the antisymmetric family and S_S is that due to the symmetric family. Although it is extremely unlikely that the two families are in fact independent at any given time, the effect in the time-averaged field may be approximately the same as if they were independent. By definition, the dispersion at the equator is caused entirely by the symmetric family. Analysis of the present geomagnetic field, which by chance has a latitudinal structure in

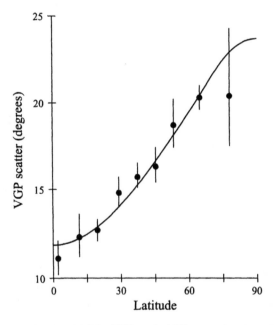

Fig. 1.17. Least squares fit of *Model G* for PSVL to the VGP scatter from lavas for the past 5 Myr. After McElhinny and McFadden (1997).

VGP scatter similar to that for the past 5 Myr, shows that the contribution from the symmetric family is effectively independent of latitude. The latitude variation comes from the antisymmetric family and up to latitude 70° the dispersion from this source is approximately proportional to latitude. Assuming a similar behavior for the paleomagnetic field, the model therefore predicts

$$S^2 = (a\lambda)^2 + b^2 \ , \qquad\qquad (1.2.12)$$

where $S_A = a\lambda$ and $S_S = b$, and a and b are constants to be determined. When this model is applied to paleomagnetic results from lavas for the past 5 Myr (McElhinny and McFadden, 1997), there is an excellent fit to the latitude variation of VGP scatter as shown in Fig. 1.17.

Chapter Two

Rock Magnetism

2.1 Basic Principles of Magnetism

2.1.1 Magnetic Fields, Remanent and Induced Magnetism

The study of magnetism originated from observation of the behavior of natural permanent magnets, the earliest known of which were used as magnetic compass needles. A permanent magnet is usually described by its "north" and "south" magnetic "poles", imagined to reside at the opposite ends of the magnet. The concept of magnetic poles has been of considerable use in analyzing the behavior of magnets, but since the discovery by Oersted in 1820 that an electric current flowing in a wire deflected a compass needle placed near it, it has been recognized that all magnetic effects are appropriately described in terms of electric currents. An immediate consequence of this is that there are no isolated magnetic poles, so the "magnetic poles" at the end of a magnet are just convenient fictions for the purpose of simple analysis. An electron in orbit around a nucleus is, in essence, a current flowing in a loop and the magnetic effects of materials can all be described in terms of such elementary current loops.

Magnetic fields are specified in terms of the two vectors $\mathbf{H}$, called the *magnetic field*, and $\mathbf{B}$, the *magnetic induction*. $\mathbf{B}$ includes the effects of the macroscopic *magnetization* $\mathbf{M}$, defined as the dipole moment per unit volume, according to the relation

$$\mathbf{B} = \mu_0(\mathbf{H} + \mathbf{M}),$$
  (2.1.1)

where μ_0 is the permeability of free space and has the value of $4\pi \times 10^{-7}$ Hm^{-1} (Henry per metre). Outside any magnetic materials $\mathbf{M} = 0$ and $\mathbf{B}$ and $\mathbf{H}$ are parallel and in this case $\mathbf{B} = \mu_0\mathbf{H}$. In geomagnetism and paleomagnetism the magnetic field of interest is almost always external to a magnetic medium, so that $\mathbf{B}$ and $\mathbf{H}$ are parallel and it is of no consequence which one is used. However, numerical conversion between the old cgs system and the current SI system of units is trivial for $\mathbf{B}$ but involves a factor of 4π (from μ_0) in the case of $\mathbf{H}$, so magnetic fields are generally specified in terms of the magnetic induction $\mathbf{B}$ rather than the magnetic field $\mathbf{H}$. Furthermore, $\mathbf{B}$ is typically referred to in an informal manner as the magnetic field $\mathbf{B}$, a usage that will often be followed in this book. Naturally, when considering magnetic fields inside a magnetic material $\mathbf{M} \neq 0$, so it becomes important to distinguish between $\mathbf{B}$ and $\mathbf{H}$ because they will not be equivalent, will not always be parallel, and can even have opposite signs. Under such circumstances it becomes important to use the magnetic field $\mathbf{H}$ (see §4.1.2, in which such a situation occurs).

A permanent magnetic and electric current loop both have a *magnetic dipole moment*, $\mathbf{m}$, associated with them. When placed in a magnetic field $\mathbf{B}$ (Fig. 2.1) each will experience a torque $\mathbf{L} = \mathbf{m} \times \mathbf{B}$ (i.e., $L = mB\sin\theta$, where θ is the angle between the long axis of the magnet and $\mathbf{B}$ or the angle between the axis drawn through the center of the current loop at right angles to the plane of the loop). Hence, the torque attempts to rotate the dipole moment into alignment with $\mathbf{B}$. In the case of the current loop the dipole moment $m = iA$ (current times the area of the loop), and in the case of the bar magnet it is $m = pl$, where p is the "pole strength" of the imaginary poles at each end of the magnet and l is the distance between the poles. Thus dipole moment is measured in units of Am^2 and the magnetization $\mathbf{M}$ (dipole moment per unit volume) is measured in Am^{-1}. Table 2.1 summarizes the basic magnetic quantities together with typical values that arise in geomagnetism and paleomagnetism.

TABLE 2.1
Various Magnetic Quantities Used in Geomagnetism and Paleomagnetism with Typical Values

Quantity	Units	Typical values
Magnetic induction, B	T(tesla)	Earth's field 10^{-4} T (0.1 mT or 100 μT)
Magnetic field, H	Am^{-1}	Earth's field $10^3/4\pi = 79.6$ Am^{-1}
Dipole moment, m	Am^2	Orbital electron 10^{-23} Am^2;
		Bar magnet 1 Am^2
Magnetization, M	Am^{-1}	Sediments $\approx 10^{-3}$ Am^{-1} ; Volcanics ≈ 1 Am^{-1}
Magnetic susceptibility, χ	Dimensionless	Magnetite ≈ 2.5
Permeability of free space, μ_0	Hm^{-1}	$4\pi \times 10^{-7}$ Hm^{-1}

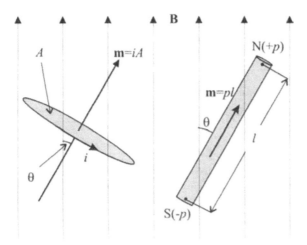

Fig. 2.1. The equivalence of a bar magnet and a current loop. When placed in a magnetic field **B** each suffers a torque according to its magnetic dipole moment **m**. The current loop has area A and current i flowing in the loop. The bar magnet has "pole strength" p and distance l between the imaginary poles at the ends of the magnet.

The magnetization of any material is generally made up of two components: the *remanent magnetization* (or simply remanence), which is that remaining in the absence of an applied field; and the *induced magnetization*, which is that induced by an applied field but which disappears after removal of that field. When dealing with rocks, the total magnetization **M** is made up of the vector sum of the remanence $\mathbf{M_n}$ and the magnetization $\mathbf{M_i}$ induced by the Earth's magnetic field, where

$$\mathbf{M} = \mathbf{M_n} + \mathbf{M_i} \ . \qquad (2.1.2)$$

In isotropic materials the induced magnetization $\mathbf{M_i}$ lies along the direction of the applied field **H** (i.e., **B**) and is proportional to the magnitude of that field, that is

$$\mathbf{M_i} = \chi \mathbf{H} = \frac{\chi \mathbf{B}}{\mu_0}, \qquad (2.1.3)$$

where χ is a constant of proportionality called the *magnetic susceptibility*. Since **H** and $\mathbf{M_i}$ have the same dimensions (2.1.1 and Table 2.1), χ is a dimensionless number. Some banded sediments, layered intrusions, and foliated metamorphic rocks are magnetically anisotropic and have greater susceptibility in the plane of layering. These are special cases and most rocks used in paleomagnetism, such as basalts, dolerites, redbeds, and limestones, are magnetically isotropic or nearly so (see also §2.3.9).

The *Koenigsberger ratio* (*Q*) has been defined as the ratio of the remanent to induced magnetization and is given by

$$Q_n = \frac{\mu_0 M_n}{\chi B} \qquad (2.1.4)$$

or

$$Q_t = \frac{\mu_0 M_t}{\chi B}. \qquad (2.1.5)$$

Q_n is the ratio of the remanence (M_n) to that induced by the Earth's magnetic field at the sampling site, whereas Q_t is the ratio of the thermoremanent magnetization (M_t) acquired in the magnetic field B to the magnetization induced by the same field at room temperature.

2.1.2 Diamagnetism and Paramagnetism

Diamagnetism is a phenomenon common to all materials. Any moving charge, including orbital electrons, experiences a force (known as the Lorentz force) in a magnetic field **B**. This Lorentz force deflects the path of electrons in such a way that they precess clockwise about **B** (when viewed along a line in the direction of **B**). This is equivalent to a current anticlockwise about **B**, which produces a negative induced magnetic moment (compare with the current direction in Fig. 2.1) known as *diamagnetism*. Thus, the susceptibility is negative (Fig. 2.2a) and very small, typically on the order of 10^{-5}.

If an atom has a resultant magnetic moment the application of a magnetic field tends to align these dipole moments along the direction of the field. Although the diamagnetic effect still occurs, it is swamped by the alignment of the atomic dipole moments. Substances that exhibit this effect are called *paramagnetics*, and the induced magnetization is in the same direction as the applied field giving a positive susceptibility (Fig. 2.2b) that lies typically between 10^{-3} and 10^{-5}. In metallic substances a further situation arises because the individual atoms and their inner orbital electrons are closer together in the solid state than the virtual radii of the valence electrons. The outer valence electrons are thus no longer associated with individual atoms and they wander freely through the metal. In an atom devoid of its valence electrons, the net atomic dipole moment is zero. The application of a magnetic field causes the "free" electrons, equal numbers of which have opposite spins, to have their spins aligned parallel to the magnetic field. The substance thus acquires a dipole moment and paramagnetism results.

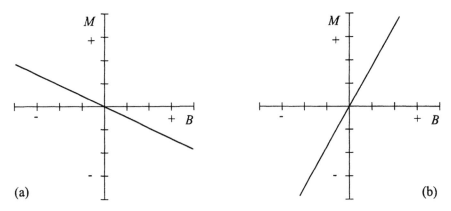

Fig. 2.2. Variation of magnetization with applied field for (a) diamagnetic and (b) paramagnetic material.

2.1.3 Ferro-, Antiferro-, and Ferrimagnetism

Diamagnetic and paramagnetic substances exhibit only weak magnetic effects because the dipole moments involved are relatively small. However, substances like iron, cobalt, and nickel exhibit strong magnetic effects resulting from a phenomenon known as *ferromagnetism*. These ferromagnetic substances are distinguished by the fact that the individual atoms, and their inner orbital electrons, are much closer together than the virtual radii of their valence electron orbits when compared with the other metallic paramagnetics. Also, there are more valence electrons available to move freely through the metal so that they become crowded together and react strongly with one another. The exchange forces between these electrons are such that their spins become aligned *even in the absence of an applied magnetic field*. Ferromagnetic substances therefore exhibit *spontaneous magnetization* because of exchange coupling between the electrons and may have a permanent dipole moment in the absence of an applied field. As the temperature is increased thermal agitation may destroy the alignment process. It becomes completely destroyed at a critical temperature for each substance called the *Curie temperature* or *Curie point*. The spontaneous magnetization reduces to zero at the Curie point and above this temperature the substance behaves like an ordinary paramagnetic. Figure 2.3 shows the variation of spontaneous magnetization, M_s, with temperature for magnetite and hematite, the two most common magnetic minerals in rocks.

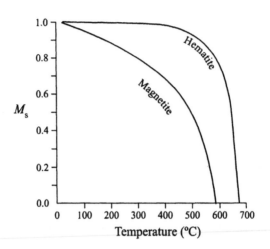

Fig. 2.3. Variation of spontaneous magnetization, M_s, with temperature for magnetite and hematite. M_s is normalized to its value at 0°C. Redrawn from Pullaiah *et al.* (1975), with permission from Elsevier Science.

Some substances are characterized by a subdivision into two sublattices (usually designated *A* and *B*). The atomic moments of *A* and *B* are each aligned but antiparallel to one another. The ferromagnetic effects cancel one another out when the moments of the two sublattices are equal (Fig. 2.4) and there is no net magnetic moment. This phenomenon is known as *antiferromagnetism*. Such

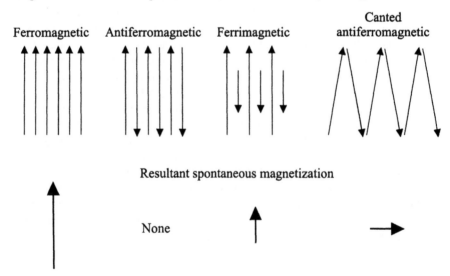

Fig. 2.4. Cartoon of the different exchange-coupled spin structures, together with the resultant spontaneous magnetization.

substances do not have a Curie temperature because there is no net ferromagnetism. In this case the ordering of the atomic moments is destroyed at a critical temperature called the *Néel temperature*, above which the substances behave like ordinary paramagnetics. If the atomic moments of the *A* and *B* sublattices are unequal, then there is a net spontaneous magnetization and a weak ferromagnetism results that is known as *ferrimagnetism*. Alternatively, the equal atomic moments in the two sublattices may not be exactly antiparallel and a small spontaneous magnetization results (Fig. 2.4). Such a substance is called a *canted antiferromagnetic*. Both the ferrimagnetic and canted antiferromagnetic substances behave as ordinary ferromagnetics; they have a Curie temperature and all the properties of ferromagnetics. The important minerals in rock magnetism are of these two types, to which the basic theories of ferromagnetism may be applied.

2.1.4 Hysteresis

When a ferromagnetic substance, initially in a demagnetized state, is placed in an applied magnetic field *B*, the specimen follows the magnetization curve from the origin as in Fig. 2.5. As *B* is increased from zero the magnetization *M* initially rises linearly as shown by the portion *a* of the curve. If *B* is reduced to zero at this point the process is reversible and *M* also falls to zero. The *initial susceptibility* ($\chi = \mu_0 M/B$ since $B = \mu_0 H$) of the ferromagnetic substance can be obtained from the slope of the *M–B* curve here. As *B* is increased further, the slope of the curve increases (in region *b*); if *B* is now reduced to zero, *M* does not fall to zero but follows the path *c*, and an *isothermal remanent magnetization* (IRM) given by M_r results. Further increases in *B* beyond point *d* on the magnetizing curve would produce no further increases in *M*, and a *saturation magnetization*, M_s, is reached at the saturating field B_{sat}. On reducing the field to zero (along portion *e*) the *saturation IRM*, or simply *saturation remanence*, M_{rs}, occurs. On applying a field in the opposite direction, the IRM is overcome and *M* is reduced to zero in a field B_c, called the *coercivity* or *coercive force*. Coercive force, however, is more usually given in terms of the equivalent field, H_c, although often quoted in units of *B*. Further increases of *B* in the negative direction causes saturation to occur in the opposite direction and repeated cycling of the field cause the magnetization to follow a *hysteresis loop* as shown in Fig. 2.5. The largest hysteresis loop occurs when the field is cycled with saturation being reached, and in this case B_c (and H_c) has its maximum value called the *maximum coercive force*. If the field is cycled without saturation being reached, then a smaller hysteresis loop results as shown.

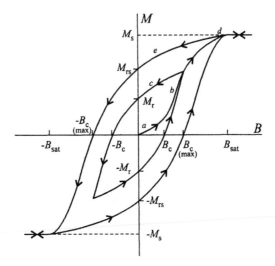

Fig. 2.5. Initial magnetization curve and hysteresis loops (*M-B* loops) for a ferromagnetic substance.

When dealing with rocks it is usually most convenient to study the IRM (M_r) since this is the value of the magnetization with no external field. Indeed, IRM curves are often used as a means of identifying the major magnetic minerals in rocks (see §3.5.2 for further discussion). If successively increasing fields are applied (and then removed) in the direction opposite to M_{rs} until the IRM is reduced to zero, then the "backfield" required to reduce this to zero is called the *coercivity of remanence* (B_{cr} or H_{cr}). It is important to understand the distinction between B_c (or H_c) and B_{cr} (or H_{cr}). If a backfield is applied and, in the presence of that backfield, the remanence is reduced to zero then that backfield is B_c (often also called the *bulk coercivity*). In contrast, if a backfield is applied and then removed and this results in zero remanence, then that backfield is B_{cr}.

2.2 Magnetic Minerals in Rocks

2.2.1 Mineralogy

The minerals that are responsible for the magnetic properties of rocks are mainly those that lie within the ternary system FeO-TiO_2-Fe_2O_3 (Fig. 2.6). For detailed descriptions of magnetic mineralogy, readers are referred to Stacey and Banerjee (1974), Lindsley (1976, 1991), O'Reilly (1984), Dunlop (1990), and Dunlop and Özdemir (1997).

In the ternary system of Fig. 2.6, it is generally sufficient to distinguish between two types of magnetic mineral. There are the strongly magnetic cubic oxides *magnetite* (Fe_3O_4) and its solid solutions with *ulvöspinel* (Fe_2TiO_4) that are known as *titanomagnetites* (see §2.2.2 for more details). The more weakly magnetic rhombohedral minerals are based on *hematite* (αFe_2O_3) and its solid solutions with *ilmenite* ($FeTiO_3$) that are known as *titanohematites* (see §2.2.3 for more details). The members of the orthorhombic *pseudobrookite* series shown in Fig. 2.6 are all paramagnetic above liquid oxygen temperatures and need not be considered further. Complete solid solution occurs only at high temperatures; at lower temperatures *exsolution* (unmixing of phases) occurs and in titanomagnetites this would result in an ulvöspinel-rich phase and a magnetite-

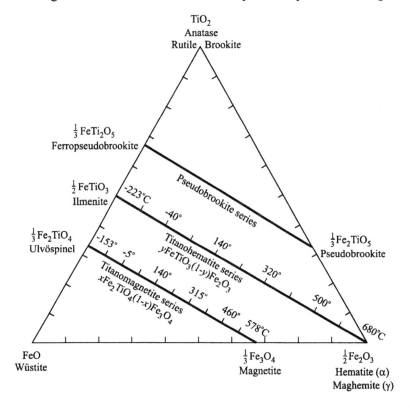

Fig. 2.6. FeO–TiO$_2$.–Fe$_2$O$_3$ ternary system showing the three principal solid solution series found in igneous rocks. Members of the pseudobrookite series are all paramagnetic above liquid oxygen temperatures and therefore have little significance for remanence. Approximate Curie (or Néel) temperatures for various values of mole function x and y (at 0.1 intervals) are indicated for the titanomagnetite and titanohematite series respectively. After Merrill and McElhinny (1983).

rich phase. However, magnetite-ulvöspinel intergrowths rarely occur in nature because of the presence of oxygen in the melt, which oxidizes the titanomagnetites. Thus, *high-temperature oxidation*, sometimes called *deuteric oxidation* if it occurs during initial cooling, commonly occurs above the Curie point in terrestrial magnetic minerals. In addition, *low-temperature oxidation* or *maghemitization* may occur at temperatures below about 200°C to produce *titanomaghemites*, a cation-deficient form of titanomagnetite that will be discussed more fully in §2.2.2.

Pyrrhotite ($Fe_{1-x}S$, where $0 < x \le \frac{1}{8}$) is a common accessory magnetic mineral in rocks, although it seldom dominates the remanence. In addition, *greigite* (Fe_3S_4) occurs quite commonly in sediments that were formed under anoxic conditions. Two important naturally occurring oxyhydroxides of iron are *goethite* ($\alpha FeOOH$) and *lepidocrocite* ($\gamma FeOOH$). Both the sulfides and oxyhydroxides are discussed further in §2.2.4. Table 2.2 summarizes the magnetic properties of some of the magnetic minerals that occur in rocks together with the properties of some typical ferromagnetic substances.

TABLE 2.2

Magnetic Properties of Some Common Minerals

Mineral	Composition	Magnetic state	M_s (10^3 Am^{-1})	T_c (°C)
Magnetite	Fe_3O_4	Ferrimagnetic	480	580
Titanomagnetite (TM60)	$Fe_{2.4}Ti_{0.6}O_4$	Ferrimagnetic	125	150
Ulvöspinel	Fe_2TiO_4	Antiferromagnetic		-153
Hematite	αFe_2O_3	Canted antiferromagnetic	≈2.5	675
Ilmenite	$FeTiO_3$	Antiferromagnetic		-233
Maghemite	γFe_2O_3	Ferrimagnetic	380	590–675
Pyrrhotite	$Fe_{1-x}S$ ($0 < x \le$ ⅛)	Ferrimagnetic	≈80	320
Greigite	Fe_3S_4	Ferrimagnetic	125	≈330
Goethite	$\alpha FeOOH$	Antiferromagnetic with defect ferromagnetism	≈2	120
Iron	Fe	Ferromagnetic	1715	765
Cobalt	Co	Ferromagnetic	1422	1131
Nickel	Ni	Ferromagnetic	484	358

2.2.2 Titanomagnetites

Titanomagnetites (generally referred to as TM) are cubic minerals within the magnetite (Fe_3O_4) – ulvöspinel (Fe_2TiO_3) series as shown in Fig. 2.6. They are members of the spinel group (inverse type). As with all spinels, the cations are located in two lattices *A* and *B*, the *A* sites in fourfold co-ordination with oxygen ions and the *B* sites in sixfold co-ordination. There are two *B* cations for each *A* cation so that the two interacting sublattices are unequal giving rise to the observed ferrimagnetism (Fig. 2.7). In normal spinels the divalent metal ion occupies the A sites, whereas the two trivalent metal ions occupy the B sites. In inverse spinels the divalent ion and one trivalent ion exchange places. There is complete solid solution at temperatures in excess of 600°C and the ionic replacement in the solid solution series takes the form

$$2Fe^{3+} \Leftrightarrow Fe^{2+} + Ti^{4+}. \qquad (2.2.1)$$

The generalized formula for titanomagnetites is usually given by the relation $xFe_2TiO_4(1-x)Fe_3O_4$ or the equivalent $Fe_{3-x}Ti_xO_4$, where x $(0 < x \leq 1)$ is the composition parameter giving the fraction of ulvöspinel represented at the point in the series or the mole fraction of Ti^{4+}. Expressing x as a percentage, the titanomagnetite composition is usually stated to lie between TM0 (magnetite with $x = 0$) and TM100 (ulvöspinel with $x = 100\%$).

The Curie point of magnetite is 580°C and it has cell dimension $a = 8.396$ Å. It is strongly magnetic with a spontaneous magnetization of 480×10^3 Am^{-1} (see Fig. 2.3 for the variation of M_s with temperature). As the proportion of ulvöspinel increases, the cell dimension increases and the Curie point decreases in a regular manner. Ulvöspinel has cell dimension $a = 8.535$ Å and is paramagnetic at room temperatures and antiferromagnetic at low temperatures with a Néel temperature of -153°C (see Table 2.2). When magnetite is cooled to about 120 K (-153°C) the unit cell is distorted slightly from cubic to monoclinic symmetry. The temperature at which the transition between these states takes place is known as the *Verwey transition temperature*, below which magnetite is an electrical insulator and above which it becomes a semiconductor. In titanomagnetites the Verwey transition is suppressed for compositions with $x > 0.1$. It will be shown later that the magnetic properties that depend on crystalline anisotropy, such as remanence, susceptibility, and coercive force (see

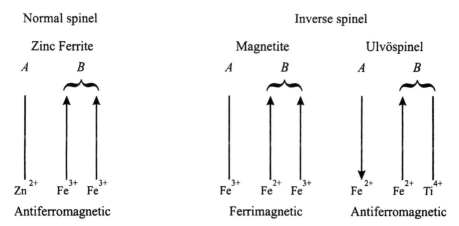

Fig. 2.7. Magnetization vectors in the two lattices A and B in normal and inverse spinels.

§2.3.2), change abruptly around the Verwey transition temperature and this property can be useful as a means of detecting magnetite in rocks (see §3.5.5).

High-Temperature (Deuteric) Oxidation

Although there is complete solid solution of the titanomagnetites at high temperatures, intermediate compositions can only be preserved as single-phase minerals at room temperature if they are cooled rapidly. For example, oceanic basalts are quenched by extrusion into seawater and their primary magnetic oxides are single-phase TM60 grains. If the basalts cool more slowly they will generally not have single-phase TM grains because in nature there is usually sufficient oxygen present to produce high-temperature (or deuteric) oxidation. The oxidation takes place at temperatures of 600–1000°C during initial cooling and is not due to weathering. See Haggerty (1976) for a discussion of this topic.

Oxidation of the TM series progresses toward the titanohematite series along lines parallel to the base of the ternary diagram of Fig. 2.6. This progressive oxidation is illustrated in the photomicrographs shown in Fig. 2.8. On cooling, the typical unoxidized TM grain (Fig. 2.8a) is restricted to compositions near the TM series end members, proceeding first with the production of ilmenite lamellae (Fig. 2.8b) within (titano)magnetite through the general reaction

$$6Fe_2TiO_4 + O_2 \Rightarrow 6FeTiO_3 + 2Fe_3O_4 \qquad (2.2.2)$$
$$\text{(ulvöspinel)} \qquad \text{(ilmenite) (magnetite)}$$

and then progressing towards the pseudobrookite series (Figs. 2.8c–2.8f) until ultimately only pseudobrookite is seen in association with (titano)hematite (the highest oxidation state shown in Fig. 2.8f) along the general lines

$$2FeTiO_3 + 2Fe_3O_4 + O_2 \Rightarrow 2Fe_2TiO_5 + 2Fe_2O_3. \qquad (2.2.3)$$
$$\text{(ilmenite) (magnetite)} \qquad \text{(pseudobrookite) (hematite)}$$

Fig. 2.8. Progressive oxidation of magnetic minerals in rocks. Photomicrographs were taken using an oil immersion objective with the scale line showing 25 μm. From Wilson and Haggerty (1966).
a–f. Progressive high-temperature oxidation of titanomagnetite in basalts.
a. A brown homogeneous titanomagnetite grain with sharp well-formed crystal faces. Titanomagnetite has a face-centered cubic structure.
b. A well-shaped titanomagnetite grain with coarse ilmenite lamellae along [111] planes. The fine needles are short lamellae of spinel exsolved along the [100] cube faces.
c. The original broad ilmenite lamellae have been oxidized to pseudobrookite (gray) and a submicroscopic intergrowth ("meta-ilmenite") of rutile and titanohematite (yellow). Brown titanomagnetite again contains dark spinel rods.
d. The smaller dark brown titanomagnetite areas have been enveloped by the streaky rutile–titanohematite intergrowth. The larger titanomagnetite areas still survive.
e. Development of pseudobrookite (gray) and titanohematite (brightish yellow) from the "meta-ilmenite" of c. Titanomagnetite contains well-formed and abundant dark exsolved spinel rods.
f. This represents the highest oxidation state in the titanomagnetite series. The pseudobrookite (gray) had been redistributed into fewer, more irregular areas in a host of titanohematite.

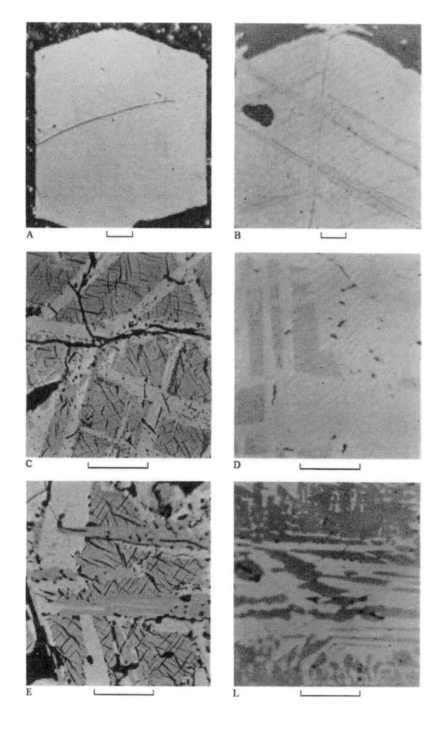

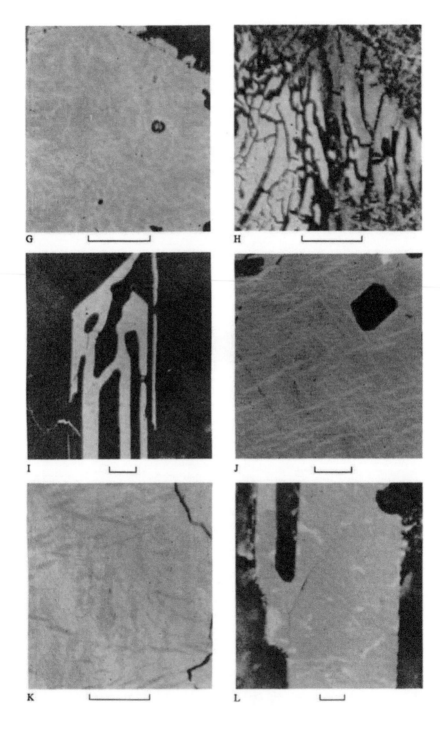

G   H

I   J

K   L

TABLE 2.3

High-Temperature (Deuteric) Oxidation Classification Scheme from Wilson and Watkins (1967)
and Watkins and Haggerty (1967)

Class	Observed characteristics
I	Homogeneous (single-phase) titanomagnetites
II	Titanomagnetites contain a few exsolved ilmenite lamellae in [111] planes
III	Abundant ilmenite lamellae with equilibrium two-phase intergrowths
IV	Ilmenite lamellae oxidized to rutile + hematite
V	Residual titanomagnetite and ilmenite oxidized to rutile + hematite
VI	Total oxidation to pseudobrookite and hematite and/or rutile

A useful empirical classification scheme for high-temperature oxidation has been developed by Wilson and Watkins (1967) and Watkins and Haggerty (1967) and is given in Table 2.3. Here the oxidation classes I–VI follow very closely the states shown successively in Figs. 2.8a–2.8f. The *oxidation index* of any sample is then defined as the weighted average of the observed oxidation states of several hundred opaque mineral grains. Watkins and Haggerty (1967) found that the highest oxidation states appear to be achieved only in very thick lava flows. Basalts with a distinctive reddish color are generally indicative of a high oxidation state.

Low-Temperature Oxidation: Titanomaghemites

Maghemite (γFe_2O_3) is the ultimate low-temperature oxidation or weathering product of magnetite. It has an inverse spinel structure similar to magnetite but has a defective lattice with one ninth of the Fe positions in the lattice being vacant with cell dimension $a = 8.337\text{Å}$. It is unstable and when heated inverts to rhombohedral hematite (αFe_2O_3) at temperatures between 250 and 750°C. The Curie temperature has only been determined indirectly since it often lies above the inversion temperature, but is estimated to be in the range 590-675°C (Table 2.2).

---

g and h. Low-temperature oxidation (maghemitization) and replacement of titanomagnetite.
g. Fine vermicular replacement of titanomagnetite (brown) by titanomaghemite (lighter color). Note renewed growth along the sharp upper edge after original crystal had become well formed.
h. The bright area is totally maghemitized. The cracks are typical at this stage. In the darker cracked area the titanomaghemite has been replaced by a fine granular, amorphous iron–titanium oxide.
i–l. Progressive high-temperature oxidation in discrete ilmenite grains.
i. Homogeneous skeletal ilmenite in the lowest oxidation state. Ilmenite crystallizes in the rhombohedral system and is typically long and lath shaped.
j. Development of fine light-colored ferrirutile blades in ilmenite. The blades have grown in the [0112] and [0001] planes.
k. Sigmoidal (or leaf-textured) rutile in titanohematite, completely replacing the original ilmenite.
l. Pseudobrookite with relic undigested hematite (bright blebs) and rutile (less bright elongated blebs) completely replacing the original ilmenite. This represents the highest oxidation state in ilmenite.

A *partially oxidized* titanomagnetite is usually referred to as titanomaghemite. Note that this is a different usage of "maghemite" from pure maghemite, which is the *fully oxidized* end member of magnetite oxidation. During low temperature oxidation the bulk composition of titanomagnetite grains follow the oxidation lines shown in Fig. 2.9, which shows the titanomaghemite field in terms of the oxidation parameter z ($0 < z \leq 1$) and the composition parameter x ($0 < x \leq 1$). The low-temperature oxidation of titanomagnetites (sometimes also referred to as *maghemitization*) converts a single phase spinel to another single phase spinel with a different structure. Note that this is different from high temperature oxidation, which results in intergrowths of spinel (near magnetite) and rhombohedral (near ilmenite) phases. Two stages of low temperature oxidation of titanomagnetite are shown in Figs. 2.8g–2.8h. Because the lattice parameters of the stochiometric titanomagnetite and the cation deficient titanomaghemite are so different, the oxidized surface is strained and frequently cracks (Fig. 2.8h). Ade-Hall *et al.* (1971) have shown how each of the deuteric oxidation states shown in Table 2.3 may change as a result of regional hydrothermal alteration.

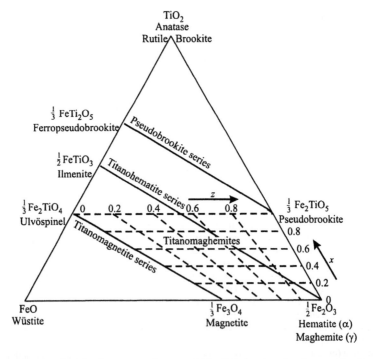

Fig. 2.9. The FeO-TiO$_2$-Fe$_2$O$_3$ ternary diagram of Fig. 2.6 showing the titanomaghemite field. During low temperature oxidation the bulk composition follows the horizontal dashed lines, where z is the oxidation parameter ($0<z\leq1$) and x is the titanomagnetite composition parameter ($0<x\leq1$). After Dunlop (1990).

Johnson and Hall (1978) developed a low-temperature oxidation classification scheme (with stages 1-5) for submarine basalts along the lines of those shown in Table 2.3.

2.2.3 Titanohematites

Titanohematites (sometimes also referred to as hemoilmenites) are rhombohedral minerals within the hematite (αFe_2O_3) – ilmenite ($FeTiO_3$) series as shown in Fig. 2.6. There is complete solid solution above 1050°C and the ionic replacement in the solid solution series follows that for the titanomagnetite series given in (2.2.1). The generalized formula for titanohematites is usually given as $yFeTiO_3(1-y)Fe_2O_3$ or the equivalent $Fe_{2-y}Ti_yO_3$, where y is the composition parameter giving the fraction of ilmenite represented in the series or the mole fraction of Ti^{4+}.

Hematite is rhombohedral with hexagonal unit cell dimensions $a_0 = 5.0345$ Å and $c_0 = 13.749$ Å. The lattice is made up of layers of cations in sixfold co-ordination parallel to the triad axis. It is a canted antiferromagnetic as illustrated in Fig. 2.4. The oppositely magnetized Fe^{3+} ions in the two sublattices A and B are canted at a small angle ($\approx 0.2°$) and this gives rise to a weak spontaneous magnetization of about 2.5×10^3 Am^{-1} (Table 2.2), only about 0.5% of that of magnetite (see Fig. 2.3 for variation of M_s with temperature). This *spin canting* phenomenon is often referred to as *parasitic ferromagnetism* because the ferromagnetism originates in and is a small fraction of the antiferromagnetism. The Curie temperature of 675°C coincides with the Néel temperature at which the antiferromagnetism disappears. As the proportion of ilmenite increases the cell dimensions increase and the Curie temperature decreases in a regular manner. Ilmenite is antiferromagnetic and has cell dimensions $a_0 = 5.0881$ Å and $c_0 = 14.080$ Å and a Néel temperature of -223°C. When hematite is cooled below about -15°C (the *Morin transition*), the intrinsic weak ferromagnetism disappears. As in the case of the Verwey transition in magnetite, this property can be useful as a means of detecting hematite in rocks (see §3.5.5).

In addition to the spin-canted ferromagnetism there is also an underlying *isotropic ferromagnetism* in hematite that probably arises from impurities or lattice defects and is often called the *defect moment*. This defect ferromagnetism is observable below the Morin transition (-15°C) and up to the Curie temperature. Because defect ferromagnetism is sensitive to structure, it can be altered by stress or heating and could thus provide spurious paleomagnetic information. However, the defect remanence of fine-grained hematite, such as occurs in red sediments, is magnetically softer than the spin-canted remanence and can be erased by partial demagnetization (see §3.4.2).

Although there is complete solid solution of the titanohematites only at high temperatures, intermediate single-phase compositions ($y \approx 0.5$–0.7) can only be preserved by rapid cooling, such as in the case of dacite pyroclastic rocks. These rocks have the property of acquiring self-reversed thermoremanence as will be discussed more fully in §4.1.2. In general, more slowly cooled rocks will contain exsolution intergrowths near the end members of the series (ilmenite and hematite). For titanohematites with compositions in the range $0 \leq y < 0.5$, the cation distribution is disordered and they are essentially antiferromagnetic with a weak parasitic ferromagnetism as in hematite. In the range $0.5 \leq y < 1$, the cation distribution becomes ordered and the titanohematite is ferrimagnetic with the maximum value of the spontaneous magnetization occurring around $y = 0.7$. However, in the range $0.5 \leq y \leq 0.7$ the Curie temperature lies between 200 and 20°C respectively and for $y > 0.7$ it is below room temperature.

Since pure hematite is already in its highest oxidation state, progressive high-temperature oxidation is observed in ilmenite, the other end member of the series. Further details can be found in Haggerty (1976). The progressive high-temperature oxidation in discrete ilmenite grains is illustrated in Figs. 2.8i–2.8l. In the highest oxidation state pseudobrookite and rutile completely replace the original ilmenite through the reaction

$$4FeTiO_3 + O_2 \Rightarrow 2Fe_2TiO_5 + 2TiO_2. \tag{2.2.4}$$
$$\text{(ilmenite)} \quad \text{(pseudobrookite) (rutile)}$$

Hematite can be formed by deuteric high temperature oxidation of titanomagnetite during cooling or through the inversion of titanomaghemite during later reheating. It also forms as the end product of the prolonged oxidation of magnetite at room temperature. In this form, pseudomorphing the original magnetite crystals, it is called *martite*. Hematite is also formed through other important secondary processes including the inversion of maghemite or the dehydration of weathering products such as goethite (see §2.2.4). In red sediments it is usually observed that hematite occurs in two forms: either as fine-grained ($<\approx 1$ μm), red pigment or cement that has been precipitated from iron-rich solutions in the pore spaces of clastic sediments giving redbeds their distinctive color or as large grains (usually >10 μm) of detrital origin often referred to as *specularite*. The relative importance of these two forms of hematite is discussed more fully in §3.4.1.

2.2.4 Iron Sulfides and Oxyhydroxides

Iron Sulfides

Until recently greigite (Fe_3S_4) was thought to be a rare mineral but it is now known to occur quite commonly in sediments formed under anoxic conditions. In the iron reduction zone in deep-sea sediments, fine-grained magnetite and other oxides tend to dissolve and reform as sulfides, especially pyrite (FeS_2). As a result, anoxic sediments are often basically nonmagnetic. However, under sulfate-reducing conditions that can occur in muds and some rapidly deposited deep-sea sediments, the magnetic sulfides greigite and pyrrhotite may be preserved. Greigite is the sulfide counterpart of magnetite and has the same inverse spinel structure (Fig. 2.7). It is ferrimagnetic with spontaneous magnetization of about 125×10^3 Am^{-1}, about 25% of that of magnetite, with Curie point of about 330°C (Table 2.2).

Pyrrhotite ($Fe_{1-x}S$, $0 < x < \frac{1}{8}$) is a reasonably common accessory mineral in rocks, especially igneous rocks formed from sulfur-rich magmas, and occurs frequently as a secondary mineral in deep-sea sediments. Natural pyrrhotite is actually a mixture of monoclinic Fe_7S_8, which is ferrimagnetic, and antiferromagnetic hexagonal phases such as Fe_9S_{10} and $Fe_{11}S_{12}$. The deficiency of Fe^{2+} gives rise to vacancies in the lattice so that monoclinic pyrrhotite is a cation-deficient ferrimagnetic like maghemite. It has spontaneous magnetization of about 80×10^3 Am^{-1} and Curie point of 320°C (Table 2.2). Above 500°C it transforms irreversibly, usually to magnetite, and at higher temperatures transforms to hematite either directly or by oxidation of magnetite.

Iron Oxyhydroxides

Iron oxyhydroxides form as weathering products that are often collectively called limonite. Orthorhombic goethite ($\alpha FeOOH$) is the most important of these minerals and is a common constituent of soils and sediments. Goethite is antiferromagnetic with a Néel temperature of 120°C, but it has a weak superimposed parasitic ferromagnetism that probably arises from a defect moment as in hematite, with which it is often intergrown. The Curie temperature of this defect moment coincides with the Néel temperature. Its spontaneous magnetization of about 2×10^3 Am^{-1} is less than that of hematite. On heating, goethite dehydrates at temperatures in the range 250–400°C to form hematite as

$$2\alpha FeOOH \Rightarrow \alpha Fe_2O_3 + H_2O. \tag{2.2.5}$$

Lepidocrocite ($\gamma FeOOH$) is only a minor constituent of soils and sediments and is antiferromagnetic with a Néel temperature well below room temperature. However, although it cannot carry any remanence, it can be important because it dehydrates to strongly magnetic maghemite when heated above 250°C as

$$2\gamma FeOOH \Rightarrow \gamma Fe_2O_3 + H_2O. \tag{2.2.6}$$

On further heating, at around 400°C, the maghemite will invert to hematite. Therefore, on heating samples containing lepidocrocite, the possible magnetic effects can be strange in that an initially nonmagnetic mineral first inverts to a strong magnetic mineral and finally to a weak magnetic mineral.

2.3 Physical Theory of Rock Magnetism

2.3.1 Magnetic Domains

When the magnetization of a body produces an external field (i.e., it exhibits a remanence), it has *magnetostatic energy* or *energy of self-demagnetization*. This arises from the shape of the body, because it is more easily magnetized in some directions than in others. The internal field tends to oppose the magnetization and is referred to as a demagnetizing field. Two extreme examples illustrate this point in Fig. 2.10. It is easier to magnetize a long thin rod along its length than across it. When magnetized along its length (Fig. 2.10a) elementary magnets can be imagined to be lined up inside so that the north pole of one lies next to the south pole of its neighbor and the magnetization is aided by their mutual attraction. All the "+" and "-" signs cancel out inside the material, but there is a "bound" magnetic "charge" at the boundaries which produces the demagnetizing field. With magnetization along the rod, the bound magnetic charge is small with the charges far apart so the demagnetizing field (and therefore the magnetostatic energy) is small. To magnetize the rod at right angles to the axis (Fig. 2.10b) requires that the elementary magnets be lined up so that north poles of neighbors are adjacent as are their south poles. Consequently, the "+" and "-" signs do not cancel and there is a large bound magnetic charge, with the charges close together. Thus, the demagnetizing field, and hence the magnetostatic energy, is large. If there were no exchange energy, the elementary magnets would rearrange themselves so that each alternate one pointed in the opposite direction (Fig. 2.10c) so that the magnetostatic energy would be zero.

The demagnetizing field is proportional to a factor called the *demagnetizing factor*, N, which is related to the shape of the body. For an arbitrarily shaped body it is usually extremely difficult to calculate the demagnetizing factors. However, for some simple shapes, such as ellipsoids, the factors are relatively easy to calculate. If N_x, N_y, and N_z are the demagnetizing factors along the three principal axes of the ellipsoid then they are related by the general relationship

$$N_x + N_y + N_z = 1. \tag{2.3.1}$$

A long thin rod (a greatly elongated ellipsoid) has $N_z = 0$ along its long axis and thus $N_x = N_y = \frac{1}{2}$ across the rod. Across a large flat plate (which is just a greatly flattened ellipsoid) $N_x = N_y = 0$ in the plane of the plate so $N_z = 1$ across it. For a sphere or a cube $N_x = N_y = N_z = \frac{1}{3}$. In the general case an ellipsoid of volume V, magnetized with magnetization M, has a magnetostatic energy, E_m, given by

$$E_m = \tfrac{1}{2}\mu_0 N V M^2, \tag{2.3.2}$$

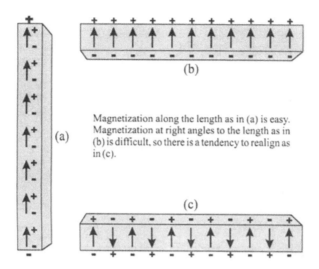

Fig. 2.10. To illustrate the ease with which a rod can be magnetized along its length compared with at right angles to its length.

where N is the demagnetizing factor in the direction of the magnetization M. Except for a sphere or a cube, where the demagnetizing factor is the same in all directions, the general situation is that it will be easiest to magnetize a body along its long axis. This effect is called *shape anisotropy*.

Suppose a ferromagnetic grain is magnetized to saturation as in Fig. 2.11a. There is a large magnetostatic energy associated with the grain. The surface poles create an internal magnetic field, H_d (in the direction opposite to that of M_S), which is usually expressed in the form

$$H_d = -NM_s, \qquad (2.3.3)$$

where N is the demagnetizing factor. If the grain is subdivided into two oppositely magnetized regions as in Fig. 2.11b, the internal field, and consequently the magnetostatic energy, is decreased. However, a boundary or wall must be formed between the two oppositely magnetized regions and magnetic energy (the wall energy) is stored in this wall. The system assumes the state of lowest total energy. The process of subdivision will continue as shown in Fig. 2.11c until the energy required for the formation of an additional boundary is greater then the consequent reduction in magnetostatic energy. The subdivided regions are called *magnetic domains* and the boundaries between them are called *domain walls*.

The change in the direction of magnetization between one domain and the next does not occur abruptly across a single atomic plane. Domain walls can have considerable variation in thickness depending on the substance and type of wall.

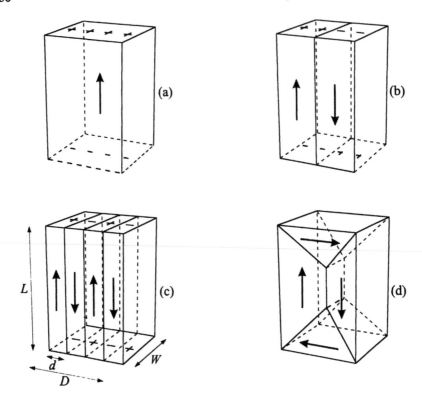

Fig. 2.11. Subdivision of a ferromagnetic grain into domains. (a) Single-domain structure with widely separated + and − poles. (b) Two-domain structure with less pole separation. (c) Four-domain state. (d) Two-domain state with closure domains. Redrawn after Dunlop and Özdemir (1997), with the permission of Cambridge University Press.

For magnetite the wall width has been calculated to be 0.28 μm or about 300 lattice spacings (Dunlop and Özdemir, 1997) and has been determined experimentally as 0.18 μm (Moskowitz *et al.*, 1988). The corresponding wall energy is approximately 10^{-3} Jm^{-2}.

The grain shown in Fig. 2.11a, where no domains occur, is referred to as a *single-domain grain* (SD grain) and at some critical size the grain will subdivide into two or more domains to form a *multidomain grain* (MD grain). Note that it is usually energetically more favorable to subdivide the grain by a wall parallel to the long axis of the grain. Another way of reducing or eliminating the magnetostatic energy is by adding closure domains to the two-domain state (Fig. 2.11d). There are now no surface poles and no external field and the wall energy is also reduced when compared with that of the four-domain state. However, the closure domains generate additional magnetoelastic energy as described later. The body and closure domains in Fig. 2.11d are strained and both are shorter than they would be in isolation.

In addition to the magnetostatic energy of a grain, arising from its shape anisotropy (i.e., it is easier to magnetize along the long axis than in other directions), there is energy from *magnetocrystalline anisotropy* that arises because it is easier for the domain magnetizations to lie along certain crystallographic axes than along others. For example, in a magnetite crystal the easy direction of magnetization is along the [111] axis and the difficult or hard direction is along the [100] axis. The magnetocrystalline anisotropy energy is the difference in magnetization energy between the hard and easy directions. Obviously, the contribution to the total energy will be a minimum when the various domain magnetizations all lie along easy directions.

The magnetization of a ferromagnetic or ferrimagnetic crystal is usually accompanied by a spontaneous change in the dimensions of the crystal, giving rise to *magnetostriction*, which is due to strain arising from magnetic interaction along the atoms forming the crystal lattice. In addition, the presence of some impurity in the crystal lattice, or the presence of dislocations, will produce internal stress that then acts as a barrier to changes in magnetization. Strain occurs when closure domains are formed as in Fig. 2.11d. The strain dependence of crystalline anisotropy is termed magnetoelasticity giving rise to *magnetoelastic energy* or *magnetostrictive strain energy*.

2.3.2 Theory for Single-Domain Grains

The theory of the magnetization of an assemblage of SD particles is due to Néel (1949, 1955). Although it appears to have wide applicability, it is based on the assumption that the grains are identical and that there are no grain interactions. These assumptions can obviously lead to shortcomings in various aspects of the theory, but the essential features of the behavior of magnetic grains over the geological time scale can be adequately described in terms of this simple theory. Consideration of grain interactions has led to the Preisach–Néel theory, which has been developed by Dunlop and West (1969), but discussion of the details of this theory is beyond the scope of this book (see also Dunlop and Özdemir, 1997).

Imagine a set of identical grains with uniaxial symmetry; that is, the magnetic moment of an individual grain may be oriented in either direction along its axis of symmetry but not in any other direction. The axes of the grains are randomly oriented so that a specimen may have zero magnetic moment if the magnetizations are directed so as to cancel out one another. On application of a magnetic field in any direction, the specimen acquires a magnetic moment because the individual grain magnetizations will be in whichever of the two directions along their axis of symmetry has a component in the direction of the external field. Thus, although individual magnetic grains are themselves magnetically anisotropic, a random assemblage of grains making up the specimen is magnetically isotropic.

The magnetic behavior of one of these grains depends on its orientation with respect to the applied field. When the field is parallel to the axis of the grain a rectangular hysteresis loop results as in Fig. 2.12a. The height is twice the saturation magnetization M_s, and the width is twice the *microscopic coercivity* $B'_c (= \mu_0 H_c = 2K/M_S$, where K is the anisotropy constant, see (2.3.10) below). At $B = +B'_c$ and $B = -B'_c$ there are discontinuities in the magnetization. At the other extreme, when the axis of the grain is perpendicular to the applied field, there is no hysteresis (Fig. 2.12b). For $B > B'_c$ and $B < -B'_c$, the magnetization is $+M_s$ and $-M_s$ respectively. As B changes from $-B'_c$ to $+B'_c$, then M_s varies linearly from $-M_s$ to $+M_s$. In a randomly oriented assemblage of grains, the average limiting hysteresis cycle is as shown in Fig. 2.12c. There is a remanence $M_r = 0.5M_s$ and a $B_c \approx 0.5B'_c$. Note that the microscopic coercivity B'_c of a single grain should be distinguished from the bulk coercivity B_c defined by the hysteresis of bulk materials as in §2.1.4.

In an SD grain the internal magnetization energy depends only on the orientation of the magnetic moment with respect to certain axes in the grain. For magnetically uniaxial grains, the energy E is given by

$$E = K v \sin^2\theta, \tag{2.3.4}$$

where v is the volume of the grain and θ is the angle between the magnetic moment and the axis. K is called the *anisotropy constant* and can arise from three factors that contribute to the magnetic anisotropy of an SD grain: shape anisotropy, magnetocrystalline anisotropy and magnetostrictive (or stress) anisotropy (see §2.3.1). The microscopic coercivity B'_c is simply related to the

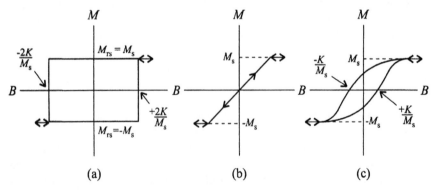

(a)   (b)   (c)

Fig. 2.12. Hysteresis loops of single domain particles. (a) Square loop produced when B and M are confined to the same axis. (b) No hysteresis produced if B and M are perpendicular. (c) Hysteresis loop produced by a set of randomly oriented grains in which the remanence $M_r = \frac{1}{2} M_s$.

anisotropy constant by the relation

$$B'_c = \mu_0 H'_c = \frac{2K}{M_s}. \qquad (2.3.5)$$

The anisotropy constant and microscopic coercivity for the three types are as follows.

Shape:
$$K = 0.5\mu_0(N_b - N_a)M_s^2, \quad B'_c = \mu_0(N_b - N_a)M_s, \qquad (2.3.6)$$

where N_b and N_a are the demagnetizing factors along the equatorial and polar axes of the prolate spheroid respectively.

Magnetocrystalline:
$$K = K_1, \qquad B'_c = \frac{2K_1}{M_s}, \qquad (2.3.7)$$

where K_1 is the first magnetocrystalline anisotropy constant.

Stress:
$$K = \tfrac{3}{2}\lambda\sigma, \qquad B'_c = \frac{3\lambda\sigma}{2M_s}, \qquad (2.3.8)$$

where λ is the average magnetostrictive coefficient and σ is the internal stress amplitude.

It should be noted that the values of M_s for magnetite (480×10^3 Am^{-1}) and hematite (2.2×10^3 Am^{-1}) are widely different and this means that different forms of anisotropy are important in the two cases. For hematite it is obvious that shape anisotropy is of no significance compared with either magneto-crystalline or stress-induced anisotropy. Hematite has high coercivities that are probably due to stress-induced anisotropy. Then $\lambda = 8 \times 10^{-6}$ and for an internal stress of $\sigma = 100$ MPa (1kb), $B'_c = 500$ mT. For magnetite the effect of mechanical stress is minor since a uniaxial stress of $\sigma = 10$ MPa (0.1 kb), which is close to the breaking strength, only gives $B'_c = 2$ mT with $\lambda = 36 \times 10^{-6}$. In the case of magnetocrystalline anisotropy, $K_1 = 1.35 \times 10^4$ Jm^{-3}, so this cannot give rise to coercivities in excess of $B'_c = 60$ mT. Shape anisotropy, however, can give rise to coercivities considerably greater than this. The theoretical maximum (for infinitely long needles) is given by $B'_c = \tfrac{1}{2}\mu_0 M_s \approx 300$ mT).

The magnetic susceptibility χ_s (initial susceptibility) of a random assemblage of SD grains (Fig. 2.12c) in which shape anisotropy predominates is given by

$$\chi_s = \frac{\mu_0 M_s^2}{3K} = \tfrac{2}{3}(N_b - N_a). \qquad (2.3.9)$$

For magnetite, χ_s therefore lies between 1.33 (infinite needles) and about 10, although this upper limit is really set by magnetocrystalline anisotropy for equidimensional grains and this can give values of χ_s as high as about 20. For hematite χ_s is four orders of magnitude smaller, with the very much lower value of M_s and the dominance of stress-induced anisotropy giving $\chi_s \approx 10^{-3}$. Model MD grains have susceptibility $1/N$ parallel to the domain magnetization and an SD-like susceptibility perpendicular. For the complicated case of many nonlinear walls, Stacey (1963) and O'Reilly (1984) give the susceptibility χ_m simply as

$$\chi_m = \frac{1}{N} .$$ (2.3.10)

2.3.3 Magnetic Viscosity

In the absence of an applied field, the magnetic moment of a uniaxial SD grain can take up two orientations of equal minimum energy, $\theta = 0$ or $\theta = 180°$ from (2.3.4). The potential barrier between these two positions is represented by the positions of maximum energy that occur at $\pm 90°$, and from (2.3.4) has value E_r given by

$$E_r = vK$$ (2.3.11)

The thermal fluctuations of energy E_t, given by

$$E_t = kT ,$$ (2.3.12)

where k is Boltzmann's constant (1.38×10^{-23} Jm^{-3}) and T is the absolute temperature, are capable of moving the magnetic moment from one minimum to the other if $E_t > E_r$, i.e., if $kT > vK$. For a given value of T there must always be some grains of volume v for which the thermal fluctuations are large enough to cause the moment to change spontaneously from one position to the other. If an assemblage of identical grains has an initial moment M_0, then under these conditions it will decay exponentially to zero according to the relation

$$M_t = M_0 \exp(-t/\tau),$$ (2.3.13)

where M_t is the moment remaining after time t and τ is the *relaxation time* of the grains. The initial moment M_0 will thus have decayed to one-half of its value ($M_t = \frac{1}{2}M_0$) after time $t = 0.693\tau$, which can be thought of as the "half-life" of the initial remanence.

The relaxation time is related to the ratio of the two energies E_r and E_t by the equation

$$\tau = \frac{1}{f} \exp\left(\frac{E_r}{E_t}\right)$$

$$= \frac{1}{f} \exp\left(\frac{vK}{kT}\right)_T \qquad (2.3.14)$$

where $f = 10^9 \, s^{-1}$ (Moskowitz *et al.*, 1997) is the frequency of successive thermal excitations and K is the value of the anisotropy constant at temperature T. From (2.3.5) the anisotropy constant K is related to the microscopic coercivity B'_c. Therefore, the relaxation time τ may alternatively be given as

$$\tau = \frac{1}{f} \exp\left(\frac{vB'_c M_s}{2kT}\right)_T . \qquad (2.3.15)$$

where B'_c and M_s are their values at temperature T. Note that τ is thus directly related to the microscopic coercivity. This has important implications relating to the magnetic stability of rocks and forms the basis of the method of "magnetic cleaning"(§3.4.1).

When the relaxation time is small, say 100 seconds, the magnetization acquired by an assemblage of grains will be lost almost as soon as it has been acquired. The grains are rendered unstable by thermal agitation and, on application of a weak field, they quickly reach equilibrium with this field. The moment so acquired is called the *equilibrium magnetization*, which, on removal of the applied field, quickly dies away at a rate determined by the relaxation time. Grains such as these are said to be *superparamagnetic* and are referred to here as SP grains. The relaxation time according to (2.3.14) becomes small when T is large (i.e., at higher temperatures) and also when v is small (small grain size). For each grain of volume v there is thus a *critical blocking temperature*, T_B, at which τ becomes small (say 100 s), but which might also be below the Curie temperature. Similarly at any given temperature T, there is a *critical blocking volume*, v_B (corresponding to a sphere of diameter d_B), at which τ becomes small.

For the moment it is convenient to neglect changes with temperature of the anisotropy constant K, although this will be considered in more detail in §2.3.8. The relaxation time τ_1 at temperature T_1 is then simply related to the relaxation time τ_2 at temperature T_2 from (2.3.14)

$$T_1 \ln(f\tau_1) \approx T_2 \ln(f\tau_2) . \qquad (2.3.16)$$

Thus, the same effect upon the remanence is obtained either by maintaining at temperature T_1 for sufficient time τ_1 or by raising to a higher temperature T_2 and maintaining this for a shorter time τ_2. For example, putting $f = 10^9 \, s^{-1}$ (2.3.16) shows that maintaining at a temperature of 150°C for 10^6 years is equivalent to maintaining at 500°C for only 1000 s. These are only approximate values; this relationship is discussed more fully in §2.3.8, in which the temperature variation of K is taken into account in relation to the acquisition of VRM. It also has implications for the thermal demagnetization of rocks as discussed in §3.4.2.

2.3.4 Critical Size for Single-Domain Grains

The theory outlined above is for SD grains, but the magnetic behavior of SD and MD grains is quite different so it is of some importance to determine which configurations are relevant to the magnetic minerals in rocks. For SD grains the saturation remanence M_{rs} (see Fig. 2.5) would ideally be expected to have a value of 0.5 of the saturation magnetization M_s so that $M_{rs}/M_s \approx 0.5$ (§2.3.2). However, for two-domain (2D) or larger grains the ratio M_{rs}/M_s would be expected to fall abruptly beyond the SD critical size to the ratio of ≤ 0.1 expected for MD grains. The data in Fig. 2.13 show that there is no such observed abrupt fall. A similar effect would be expected in the ability of grains to acquire thermoremanent magnetization (TRM; see §2.3.5), but again there is no such observed effect. Stacey (1962) therefore proposed that there are MD grains, containing only a small number of domains, that act akin to SD grains. He termed these *pseudo-single-domain* (PSD) grains. It is now recognized that the PSD size range includes most of the magnetite or titanomagnetite carrying stable TRM in igneous rocks.

The mechanism of PSD behavior remains far from certain, but Stacey (1962) preferred *Barkhausen discreteness*. In MD grains the potential barriers opposing changes in remanence arise from crystal imperfections that cause local stresses and variations in spontaneous magnetization. The energy of a domain wall may thus be a minimum at several discrete positions only. Changes in magnetization then take place when the domain wall moves in jumps, so-called Barkhausen jumps, from one minimum in potential to the next. In a large MD grain the domain walls can generally find suitable positions so that the magnetostatic energy is zero. However, in small MD grains the Barkhausen discreteness of the positions of the domain walls prevents them from occupying the precise positions necessary to give the grain zero magnetic moment. In such a model the PSD moment is not independent of MD processes. For example, the moment can only reverse if domain walls are displaced. Conversely, typically they cannot be demagnetized. The behavior of such grains is, therefore, somewhere between that of SD and MD grains.

Recent models, based on the domain wall moment and on metastable SD grains, have superseded the Stacey model. However, these models are viable only over the lower part of the observed PSD range. A domain wall has a net moment perpendicular to the domain magnetization, with a choice of two

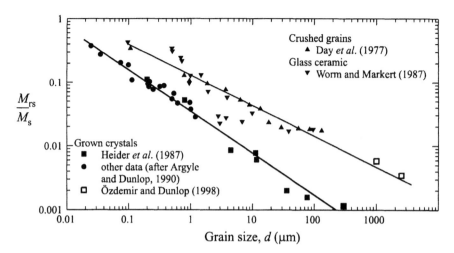

$\dfrac{M_{\mathrm{rs}}}{M_{\mathrm{s}}}$

Grain size, d (μm)

Fig. 2.13. Grain size dependence of the saturation remanence ratio $M_{\mathrm{rs}}/M_{\mathrm{s}}$ in magnetite grains. The continuous decrease observed is at odds with the abrupt decrease above the critical SD size predicted by multidomain theories. After Dunlop (1995).

orientations depending on the sense of spin rotation across the wall. Wall moments have truly SD behavior in that they are perpendicular to and can reverse independently of the domain magnetizations (Dunlop, 1977, 1981). Halgedahl and Fuller (1980, 1983) noted that the magnetic structure (e.g., SD or MD) depended not only on the size of the grain but also on its history. Thus, if the magnetic structure of a grain is derived rather than prescribed, there is usually no sharp SD to MD transition. Using this observation, Halgedahl and Fuller developed a PSD model on the idea that wall nucleation sites are randomly distributed and that nucleations are (in a statistical sense) rare events. This model predicts that the number of walls is Poisson distributed, which provides a good match with observation. This implies that metastable SD grains (i.e. no walls) account for most of the remanence. Moon and Merrill (1984, 1985) provided the first understanding using micromagnetic theory that a grain magnetic structure could depend on its history. It appears that there is an intermediate structure, referred to as a vortex state (Newell *et al.*, 1993), that is the lowest energy state for sizes slightly larger than SD but smaller than MD.

The domains illustrated in Fig. 2.11 show that, in the absence of an applied external field, the magnetostatic energy decreases as the number of domains n increases. In classical domain theory the critical size d_0 for the transition between the SD and 2D states arises when the decrease in magnetostatic energy exactly balances the wall energy, E_{w} (i.e., the energies of SD and 2D structures are equal). Dunlop and Özdemir (1997) give the useful approximation for the magnetostatic energy E_{nD} of a crystal with n domains in terms of its SD counterpart E_{SD} as

$$E_{nD} \approx \frac{1}{n} E_{SD} \qquad (2.3.17)$$

so that

$$E_{SD} = E_{2D} + E_w , \qquad (2.3.18)$$

where

$$E_{SD} = \frac{1}{2} \mu_0 N_{SD} M_s^2 d_0 LW = 2E_{2D} \qquad (2.3.19)$$

and

$$E_w = \gamma_w LW \qquad (2.3.20)$$

where γ_w is the wall energy per unit area (see Fig. 2.11 for definitions of L and W). Hence, combining (2.3.18), (2.3.19) and (2.3.20)

$$d_0 = \frac{4\gamma_w}{\mu_0 N_{SD} M_s^2} . \qquad (2.3.21)$$

Substituting values for magnetite of $\gamma_w = 10^{-3}$ Jm^{-2} and $M_s = 480 \times 10^3$ Am^{-1}, then $d_0 \approx 0.04$ μm for cubes in which $N_{SD} = 0.33$. This is only 50 lattice spacings and considerably less than the wall width for magnetite of ~0.1–0.2 μm. This apparent contradiction was first pointed out by Néel (1947). One would expect the critical SD size to be at least this large to accommodate the domain wall. This highlights the fact that the thin-wall model of Fig. 2.11 is inappropriate and (2.3.21) gives a significant underestimate of d_0. However, the structure of (2.3.21) provides interesting insight. Since the critical size $d_0 \propto 1/M_s^2$ and M_s decreases with increasing temperature, grains that are 2D at room temperature may transform to SD at high temperatures, particularly near the Curie point. For titanomagnetites, where the value of M_s decreases with increasing x, the critical SD size can be expected to be larger than that for pure magnetite. In the case of hematite $M_s = 2.2 \times 10^3$ Am^{-1} and so d_0 is much larger, with a value of 15 μm determined experimentally for equidimensional grains (Dunlop and Özdemir, 1997). So, for most purposes hematite grains can be considered to be SD grains.

In practice, grains are rarely equidimensional and elongated grains of magnetite, with much smaller N_{SD}, will support single domains to much larger sizes. Evans and McElhinny (1969) generalized the calculations to the case of prolate ellipsoids (N can be obtained from standard tables). They showed that elongated grains of magnetite of up to several micrometres in length could act as SD grains. Butler and Banerjee (1975) argued that the faceted crystals commonly found in rocks were better represented by cuboids (rectangular parallelepipeds) and recalculated the possible SD size range with results similar to those of Evans and McElhinny (1969). Newell and Merrill (1999) applied

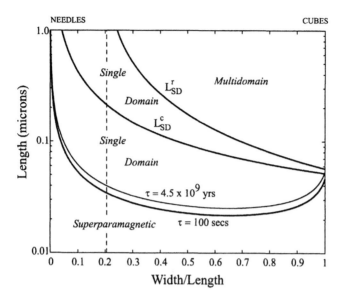

NEEDLES   CUBES

Fig. 2.14. Superparamagnetic and SD fields for rectangular parallelepiped magnetite grains. Compiled from Newell and Merrill (1999).

micromagnetic theory to the problem. In this theory the energy terms are the same as in domain theory, but the magnetization is allowed to vary continuously and the lowest energy state is then derived rather than assumed.

The calculations made by Newell and Merrill (1999) using micromagnetic theory include the effects of shape, crystallographic orientations, and stress. Their results are illustrated in Fig. 2.14 for the case of rectangular parallelepipeds of varying width/length ratio. There are two critical sizes, L_{SD}^c and L_{SD}^r, that correspond to maxima in coercivity and remanence respectively where L_{SD}^r is always larger than L_{SD}^c. The region of superparamagnetic behavior is also shown in Fig. 2.14, being defined for relaxation times $\tau \leq 100$ s. The boundary for grains with relaxation time 4×10^9 years lies close to the superparamagnetic boundary, so that the transition region is sharp. Figure 2.14 indicates that there is a wide range of shapes and sizes over which SD behavior can be expected in magnetite. The SD boundary for L_{SD}^r approaches infinity for particle elongations of 5:1 so that large, highly elongated particles of magnetite can have SD-like remanence.

Throughout almost the whole of the SD field in Fig. 2.14 the relaxation time is very much greater than the age of the Earth, so that these grains are of considerable importance to paleomagnetism, being capable of retaining their initial remanence over the whole of geological time. The presence of small magnetite grains in the appropriate size–shape range has been confirmed by

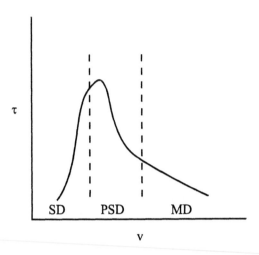

Fig. 2.15. Hypothetical change in relaxation time τ as a function of grain volume v, all other factors being held constant. SD, single domain; PSD, pseudo-single-domain; MD, multidomain. From Merrill and McElhinny (1983).

electron microscopy (Evans and Wayman, 1970, 1974). It should be noted that the deuteric oxidation of titanomagnetites can cause the growth of exsolution lamellae of ilmenite that effectively subdivides the crystal into sheet-like or rod-like subgrains of magnetite. These elongated subgrains can be of suitable size to be interacting SD particles with high shape anisotropy and high coercivity (Strangway *et al.*, 1968). Experimental work by Davis and Evans (1976) confirms that this is the case.

For SD grains the relaxation time increases with increase in volume v as indicated by (2.3.14). The way in which the relaxation time varies in the PSD and MD region is not well known, but theory and observation suggest the hypothetical variation with grain size given in Fig. 2.15. It appears to have a single maximum either in the SD size range or in the PSD size range.

2.3.5 Thermoremanent Magnetization

The remanence acquired by a rock specimen during cooling from the Curie point to room temperature is called the *total* TRM. Upon cooling from a high temperature, spontaneous magnetization appears at the Curie point T_c, and this assumes an equilibrium magnetization in the presence of an applied field. Grains of different volumes v will each have different blocking temperatures T_B. As the temperature cools below T_c and then passes through each T_B, the relaxation time of each of these grains increases very rapidly. The equilibrium magnetization becomes "frozen in" and subsequent changes in the field direction occurring at

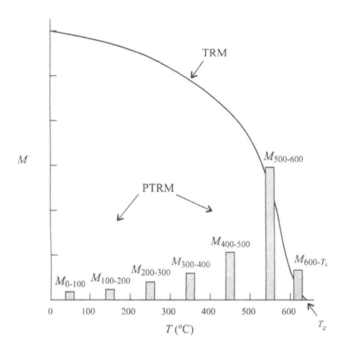

Fig. 2.16. The acquisition of TRM. The PTRMs acquired over successive temperatures add up to give the total TRM curve.

temperatures below T_B are ineffective in changing the direction of magnetization. From (2.3.16), using a value of $\tau = 100$ s at the blocking temperature, a grain with $T_B = 530°C$ (800K) has relaxation time of 10^{13} years on cooling to room temperature (30°C). Even grains with relatively low blocking temperatures of 330°C (600 K) have a relaxation time of 10^8 years on cooling to room temperature. This is one of the basic appeals of paleomagnetism; TRM is essentially stable on the geological time scale.

The TRM is not all acquired at the Curie point T_c, but over a range of blocking temperatures from the Curie point down to room temperature (Fig. 2.16). The total TRM may be considered to have been acquired in steps within successively cooler temperature intervals $(T_c\text{-}T_1)$, $(T_1\text{-}T_2)$, and so on. The fraction of the total TRM acquired in these temperature intervals is called the *partial* TRM for each temperature interval. The PTRM acquired in any interval is not affected by the field applied at subsequent intervals on cooling. Thus, the total TRM is equal to the sum of the PTRMs acquired in each consecutive temperature interval between the Curie temperature and room temperature (Fig. 2.16). This is known as the *law of additivity* of PTRM (Thellier, 1938). Conversely, on reheating to any temperature $T < T_c$, the original magnetization of all grains with blocking temperatures less than T is destroyed. This has important implications relating to the magnetic stability of rocks and forms the basis of the method of "thermal cleaning" (§3.4.1).

Néel (1949) considered the TRM of uniaxial grains of volume v aligned with the direction of an applied field $B = \mu_0 H$. He assumed that there is a blocking temperature (§2.3.4) above which the grain assembly is in equilibrium, but below which the magnetization is locked into the sample. The room temperature TRM is then given by

$$M_{TRM}^{SD} = M_s \tanh\left(\frac{vM_{sB}B}{kT_B}\right), \tag{2.3.22}$$

where M_s is the saturation magnetization at room temperature and M_{sB} that at the blocking temperature T_B. This is the basic equation for the theory of TRM. The general case of a randomly oriented assembly of grains can be solved for small fields (0–0.1 mT), as in the case of the Earth's field, and from Stacey (1963) is given by

$$M_{TRM}^{SD} = \frac{M_s vM_{sB}B}{3kT_B}. \tag{2.3.23}$$

In this case TRM is proportional to the applied field. The theory of course applies strictly only to noninteracting SD grains. It applies in this case to hematite grains, but only to equidimensional magnetite grains of up to about 0.2 μm and more elongated grains of up to a few micrometres. Néel's theory has been criticized as being too simplistic; however, more rigorous treatments give essentially the same results.

For true MD grains the theory becomes complex and readers are referred to Dunlop and Özdemir (1997) for a complete treatment. However, a simplified treatment by Stacey (1958) gives the relation

$$M_{TRM}^{MD} = \frac{M_s B}{\mu_0 NM_{sB}}. \tag{2.3.24}$$

This suggests that the TRM of MD grains is proportional to the applied field and is roughly independent of grain size.

Figure 2.17 shows experimental determinations of the grain-size dependence of weak-field (0.1 mT) TRM in magnetite. Néel's theory (2.3.23) predicts an increase in TRM as v increases. This is exactly as observed for grain sizes up to the critical equidimensional size d_0 for magnetite of about 0.1–0.2 μm (§2.3.4). Beyond the SD range, TRM decreases with increasing grain size to the value for true MD grains (2.3.24). In the Halgedahl and Fuller (1980, 1983) theory for PSD magnetization that might apply in the intermediate range, the TRM of

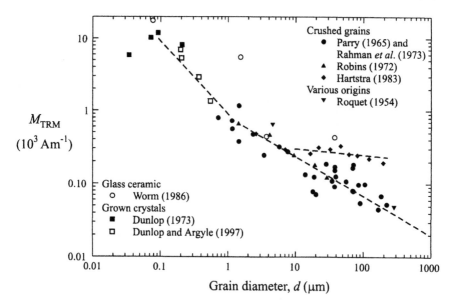

Fig. 2.17. The grain size dependence of the intensity of weak-field (0.1 mT) TRM in magnetite. After Dunlop and Argyle (1997).

metastable SD grains depends on the probability that no domain walls will have nucleated and therefore becomes a fraction of the SD remanence. For a given grain size some of the grains will have no domain walls and others will have differing numbers of walls with an *average* value of $\overline{w}$ for that grain size. The TRM for any given grain size will therefore be a combination of the TRM of the metastable SD grains (no domain walls) and the multidomain TRM from the remaining grains. Fuller (1984) suggests the following relationship

$$M_{TRM} = e^{-\overline{w}} M_{TRM}^{SD} + (1 - e^{-\overline{w}}) M_{TRM}^{MD} \quad . \tag{2.3.25}$$

For large values of $\overline{w}$ this gives the MD value and for $\overline{w} = 0$ the SD value. The appropriate grain size relationship of $\overline{w}$ for the TRM after weak-field cooling is not known for magnetite or titanomagnetite but $\overline{w} \approx (r^{\frac{1}{2}} - 1)$ will be a rough approximation, where r is the ratio of the grain size to the critical SD grain size. As shown by Fuller (1984) the TRM given by (2.3.25) predicts the fall in TRM in the PSD range shown by the experimental data of Fig. 2.17.

It is worth noting that (2.3.24) and (2.3.10) enable the Koenigsberger ratio Q_t of (2.1.5) to be calculated for MD grains of magnetite, since Q_n for the NRM is frequently quoted in paleomagnetic studies. After allowing for the effect of the internal demagnetizing field, Dunlop and Özdemir (1997) calculate $Q_t = 0.6$ that is close to the experimental value for MD magnetite. For SD grains of magnetite the TRM is very much higher by at least an order of magnitude (Fig. 2.17), so that for these grains usually $Q_t > 10$. Most rocks used for paleomagnetism have $Q_n > 1$ so that pure multidomain TRM typically makes little contribution to the total TRM.

When carrying out laboratory tests relating to TRM in rocks, the very act of reheating the rock sample can cause chemical changes that can make the test ineffective. The acquisition of TRM involves the thermal activation of grains so that the energy barriers preventing the resetting of the magnetization in grains can be overcome. Anhysteretic remanent magnetization (ARM) is often used as an analog of TRM for laboratory testing. ARM is that remanence acquired when a sample is subjected to a decreasing alternating magnetic field in the presence of a small steady magnetic field (see Table 1.2). The alternating field must initially be of sufficient strength to be able to saturate the magnetic grains in the sample. The alternating magnetic field then becomes the low-temperature analog of thermal activation. Weak-field ARM and TRM exhibit similar alternating field demagnetization characteristics so heating of the sample can be avoided (see §3.5.3).

2.3.6 Crystallization (or Chemical) Remanent Magnetization (CRM)

CRM results from the formation of a magnetic mineral at low temperatures (below the Curie point) in the presence of an applied field. This may take the form of single-phase (or grain-growth) CRM by nucleation and growth through the critical blocking diameter d_B (or corresponding volume v_B) as defined in §2.3.4. Alternatively, it may take the form of two-phase (or parent–daughter) CRM through the alteration of an existing magnetic phase. CRM is often referred to as chemical remanent magnetization but this is not always strictly correct. For example, the transformation of maghemite (γFe_2O_3) to hematite (αFe_2O_3) is one from the spinel to the rhombohedral structure that occurs with no chemical change, only a restacking of the lattice.

Unfortunately, it is not always easy to recognize CRM because its unblocking temperatures and coercivities overlap those of TRM (§2.3.5) and DRM (§2.3.7). Here, three major aspects of CRM that are significant in continental studies will be considered: CRM acquisition through single-phase or grain-growth at constant temperature, CRM as it applies to redbeds, and CRM caused by the alteration of carbonates. The problem of CRM acquired through the alteration of titanomagnetites in oceanic basalts will be considered in the discussion of oceanic paleomagnetism in chapter 5.

Single-Phase or Grain-Growth CRM
Suppose a small grain is superparamagnetic at room temperature and therefore has no remanence since the thermal fluctuations are too great and the grain is magnetically unstable (§2.3.4). When such a grain nucleates and grows in a

weak magnetic field, it may grow to a sufficient size to pass through the critical blocking volume v_B, when the relaxation time of the grain increases very rapidly. The equilibrium magnetization then becomes "frozen in" and, as the grain grows further, subsequent changes in the field direction have no effect on the direction of magnetization. The process is analogous to the acquisition of TRM.

From (2.3.16), as a superparamagnetic grain grows at temperature T, its magnetization becomes stabilized at some volume v_B when

$$v_B = \frac{kT \ln(f\tau)}{K},$$
(2.3.26)

where $\tau = 100$ s and $f = 10^9$ s^{-1}. Hematite grains, for which $K = 1.2 \times 10^4$ Jm^{-3}, are of particular interest here. Assuming the grains are spheres, then at a temperature $T = 300$ K (27°C), the critical blocking diameter $d_B \approx 0.1$ μm. Experimental determinations suggest a slightly different value of $d_B \approx 0.2$–0.3 μm (Dunlop and Özdemir, 1997). Note that when the grains have grown from a diameter of 0.10 to 0.13 μm, the relaxation time has already increased to 10^9 years! Again, this is one of the basic appeals of paleomagnetism; that original CRM by grain growth can be stable over the geological time scale.

Stacey (1963) considered the CRM acquisition of SD grains, whose anisotropies are aligned in the direction of the applied field B in which they are growing at temperature T. The magnetization is given by

$$M_{CRM} = M_s \tanh\left(\frac{v_B M_s B}{kT}\right).$$
(2.3.27)

In small fields (0–0.1 mT), a random assemblage of such grains produces a CRM given by

$$M_{CRM} = \frac{v_B M_s^2 B}{3kT}.$$
(2.3.28)

Substituting values for v_B from (2.3.26) for hematite gives

$$M_{CRM} = 4 \times 10^3 \, B.$$
(2.3.29)

This suggests that the CRM of noninteracting grains of hematite is independent of the size to which the grains have grown, providing they exceed the critical volume v_B and that they remain SD.

The magnetic characteristics of CRM are similar to those of TRM. Experimental work confirms that CRM magnetization is proportional to the applied field as predicted from theory. In general, the magnitude of CRM is less than that of TRM, but this depends on several factors, as studied in detail by McClelland (1996).

CRM in Redbeds

Large black crystalline hematite grains (specularite) of detrital origin are found in both red and non-red sediments. The fine-grained hematite pigment, which gives redbeds their distinctive color, may be derived from the alteration of an existing magnetic phase in three possible ways – by the oxidation of magnetite to hematite, the inversion of maghemite to hematite, or the dehydration of goethite to hematite – according to the reactions

$$\left. \begin{array}{r} 4Fe_3O_4 + O_2 \Rightarrow 6\alpha Fe_2O_3 \\ \gamma Fe_2O_3 \Rightarrow \alpha Fe_2O_3 \\ 2\alpha FeOOH \Rightarrow \alpha Fe_2O_3 + H_2O \end{array} \right\} . \qquad (2.3.30)$$

In the presence of an applied field, each of the reactions results in a two-phase or daughter–parent CRM. All the reactions are slow at room temperature, but accelerate with mild heating such as during burial in a sedimentary sequence.

In all cases the lattices of the parent and daughter phases are incompatible. Therefore, a growth CRM, controlled by the applied field, would be expected to accompany each reaction. However, the growing daughter phase may also be influenced by its magnetic parent phase due to magnetostatic or exchange coupling of varying degree. In fact most experimental studies, with minor exceptions, suggest that grain-growth CRM is indeed essentially controlled by the applied field. In the case of the goethite dehydration process, all traces of goethite would have disappeared before the resulting hematite grains have grown to the critical SD size so that all memory of any previous magnetization in the goethite is lost. With continued growth therefore, the hematite pigment derived in this way acquires a true grain-growth CRM.

From (2.3.29), it was seen that the magnetization of grain-growth CRM in hematite is independent of the size to which the grains have grown, provided it exceeds the critical size. Thus, the CRM will not necessarily be related to the amount of hematite present, even if all the grains are above the critical size. The maximum CRM for a given amount of hematite will be observed when all the grains are just above the critical size. A rock specimen typically containing 1% of hematite by volume produced by grain growth in the Earth's magnetic field (0.05 mT) should thus be capable of acquiring a maximum value $M_{CRM} \approx 0.1$ Am^{-1}. This is commonly the upper limit of magnetizations (10^{-3}–10^{-1} Am^{-1}) observed in redbeds. This is stronger than the typical detrital or post-depositional magnetization of sediments (DRM or PDRM – see §2.3.7). The problem in paleomagnetic studies of redbeds is to determine how much of the magnetization is CRM carried by the hematite pigment and to discover how long after the rock formed that the pigment developed. This is discussed further in §3.4.1.

CRM in Altered Carbonates

Extensive investigations of North American Paleozoic sedimentary rocks have established that both carbonates and redbeds were remagnetized during the Late Carboniferous. Useful reviews of this topic are given by McCabe and Elmore (1989) and Elmore and McCabe (1991). Such remagnetization is widespread on both sides of the Atlantic, not only in the Appalachians and Hercynian belts but also in stable platform areas. Particularly interesting from the paleomagnetic viewpoint is the low-temperature CRM carried by authigenic or diagenetically

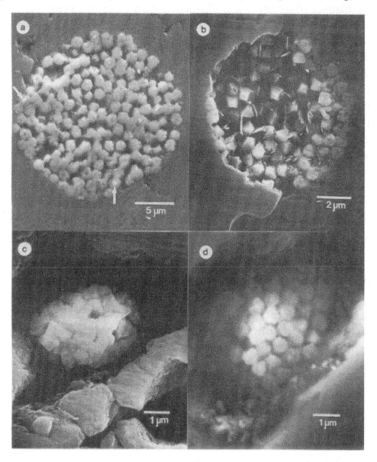

Fig. 2.18. Scanning electron microscope images of small magnetite grains in thin sections of carbonates from New York State. (a) Spheroidal aggregates of mostly magnetite but with occasional bright cores (arrow) where relict pyrite is found. The interpretation is that magnetite is replacing the pyrite. (b–d) Similar spheroidal aggregates in voids or cracks. Note the octahedral crystal shape. The matrix is calcite. The scale is indicated by the bars. From Suk *et al.* (1990), reproduced with permission from *Nature*.

altered magnetites in undeformed and only mildly heated platform carbonates. This is thought to have been caused by the migration of chemically active and perhaps hot fluids during plate convergence and subsequent mountain building.

Not all platform carbonates are remagnetized and the problem has been to distinguish between the remagnetized and the unremagnetized. Originally the mere presence of magnetite in a sedimentary rock was usually interpreted as evidence of detrital origin and early acquisition of the remanence. Because of the low concentration of magnetite in remagnetized carbonates (typically about 10 ppm), distinct observation and characterization of the magnetic carriers has been difficult. However, magnetic extracts from remagnetized carbonates have been studied by scanning electron microscopy and show spheroidal and botryoidal morphologies that are consistent with a diagenetic origin (Fig. 2.18). In addition, the magnetic properties of remagnetized and unremagnetized carbonates are distinctly different (Channell and McCabe, 1994). The remagnetized carbonates contain fine-grained high-coercivity SD magnetite with a high proportion of SP magnetite. Magnetite in unremagnetized carbonates appears to be concentrated in PSD grains. These differences are explained in more detail in §3.5.4.

A direct connection between CRM acquisition and alteration by fluids has also been established around mineralized veins. Several carbonate units that are hydrocarbon bearing also contain CRM carried by authigenic magnetite. It seems that the chemical conditions created by the hydrocarbons caused the precipitation of the authigenic magnetite and the acquisition of the associated CRM. This could have significance for oil exploration.

2.3.7 Detrital and Post-Depositional Remanent Magnetization

The process of alignment of magnetic particles by an applied magnetic field as they fall through water and then settle on the water sediment interface at the bottom is termed *detrital* (or *depositional*) *remanent magnetization* (DRM). However, DRM is not finally set in orientation until the sediment has been compacted by the weight of later deposits and the water has been excluded in the consolidation process. After deposition, wet unconsolidated sediments are often disturbed through bioturbation and slumping so that DRM should lose most of its directional coherence. Irving and Major (1964) proposed that magnetic particles would still remain free to rotate in the water-filled interstitial holes of a water saturated sediment until compaction and reduction of the water content eventually restricted their movement. This process is termed *post-depositional remanent magnetization* (PDRM).

Detrital Remanent Magnetization

Suppose a spherical or near-spherical grain of diameter d (volume $\pi d^3/6$) and remanence M is falling through water with viscosity η ($\approx 10^{-3}$ Pa s at room temperature) in the presence of an applied field B. There is a couple L, turning the magnetic moment of the grain toward the field direction (§2.1.1, Fig. 2.1), given by

$$L = -A\sin\theta = -\left(\frac{\pi d^3}{6}\right)MB\sin\theta, \qquad (2.3.31)$$

where θ is the angle between M and B. The motion of such grains is likely to be highly damped so that inertia can be neglected. Under these conditions the couple can be equated to the viscous drag on the rotation of the grain so that

$$L = C\frac{d\theta}{dt} = \pi d^3 \eta \frac{d\theta}{dt} \qquad (2.3.32)$$

If the angle θ is given by θ_0 at time $t = 0$, then for small angles

$$\theta = \theta_0 \exp(-t/t_0), \qquad (2.3.33)$$

where
$$t_0 = \frac{C}{A} = \frac{6\eta}{MB} \qquad (2.3.34)$$

and is the "time constant" of the rotation, or the time taken for the initial angle θ_0 to be reduced to $1/e$ of its value.

If the alignment process is to be reasonably complete, the particle must fall through water for at least time t_0. From Stokes' law, Stacey (1963) has shown that the time of fall t of a spherical grain through a depth h of water is given by

$$t = \frac{18\eta h}{d^2(\rho - \rho_0)g}, \qquad (2.3.35)$$

where $(\rho-\rho_0)$ is the density difference between the grain and the water and g is the acceleration due to gravity. There is thus some critical height h_0 through which the grain must fall for the time t_0 to have elapsed. Equating (2.3.34) and (2.3.35), h_0 is given by

$$h_0 = \frac{d^2(\rho-\rho_0)g}{3MB}. \qquad (2.3.36)$$

Typical conditions that might exist are $B = 0.05$ mT, $(\rho-\rho_0) = 4 \times 10^3$ kg m^{-3} and $M = 10^3$ Am^{-1} for PSD grains of diameter 10 μm. This gives $h_0 \approx 20$ μm and then $t_0 \approx 0.1$ s, so that complete alignment is almost instantaneous. Both the height and the time will be even less for smaller grains. Shive (1985) has demonstrated that perfect alignment is achieved in fields of 0.03–0.1 mT as the theory above suggests.

When magnetized grains settle on the bottom there will be a mechanical torque exerted on the grain by the surface on which it settles (it will roll or fall into the position of least potential energy). This torque will misalign M with B and, for grains larger than 10 μm, these mechanical torques become stronger than the magnetic aligning force (Dunlop and Özdemir, 1997). For increasing grain sizes the magnetic alignment achieved during fall will be completely destroyed on settling. Models of these mechanical torques suggest that grain rotations essentially have a random effect on the observed declination D, but they result in a systematic decrease in the inclination I, known as the *inclination error*. Laboratory studies of these effects show that the sediment inclination I_S will invariably be less than the applied field inclination I_B, where

$$\tan I_S = f \tan I_B \qquad (2.3.37)$$

and f is normally about 0.4 (King, 1955; Griffiths *et al.*, 1960). Because of the randomizing effects on the declination D, the DRM intensity will also be much weakened.

Although such inclination errors have been observed in some sediments, studies of deep-sea sediments suggest that they do record the field direction without significant inclination error (Fig. 2.19). This is almost certainly due to the effectiveness of PDRM as described below. The effect of compaction may also cause a shallowing of inclination because grains tend to be rotated into the bedding plane (Blow and Hamilton, 1978; Anson and Kodama, 1987) and inclination errors, where observed, are more likely to be caused by this effect than those due to the mechanical torques that are applied when the grains settle.

Post-Depositional Remanent Magnetization

When magnetic grains settle on the water–sediment interface and then acquire the possible inclination errors described above, they generally fall into water-filled voids or interstitial holes where they are still free to rotate (Irving and Major, 1964). As the general theory of DRM shows, the magnetic alignment of the magnetic particles is virtually instantaneous, so that the suspended grains tend to become realigned again in the direction of the applied field B. PDRM is most efficient if the magnetic grains are significantly finer than the silicate grains and they can rotate readily in pore spaces. The density differential during sediment transport seems to ensure that this situation generally holds.

Tucker (1980) has examined the time dependence of PDRM following reorientation of the applied field in an artificial magnetic slurry and Hamano (1980) calculated theoretical time constants for the realignment of PDRM in

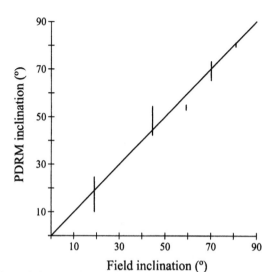

Fig. 2.19. Experimental determination of PDRM inclination in laboratory redeposited deep-sea sediments plotted as a function of the applied field inclination. Vertical bars indicate the spread of replicate measurements. From Kent (1973), reproduced with permission from *Nature*.

sediments with various void ratios and axial ratios of magnetic particles. These studies all confirm that PDRM is a viable and natural process and realignment readily occurs within a very short time.

The main factor in the PDRM process is the water content. When water content drops below some critical value the remagnetization process ceases and the particle rotations are blocked. Slow deposition, such as occurs in deep-sea sediments, and fine sediment particle size favor PDRM because they promote high water content and delay compaction. PDRM experiments on redeposited deep-sea sediments (Fig. 2.19) confirm that they record the field direction without significant inclination error (Kent, 1973). Opdyke and Henry (1969) and Schneider and Kent (1990) have shown that the inclinations observed in deep-sea sediments worldwide follow the expected latitude dependence for a geocentric axial dipole field (§1.2.3) with only second-order variations over the past several hundred thousand years.

2.3.8 Viscous and Thermoviscous Remanent Magnetization

After acquiring their primary remanent magnetization on formation, rocks are continually exposed to the Earth's magnetic field throughout their history. Because of the effect of magnetic viscosity (§2.3.3), those grains that have shorter relaxation times τ given by (2.3.14) or (2.3.15) can acquire a secondary magnetization long after the formation of the rock. Such a secondary magnetization is termed *viscous remanent magnetization* (VRM) and involves the realignment of the magnetic moments of those grains having values of τ less than the age of the rock. This can usually be erased by using suitable demagnetization techniques discussed in §3.4.

The VRM, M_{VRM}, at a given temperature is acquired according to the relation

$$M_{VRM} = S \log t, \tag{2.3.38}$$

where t is the time (in seconds) over which the VRM is acquired and S is known as the viscosity coefficient. Because of the logarithmic growth of VRM with time, VRM is usually dominated by that acquired in the most recent field to which the rock has been exposed. Rocks with large VRM components generally have their NRM aligned in the direction of the present geomagnetic field.

Since their time of formation rocks may be subjected to heating either from deep burial and subsequent uplift or from the effects of metamorphism. In §2.3.3 it was shown from the approximate relationship given by (2.3.16) that the effect of maintaining a rock for a short time at a higher temperature is equivalent to maintaining it for a much longer time at a lower temperature. Rocks that have been heated to temperatures below the Curie temperature of their magnetic minerals for a (geological) short time during their history, and then subsequently cooled again in the prevailing magnetic field, will acquire a *thermoviscous remanent magnetization* (TVRM), a terminology used by Butler (1992), although Chamalaun (1964) and Briden (1965) originally referred to it as a *viscous partial thermoremanent magnetization* (viscous PTRM).

Pullaiah *et al.* (1975) developed a blocking temperature diagram approach to the acquisition of TVRM for magnetite and hematite bearing rocks that makes use of the more precise form of (2.3.15) in the following way. Because both B'_c and M_s are functions of temperature, (2.3.15) can be rewritten by replacing B'_c with $B'_c[T]$ and M_s with $M_s[T]$ to indicate their values at the temperature T so that (2.3.16) can be replaced more precisely with

$$\frac{T_1 \ln(f\tau_1)}{B'_c[T_1]M_s[T_1]} = \frac{T_2 \ln(f\tau_2)}{B'_c[T_2]M_s[T_2]}. \tag{2.3.39}$$

For SD magnetite the microscopic coercivity is dominated by shape anisotropy, so that B'_c is then given by (2.3.6) and $B'_c[T] \propto M_s[T]$. For SD hematite the microscopic coercivity is caused by magnetoelastic effects whose temperature variation is not well known. Pullaiah *et al.* (1975) suggested the relationship $B'_c[T] \propto M_s^3$ holds so that in these two cases (2.3.39) becomes

Magnetite: $$\frac{T_1 \ln(f\tau_1)}{M_s^2[T_1]} = \frac{T_2 \ln(f\tau_2)}{M_s^2[T_2]} ; \tag{2.3.40}$$

Hematite:
$$\frac{T_1 \ln(f\tau_1)}{M_s^4[T_1]} = \frac{T_2 \ln(f\tau_2)}{M_s^4[T_2]} \quad . \qquad (2.3.41)$$

Using the known temperature dependence of M_s for magnetite and hematite (Fig. 2.3), the above relationships are displayed in the blocking temperature (T_B) versus relaxation time (τ) curves of Fig. 2.20. These plots show the locus of the points in τ-T_B space that reset the same grains and enable the prediction of time–temperature stabilities. For example, point 1 in Fig. 2.20a corresponds to SD

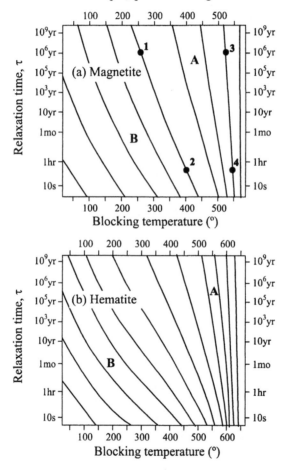

Fig. 2.20. Blocking temperature versus relaxation time diagrams for (a) magnetite and (b) hematite. Lines on each diagram represent the locus of points where the magnetization in a given population of grains will be reset. Grains in region A have sharply defined blocking temperatures within 100°C of the Curie temperature. Grains in region B have blocking temperatures on the laboratory time scale at least 100°C below the Curie temperature. From Pullaiah *et al.* (1975), with permission from Elsevier Science.

grains of magnetite with relaxation time of 10 Myr at 260°C. Such grains can be expected to acquire a substantial TVRM if held at 260°C for 10 Myr and then cooled to 0°C. However, the same grains at point 2 have τ = 30 mins at 400°C. Therefore by heating these grains to 400°C in the laboratory for 30 mins in zero magnetic field, the TVRM acquired at 260°C for 10 Myr can be unblocked and reset to zero. Point 3 in Fig. 2.20a also corresponds to SD grains of magnetite with relaxation time of 10 Myr but now at 520°C. The same grains at point 4 have τ = 30 mins at 550°C. Therefore the TVRM acquired over 10 Myr at 520°C can be unblocked by heating to a slightly higher temperature of 550°C for 30 mins in zero magnetic field.

Two regions labeled A and B are indicated in each of the diagrams in Fig. 2.20. Region A refers to those grains that have sharply defined blocking temperatures within 100°C of the Curie temperature. Such grains are resistant to resetting their magnetization except by heating to temperatures approaching the Curie temperature. Region B includes those grains that have blocking temperatures at least 100°C below the Curie temperature on laboratory time scales (30 mins). They are capable of acquiring TVRM at moderate temperatures (~300°C) if exposed to those temperatures for geologically reasonable lengths of time (~10 Myr). Thus, grains in region B are not good carriers of the primary magnetization and can likely acquire TVRM or VRM. Figure 2.20 indicates that the primary NRM in rocks can survive heating to the greenschist facies (300-500°C) but not to the amphibolite facies (550–750°C).

2.3.9 Stress Effects and Anisotropy

Most rocks are subjected to stress during their history, either from deep burial or from tectonism. Magnetocrystalline anisotropy is stress dependent so that application of stress can cause a change in the magnetization of a grain, the effect being referred to as magnetostriction (see also §2.3.1). It should be noted that stress alone cannot induce magnetic moments; the application of stress to an initially isotropic material causes stress-induced anisotropy, which may change the state of magnetization.

In the early days of paleomagnetism it was thought that the simple, reversible application of elastic stress to a rock would cause a substantial deflection of its remanent magnetization. Thus, although a rock that cooled under stress would acquire a magnetization in the direction of the field in which it cooled, upon release of the stress before measurement, changes in magnetization would occur that would make the original field direction impossible to determine. However, experimental results (Stott and Stacey, 1960; Kern, 1961) showed clearly that for isotropic rocks the TRM was always acquired parallel to the applied field. It appears that when an intrinsically isotropic rock is subjected to stress while cooling in a magnetic field, it acquires a TRM in a direction deflected away from the field by such an angle that it returns to the field direction precisely when unloaded.

The development of folds in rock formations relieves compressive tectonic stress and involves large internal strains. It is possible to "destrain" the natural remanence in deformed rocks if a detailed knowledge of both the strain mechanism and the response of the magnetic grains and their remanence vectors to the strain can be measured (e.g. Cogné and Perroud, 1987; Borradaile, 1993). This will be discussed in more detail in §3.3.2 with respect to the fold test in paleomagnetism.

If rocks have an intrinsic anisotropy, the TRM may be deflected away from the direction of the applied field towards the direction of easy magnetization (Fig. 2.21). Anisotropy is measured by the variation in values of susceptibility, saturation magnetization, IRM, or TRM in different directions in a rock specimen. Normally, anisotropy of susceptibility is considered, since this is the easiest to measure and the procedure does not generally alter the state of magnetization of the specimen. The degree of anisotropy A_n is then expressed as the ratio of maximum to minimum susceptibility

$$A_n = \frac{\chi_{max}}{\chi_{min}}. \tag{2.3.42}$$

A value of A_n of 1.25 means that the maximum susceptibility exceeds the minimum by 25%. Such a specimen is often referred to as having "25 percent anisotropy".

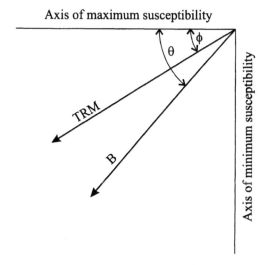

Fig. 2.21. Deflection of TRM from an applied field B due to anisotropy.

The main issue that arises in a paleomagnetic context is the degree of anisotropy that can be tolerated in a specimen before the TRM is deflected through a significant angle. It is simplest to consider the most favorable case for the deflection of TRM, which occurs when the direction of the applied field lies in the plane containing the maximum and minimum susceptibility directions. Suppose the applied field B makes an angle θ with the direction of maximum susceptibility (Fig. 2.21). Since TRM is proportional to the applied field in low fields ($M = cB$), the specimen will acquire TRM components in the maximum and minimum directions given by

$$M_{max} = c_{max}B\cos\theta$$
$$M_{min} = c_{min}B\sin\theta, \tag{2.3.43}$$

where c_{max} and c_{min} are the constants of proportionality in the maximum and minimum directions respectively. The angle ϕ that the resulting TRM will make with the direction of maximum susceptibility is then given by

$$\tan\phi = \frac{M_{min}}{M_{max}} = \frac{c_{min}B\sin\theta}{c_{max}B\cos\theta}$$
$$= \frac{\tan\theta}{A_t}, \tag{2.3.44}$$

where $A_t = c_{max}/c_{min}$ and is the degree of anisotropy of TRM. The TRM will be deflected through the angle ($\theta - \phi$), which will have its maximum value when $\theta = \tan^{-1}\sqrt{A_t}$ so that

$$(\theta - \phi)_{max} = \tan^{-1}\left(\frac{A_t - 1}{2\sqrt{A_t}}\right). \tag{2.3.45}$$

For multidomain magnetite and other strongly magnetic minerals $A_t \approx A_n^2$ from both theory (Dunlop and Özdemir, 1997) and experiment (Cogné, 1987) so that, from (2.3.42)

$$A_t \approx A_n^2 \approx 1 + \frac{2\Delta\chi}{\chi_{av}} \tag{2.3.46}$$

where $\Delta\chi = (\chi_{max} - \chi_{min})$ and $\chi_{av} = (\chi_{max} + \chi_{min})/2$. Therefore, the percentage anisotropy of TRM is approximately twice the percentage anisotropy of susceptibility. Also, from (2.3.45)

$$(\theta - \phi)_{max} = \tan^{-1}\left(\frac{A_n^2 - 1}{2A_n}\right). \tag{2.3.47}$$

For an anisotropy of susceptibility of 5%, $A_n = 1.05$ and the maximum TRM deflection is 2.7°, for 10% it is 5.2° and for 20% it is 10.0°. Thus 10% anisotropy can be tolerated without seriously deflecting the TRM. The rocks most commonly used in paleomagnetism, such as basic igneous rocks, redbeds and unmetamorphosed sediments, rarely have anisotropies exceeding a few percent, so the effect is only likely to be significant in some metamorphic or strongly foliated rocks.

Chapter Three

Methods and Techniques

3.1 Sampling and Measurement

3.1.1 Sample Collection in the Field

The typical sampling scheme for a land-based paleomagnetic study is hierarchical, consisting of *specimens* cut from independently oriented *samples* collected at different *sites* distributed throughout a *rock unit*. There are no hard and fast rules; the details of any specific scheme are dictated by the goals of the study, the rock type, accessibility of exposure, and the statistical properties of the magnetization (see §3.2). The rock unit would be a set of sedimentary beds, lava flows, or intrusives that have been recognized as units on geological grounds.

Conceptually, a typical paleomagnetic *site* should represent, as near as possible, a single point in time relative to the time frame over which the paleomagnetic field changed. For studies in tectonics the ideal goal is to obtain a reliable and acceptably precise estimate of the paleomagnetic field direction at the point in time represented by that site. It is necessary to have a sufficient number of sites throughout the rock unit to be able to average out the ancient secular variation and obtain a good estimate of the paleomagnetic pole position. For example, a single lava flow cools rapidly so it is unlikely to represent more than an instant in time. In general, therefore, no matter how many different localities are sampled in a single flow, they will all represent the same site in paleomagnetic terms. In sedimentary rocks, however, just a few metres thickness of sediment may well represent a considerable interval of time so spot readings of the field can only really be made by sampling over a limited thickness. In this case it might be possible to sample several sites all at a single geographical locality just by sampling different horizons several metres apart stratigraphically.

The basic collection unit within a site is known as a *sample*, which is usually considered to be a separately oriented piece of rock. The vagaries of the magnetization process and the errors involved in sample orientation mean that different samples from a single site will give different directions (typically known as within-site or between-sample scatter). Consequently, several samples, which should be independent observations of the same spot reading of the paleomagnetic field, are usually collected from each site so as to be able to average out this variation and allow estimation of its statistical characteristics.

On land, collection of samples from an outcrop is usually carried out by gathering oriented cores with the aid of a portable rock drill or by hand sampling oriented blocks. The basic objection to block sampling is that the most convenient samples are those associated with cracks or joints and so such samples tend to be more weathered. The orientation of the cores or hand samples is usually carried out with the aid of a sun compass that has been designed for this purpose, although orientation using a magnetic compass can be acceptable if the magnetization of the rocks is not high enough to affect the compass. See Collinson *et al.* (1967) and Collinson (1983) for descriptions of equipment that has been designed specifically for this purpose. Finally, several *specimens* are then cut and/or drilled and sliced from each sample. These specimens are the objects that are actually individually measured. If several specimens from a sample are measured then these observations are combined to provide a sample observation.

The variation from site to site (typically referred to as between-site scatter) gives useful information about the ancient secular variation and so it is important to design the sampling scheme so that the statistical properties of this variation can be estimated and the variation averaged out. However, the age spread of the sites should not be so large as to encompass movement of the pole relative to the sampling locality. At each site it is necessary to be sure that the number of samples is sufficient for the within-site (between-sample) scatter to be averaged out. For this purpose one would set a minimum of at least four samples.

As a particular example, for studies of paleosecular variation from lavas (§1.2.6), the parameter of interest is the between-site scatter; thus it is important to determine the typical within-site scatter so as to be able to correct for this in the overall scatter. Therefore, in this case eight samples per site are often collected. For studies in magnetostratigraphy the determination of polarity is the major objective at each site. Opdyke and Channell (1996) suggest three or more samples are required for an unambiguous determination of polarity at each horizon (site).

In some areas, such as in tropical countries, the incidence of lightning strikes to ground is high. Lightning strikes to ground consist of currents of the order of 10^4–10^5 amp that can travel along the surface of an outcrop. In the immediate vicinity of the current the magnetic field created is large and can substantially remagnetize the outcrop adjacent to its path. Assuming a lightning strike is equivalent to an infinite conductor carrying a current i, then the magnitude B of the magnetic field at a distance r from the current is given by

$$B = \frac{\mu_0 i}{2 \pi r} \ . \tag{3.1.1}$$

At a distance of 0.1 m, a current of 5×10^4 amp produces a field of 100 mT, which is sufficient to induce a substantial IRM. At a distance of 1 m, the field is reduced to 10 mT. Cox (1961) and Graham (1961) carried out experiments on outcrops that had been struck by lightning and showed that the IRM produced by fields of this order can be selectively removed by alternating field demagnetization (§3.4.1) in peak fields of the same order. If samples are always collected several metres apart then the chances of their having been taken too near the path of a strike are minimized. The occasional spurious sample can then be identified and rejected. Igneous rocks that have been struck by lightning become strongly magnetized, so a magnetic compass, held in fixed azimuth and passed over the outcrop being sampled at a distance of 10–15 cm, will record significant deflections ($\geq 5°$) over the region of a lightning strike. This technique may then be used to avoid the lightning affected regions. Doell and Cox (1967a) describe a simple fluxgate magnetometer for the measurement of NRM at the outcrop that can also be used for identifying and thus avoiding lightning-affected regions.

In the oceans, deep-sea sediments were initially collected using conventional piston cores dropped vertically from a ship. The maximum length of core that could be obtained by this method was about 25 m and this limited the time span covered by each core. The development of the hydraulic piston corer and advanced piston corer by the Deep-Sea Drilling Project (DSDP) and Ocean Drilling Project (ODP) enabled the collection of cores as long as 250 m. Single samples are taken from the core at close spacing, although an initial continuous measurement of "u-channels" is often made. A "u-channel" is a rectangular channel whose cross-section is the shape of a "u", typically about 1.5 m long with a cross-section of about 4 cm^2 that is pressed into a core to produce a long continuous sample (Tauxe *et al.*, 1983a; Nagy and Valet, 1993). These are then measured through the use of special magnetometers (see §3.1.2 below). The cores are not oriented so only the inclination of the magnetization, and sometimes the declination relative to some arbitrary mark, can be determined at each point.

3.1.2 Sample Measurement

In the early days of paleomagnetic investigations, the astatic magnetometer was developed as the main instrument used for measurement of the remanent magnetization, even of some weakly magnetized sediments (Blackett, 1952). The instrument consisted of two small magnets separated by a short distance and set with their magnetic moments in opposition to one another. The magnets were suspended by a torsion fiber and so acted as a magnetic gradiometer, allowing measurement of weak local magnetic fields. The measurement limit of these magnetometers was about 10^{-3} Am^{-1}. In the 1960s the spinner magnetometer became the popular measuring instrument. Many different varieties have been developed, but they all involve spinning a rock specimen within or adjacent to a magnetic field sensor (typically a conductive coil with many turns) that produces an oscillating signal capable of being amplified electronically (e.g., Gough, 1964). Despite the electronic amplification, most such spinners rotate the specimen rapidly in order to generate an acceptable signal. This rapid rotation places significant demands on the physical competence of specimens. The sensitivity of these magnetometers improved greatly with new developments in electronic circuitry and they are still in use in many laboratories today. The most recent models that are available commercially can measure magnetizations less than 10^{-5} Am^{-1}.

The development of the fluxgate magnetometer as an instrument for measuring small magnetic fields provided an alternative and simple method for measuring the magnetization of rock samples. By placing two probes in opposition a fluxgate gradiometer is produced and can be used in much the same way as an astatic magnetometer (Helbig, 1965). Foster (1966) combined a fluxgate magnetometer with a slow speed spinner that enabled samples of deep-sea sediments to be measured without being subjected to the high rotation speeds required of traditional spinner magnetometers. A detailed account of these various instruments is given by Collinson (1983) and Fuller (1987).

In the 1970s cryogenic magnetometers were developed with much greater sensitivity and speed of measurement than spinner magnetometers (Goree and Fuller, 1976; Collinson, 1983; Fuller, 1987). This enabled very weakly magnetized sediments, such as limestones, to be measured quickly and easily for the first time and opened up a new era in paleomagnetic studies. Cryogenic magnetometers use a detector called a SQUID (superconducting quantum interference device) that is based on the properties of a superconducting ring. The flux lines ϕ threading a superconducting ring are quantized ($\phi = n\phi_0$). They can only change in integral multiples of a basic flux unit $\phi_0 = h/2e = 2.07 \times 10^{-15}$ weber, where h is Plancks's constant and e is the charge on the electron.

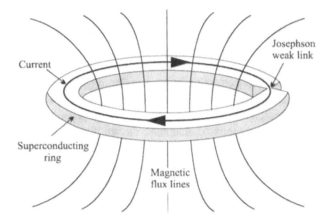

Fig. 3.1. Principle of the cryogenic magnetometer. The flux threading a superconducting ring is an integral multiple of the flux unit $\phi_0 = h/2e$. Changes in the external magnetic field cause changes in the current i in the ring. The critical value of this current can be reduced by narrowing the ring at one point, called the *Josephson weak link*.

When the ring is cooled through the critical temperature and becomes superconducting in the presence of a magnetic field B along the axis of the ring, flux is expelled from the material of the ring and this makes the superconductor a perfect diamagnet, a phenomenon known as the *Meissner effect*. This flux expulsion is achieved by an induced current i circulating in the ring (Fig. 3.1). However, the field in which the ring was originally cooled is effectively trapped and maintained in the area enclosed by the ring. This effect is exploited in a magnetometer by trapping a zero field within the ring and using this volume for measurement. If the current i exceeds some critical value i_c, the superconductivity is suppressed, the current i decreases because the ring becomes resistive, the flux linking the ring changes sharply, and the superconducting state is regained. This is sometimes referred to as resistive switching. The ring is driven by an external radio frequency field just large enough to cause the circulating current to exceed i_c in each cycle. At each point in the cycle where this happens a quantum of flux enters the ring and the rapid change in flux linkage can be detected. If a magnetic sample is introduced then the base level of the circulating current is changed to exactly cancel the introduced change in flux. This then changes the point in each cycle at which the resistive switching occurs, so the presence of the introduced field may be detected.

The sensitivity of the SQUID detector relies on the detection of extremely small current changes. To achieve the required sensitivity the cross-section at a point in the circumference of the ring is reduced to a very small diameter of ~1 μm. This reduced cross-section point is called the *Josephson weak link*

(Fig. 3.1). In this way a tiny change in the applied flux (of the order of ϕ_0) can cause the critical current density to be exceeded at the junction. Cryogenic magnetometers can measure magnetizations of 10^{-6} Am^{-1}.

Pass-through cryogenic magnetometers are now available that can be used to measure either whole cores or split half-cores without the necessity of physically sampling the cores. Weeks *et al.* (1993) described a magnetometer in which the SQUID sensing coils are arranged for high-resolution measurement of magnetization from continuous "u-channels" passed through the sensing region. The width of the response function of the sensing coils is about 3.5 cm and deconvolution of the signal in the time-domain can give comparable resolution to that achieved by back-to-back discrete sampling along the core (Constable and Parker, 1991). These instruments include the capability of carrying out alternating field demagnetization. The main drawback of such continuous measurement techniques is that it is often desirable to carry out thermal demagnetization or to do progressive demagnetization in higher alternating fields than are available in such instruments.

3.2 Statistical Methods

3.2.1 Some Statistical Concepts

The usual sampling scheme in paleomagnetic studies is hierarchical, comprising several levels. The directions of magnetization of specimens are combined first to give a sample mean. Sample means are then in turn combined to give the site means, which are then combined to give the overall formation or unit mean. To be able to calculate such means and the associated errors and to perform statistical tests it is necessary to develop a statistical framework that gives a reasonable description of the observed magnetic directions.

Probably the best known statistical distribution is the *normal distribution*, sometimes referred to as a *Gaussian distribution*. It provides a useful description of the random variation in a single variable (e.g., the heights of people in the human population) for many statistical populations. Because it is a description of the variation in a single variable it is referred to as a *univariate distribution*. Magnetic directions, however, need two variables (e.g., declination and inclination) to describe their distribution in space. Consequently, it is necessary to have a single statistical distribution that provides a joint description of the variation of both variables. Such a distribution is referred to as a *bivariate distribution*. The statistical distribution most commonly used in paleomagnetism is, in most instances, approximated by a bivariate normal distribution. This distribution is the combined description of two orthogonal normally distributed variables. Thus it is useful first to define the concepts and common parameters within the framework of the normal distribution.

Suppose that x_i represents the ith value obtained from a sample of n observations drawn randomly from some population that has a normal distribution. Note that in this context "sample" refers to an independent set of observations. In describing the variation of a single variable x, the normal distribution has two *parameters*: the population *mean* μ and the *variance* σ^2. A typical goal then, is to use the observations x_i to obtain the best estimates of the parameters μ and σ^2.

The mean μ is best estimated by the arithmetic mean, $\bar{x}$, given by

$$\bar{x} = \frac{1}{n} \sum_{i=1}^{n} x_i \; . \tag{3.2.1}$$

The phrase "best estimate" usually implies that the estimate should be *unbiased* and *consistent*. If $\bar{x}$ is calculated for a large number of independent samples each of size n observations, and the arithmetic mean of these $\bar{x}$ values gives μ, then $\bar{x}$ is an unbiased estimator for μ. If $\bar{x}$ is equal to μ for n very large, then $\bar{x}$ is a consistent estimator for μ.

The scatter of the population about μ is described by the variance σ^2. In a sample of size n this parameter is best estimated by s^2, where

$$s^2 = \frac{1}{n-1} \sum_{i=1}^{n} (x_i - \bar{x})^2 \; . \tag{3.2.2}$$

However, the variance does not have the same units as those of the original observations, so the width of the distribution is described by the *standard deviation* σ, which may be estimated by s, where

$$s = \sqrt{\frac{1}{n-1} \sum_{i=1}^{n} (x_i - \bar{x})^2} \; . \tag{3.2.3}$$

The standard deviation gives the deviation from the mean within which 63% of the values lie. Therefore, the probability that a given value will deviate from $\bar{x}$ by more than s is about 37% ("about" because $\bar{x}$ and s are only the estimates of μ and σ). A useful parameter is the *95% deviation*, x_{95}, which is the deviation from the mean beyond which only 5% of values will lie, given approximately by

$$x_{95} = 1.95s \; . \tag{3.2.4}$$

Note that the above parameters refer to a single sample of n observations drawn from a population. The standard deviation is then often referred to as the sample standard deviation. One can also determine the mean $\bar{x}$ many times with independent samples, each of size n, from the same population. In that case the means themselves will have a normal distribution, with the same population mean μ but a much smaller variance σ^2/n. The standard deviation, $\sigma/\sqrt{n}$, of this population of means is called the *standard deviation of the mean* and is usually estimated by the *standard error of the mean*, s_m, given by

$$s_m = \frac{s}{\sqrt{n}} \; . \tag{3.2.5}$$

Thus the standard error of the mean can be calculated quite simply from the parameters used in determining a single mean.

In paleomagnetism, as with most observational studies, it is often necessary to perform statistical tests of hypotheses. These hypotheses might be, for example, that two independent samples were obtained from a common parent population, or that a sample of observations is consistent with some model (e.g., a standard statistical distribution). The *chi-square* (χ^2) *distribution* and the *F distribution*, both derived from the parameters of the normal distribution, are particularly useful in making decisions about such hypotheses. For a normal distribution the statistic $(n-1)s^2$, for example, has a chi-square distribution with $\nu = n-1$ degrees of freedom, denoted as χ_ν^2. The ratio of two independent chi-square distributed variables, each divided by its number of degrees of freedom (ν_1 and ν_2), will be F distributed, denoted as $F[\nu_1,\nu_2]$.

When performing a statistical test the general process is to determine whether the data are consistent with some model. If the data conform closely with some model, then the model is acceptable. However, if the probability that the data conform with the model is small (usually taken as less than 5%) then the model is rejected. To test if a sample of observations is consistent with a particular model, the *chi-square test* is often used. The test variable X^2 is defined as

$$X^2 = \sum_{i=1}^{n} \frac{(o_i - e_i)^2}{e_i} \; , \tag{3.2.6}$$

where n is the number of cells (groupings of data) used for the comparison and o_i and e_i represent respectively the observed and expected frequencies for the ith cell. This variable X^2 will have some distribution as the observed values vary randomly. If $X^2 = 0$ there is perfect agreement with expectation, but if X^2 is large then the differences from expectation are large. The problem then is to determine the critical value, X_c^2, for X^2 so that for $X_c^2 \leq X^2$ it will be concluded that the observations are consistent with the model, but for $X_c^2 \geq X^2$ it will be concluded that the observations are not consistent with the model. Under the hypothesis that the observations are consistent with the model (often referred to as the *null hypothesis*), the variable X^2 will be chi-square distributed with ν degrees of freedom. This number ν depends on the number of cells and the number of parameters in the model that had to be estimated to calculate the expected values e_i. Knowing this, the critical value of X_c^2 may then be set equal to $\chi_\nu^2(1-P)$, which is the value that a chi-square distribution with ν degrees of freedom will exceed with probability P. Typically, the 5% critical value is used corresponding to probability 0.05. There is then less than 5% chance that an observed value of X^2 will exceed the critical value if the observations are in fact drawn from a population described by the model. Thus, if X^2 does exceed the critical value it seems unlikely that the observations are consistent with the model and so the null hypothesis is rejected "at the 95% confidence level". Conversely, if X^2 does not exceed the critical value, the model is accepted.

As another example, if independent variance estimates s_1^2 and s_2^2 are obtained from samples drawn from some parent population, then the ratio

$$f = \frac{s_1^2}{s_2^2}, \tag{3.2.7}$$

will be F distributed with ν_1 and ν_2 degrees of freedom and should have a value close to 1. Thus, if f exceeds the 5% critical value of $F[\nu_1,\nu_2]$, the null hypothesis of the common variance may be rejected at the 95% confidence level. The common phraseology is that the observed statistics are "significantly different" or, perhaps better, that the observed statistics are "discernibly different at the 95% confidence level".

3.2.2 The Fisher Distribution

The magnitude of the magnetization of a specimen is dependent on many factors, and within a given rock unit there is not necessarily an obvious link between this magnitude and the reliability of the direction recorded by the specimen. Consequently, in the development of a statistical framework it has been considered appropriate to give unit weight to each individual directional observation regardless of its magnitude. Thus, the directional observations are dealt with as if they were unit vectors. The ends of these vectors are represented as points on the surface of the unit sphere.

Fisher (1953) suggested a distribution in which the concentration of points on the unit sphere about the true mean direction is proportional to $\exp(\kappa\cos\psi)$, where κ is a *precision* (or concentration) parameter and ψ is the angle between an observation and the true mean direction. This distribution is the unit sphere analog of a normal distribution and has come to be known as the *Fisher distribution.* If the precision parameter, κ, is large (i.e., greater than about 4), then a Fisher distribution is well approximated by a bivariate normal distribution and κ is in effect the *invariance* or the reciprocal of the variance in all directions. This is particularly convenient because it means that several tests developed for the normal distribution provide excellent approximations as tests for the Fisher distribution. Watson (1956a,b) and Watson and Irving (1957) made use of this to provide workers with ready access to a range of important tests.

In a *Fisher distribution* (Fisher, 1953) the density $P_{\delta A}$ of points on the unit sphere is given by

$$P_{\delta A} = \frac{\kappa}{4\pi \sinh \kappa} \exp(\kappa \cos \psi) \ , \qquad\qquad (3.2.8)$$

where $P_{\delta A}\delta A$ is the probability of an observation falling in a small element of area δA at an angular distance ψ from the true mean direction. The mean direction occurs at $\psi = 0$ and here the density is a maximum. The precision parameter κ describes the dispersion of the points. If $\kappa = 0$ the points are uniformly distributed (there is no preferred direction) and when κ is large the

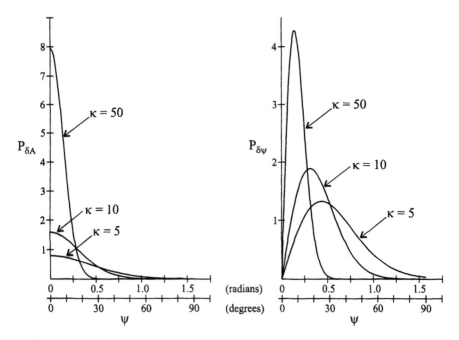

Fig. 3.2. The Fisher distribution illustrated for $\kappa = 5$, 10, and 50. (a) The function $P_{\delta A}$, where $P_{\delta A}\delta A$ is the probability of finding a direction within a small area δA at an angle ψ from the mean. (b) The probability density $P_{\delta \psi}$, where $P_{\delta \psi}\delta \psi$ is the probability of finding a direction within a band of width $\delta \psi$ between angles ψ and $\psi + \delta \psi$ from the mean.

points cluster tightly about the true mean direction. The constant factor, $(\kappa/4\pi\sinh\kappa)$, in (3.2.8) ensures that the density adds up to unity over the whole sphere.

The element of area δA is given by $\sin\psi\,d\psi\,d\phi$, where ϕ is the azimuth (or longitude) of the observation about the true mean direction and ψ is the colatitude. The azimuthal angle ϕ is uniformly distributed (i.e., there is no preferred value) so from (3.2.8) the probability density $P_{\delta\psi}$ is given by

$$P_{\delta\psi} = \frac{\kappa}{2\sinh\kappa}\exp(\kappa\cos\psi)\sin\psi\,, \qquad (3.2.9)$$

where $P_{\delta\psi}\delta\psi$ is the probability of finding a direction in a band of width $\delta\psi$ between angles ψ and $\psi+\delta\psi$ from the mean direction. The functions of (3.2.8) and (3.2.9) are illustrated in Fig. 3.2 for $\kappa = 5$, 10, and 50. Note that when calculating probabilities the angles $\delta\psi$ and $\delta\phi$ must be in radians.

In paleomagnetic studies the direction of magnetization of a rock sample is specified by the declination D, measured clockwise from true north, and the inclination I, measured positively downwards from the horizontal. The unit vector of the ith direction is specified by its direction cosines (l_i, m_i, n_i) as

$$\left.\begin{array}{ll} \text{North component:} & l_i = \cos D_i \cos I_i \\ \text{East component:} & m_i = \sin D_i \cos I_i \\ \text{Down component:} & n_i = \sin I_i \end{array}\right\}. \qquad (3.2.10)$$

Given a sample of N points drawn from a single Fisher distribution, Fisher (1953) showed that the best estimate (X,Y,Z) of the true mean direction is the direction of the vector sum (resultant) of the N individual unit vectors; therefore,

$$X = \frac{1}{R}\sum_{i=1}^{N}l_i\,, \quad Y = \frac{1}{R}\sum_{i=1}^{N}m_i\,, \quad Z = \frac{1}{R}\sum_{i=1}^{N}n_i\,, \qquad (3.2.11)$$

where $R\ (\leq N)$ is the length of the resultant vector given by

$$R^2 = \left(\sum l_i\right)^2 + \left(\sum m_i\right)^2 + \left(\sum n_i\right)^2. \qquad (3.2.12)$$

The declination D_m and inclination I_m of this mean direction are given by

$$\tan D_m = \frac{\sum m_i}{\sum l_i} \qquad (3.2.13)$$

and

$$\sin I_m = \frac{1}{R}\sum n_i\,. \qquad (3.2.14)$$

The best estimate k of the precision parameter κ, for $\kappa > 3$, is given by

$$k = \frac{N-1}{N-R}.$$

(3.2.15)

When the true mean direction is known but the precision is not, the best estimate k' is then given by

$$k' = \frac{N}{N - R\cos\zeta},$$

(3.2.16)

where ζ is the angle between the true mean direction and the resultant vector of the N observations.

In some rare instances, for example, when testing whether a set of observations is consistent with a Fisher distribution (§3.2.3), the appropriate estimate is the maximum likelihood estimate $\hat{k}$ of κ, given by

$$\hat{k} = \frac{N}{N-R}.$$

(3.2.17)

McFadden (1980a) provides a discussion on the best estimates of the precision parameter κ.

The four parameters needed to characterize a set of observations drawn from a Fisher distribution are then the mean direction given by D_m and I_m, the number of observations N, and the length R of the resultant vector.

The angle $\psi_{(1-P)}$ that will be exceeded with probability P has been given by McFadden (1980b) as

$$\cos\psi_{(1-P)} = 1 - (N-R)\left[\left(\frac{1}{P}\right)^{\frac{1}{N-1}} - 1\right].$$

(3.2.18)

The following approximate relations hold for analogs of the normal distribution:

Probable error: $\psi_{50} = \dfrac{67.5°}{\sqrt{k}}$ (3.2.19a)

Standard deviation: $\psi_{63} = \dfrac{81°}{\sqrt{k}}$ (3.2.19b)

95% deviation: $\psi_{95} = \dfrac{140°}{\sqrt{k}}$ (3.2.19c)

The angle ψ_{63} of (3.2.19b) is often referred to as the *angular standard deviation* or *angular dispersion* and ψ_{95} represents the angle from the mean direction beyond which only 5% of the directions lie.

3.2.3 Statistical Tests

Estimate of Accuracy
Fisher (1953) has shown that if N observations are drawn from a Fisher distribution with precision parameter κ and the length of the resultant vector is R, then the direction of this resultant vector (i.e., the mean direction) will also be Fisher distributed about the true mean direction but with precision κR. From this he showed that for $\kappa > 3$ the true mean direction of the population will, with probability $(1-P)$, lie within a circular cone of semi-angle $\alpha_{(1-P)}$ about the resultant vector R, where

$$\cos \alpha_{(1-P)} = 1 - \frac{N-R}{R}\left[\left(\frac{1}{P}\right)^{\frac{1}{N-1}} - 1\right]. \qquad (3.2.20)$$

Normally, P is taken as 0.05 to give a circle of 95% confidence about the mean. When α is small the following approximate relations may be used

$$\text{Circular standard error of the mean,} \quad \alpha_{63} = \frac{81°}{\sqrt{kR}}$$

$$\text{Circle of 95\% confidence,} \quad \alpha_{95} = \frac{140°}{\sqrt{kR}}. \qquad (3.2.21)$$

Note that previous texts (e.g., Irving, 1964; McElhinny, 1973a; Butler, 1992) have used $\sqrt{kN}$ instead of $\sqrt{kR}$ in (3.2.21). However, since the mean is distributed with precision κR (Fisher, 1953), $\sqrt{kR}$ is a more appropriate choice and gives a better approximation to the correct value given by (3.2.20). For large values of k (e.g., >50), R approaches N and it matters little which is used.

If some known direction, such as that of the present Earth's field at the sampling site, falls within α_{95}, then statistically there is no reason to suppose that the observations were drawn from a distribution with a true mean direction that differs from the known direction. However, if the known direction falls outside α_{95}, then the hypothesis that the known direction is the true mean direction of the observations can be rejected at the 95% confidence level.

When designing sampling schemes it is important that different statistical properties of results from different rock types be taken into account when calculating a mean formation direction. Suppose that n_i samples are collected at the ith site leading to an estimate k_{wi} of the within-site precision. The site mean direction will then be Fisher distributed with a precision of about $k_{wi}n_i$ (which is easier to use here than $k_{wi}R_i$), so it is important to keep the product $k_{wi}n_i$ roughly constant at each site, thereby maintaining an approximately constant error in each site mean direction (3.2.20). Thus, if the rock types being collected are much the same with similar within-site precisions (k_{wi} constant), the number of samples being collected at each site can be the same. If k_{wi} is not constant at each site because different rock types are being sampled, then the number of samples at each site needs to be varied to accommodate the varying k_{wi}. Under these conditions the mean formation direction can be calculated in the usual way as the mean of the site mean directions.

Discordant Observations

One problem that arises in paleomagnetic investigations is determining whether an apparent outlier is truly discordant with other observations and, if so, whether it should be rejected. Consider a set of $(N+1)$ observations in which N of the observations seem to group together but the $(N+1)^{th}$ observation is an outlier. McFadden (1982) has shown that, given the observed grouping of the N concordant observations, with resultant vector of length R, there is a probability P that an outlier from the same distribution will exceed an angle $\gamma_{(1-P)}$ from the mean of the concordant group, where

$$\cos\gamma_{(1-P)} = 1 - \frac{(R+1)(N-R)}{R}\left[\left(\frac{1}{1-(1-P)^{1/(N+1)}}\right)^{1/(N-1)} - 1\right]. \quad (3.2.22)$$

Thus, with $P = 0.05$, if the outlier lies further than $\gamma_{0.95}$ from the mean of the other N observations then it may be concluded with 95% confidence that the outlier is discordant with the other observations. This does not automatically mean that the observation was drawn from a different population (i.e., it is in error) and McFadden (1982) stresses that further investigation is required to determine whether the observation is actually an error and justifies correction or rejection. Fisher *et al.* (1981), using a different philosophical approach, developed a discordancy test using a different statistic. To several decimal points their test gives exactly the same values as the one described here.

Comparison of Precisions

It is frequently necessary to determine whether two sets of observations could have been drawn from populations sharing a common precision κ. If κ is not too small, the ordinary methods for the comparison of variances may be used to test this. Watson (1956a) showed that if samples N_1 and N_2 are drawn from populations with a common κ and gave precision estimates k_1 and k_2 respectively, then the ratio k_1/k_2 is given by

$$\frac{k_1}{k_2} = \left(\frac{N_1-1}{N_1-R_1}\right)\left(\frac{N_2-R_2}{N_2-1}\right) = h \sim F[2(N_2-1), 2(N_1-1)], \quad (3.2.23)$$

where the symbol "~" is to be read as "is distributed as" and $F[v_1,v_2]$ is the F distribution with v_1 and v_2 degrees of freedom as in (3.2.7). Values of $h = k_1/k_2$ far from unity strongly suggest the two populations do not have the same precision. The critical values of h at a given confidence level may be determined from tables (or computer code) of the F distribution.

If more than two populations are involved and the sample sizes are all the same, then the test may be performed as follows. The largest observed value of k is tested against the smallest value and if these two values could have been obtained by random sampling from populations having a common precision then so could all the intermediate values. If the result is marginal, or if the sample sizes vary, the test is not quite so simple. Bartlett (1937) presented a test for multiple observed variance estimates and its application to the Fisher distribution has been discussed by Stephens (1969), Mardia (1972), and McFadden and Lowes (1981).

Comparison of Mean Directions

It is often necessary or desirable to determine whether two or more sets of paleomagnetic observations could have been drawn from a common distribution; that is, do they have a common true mean direction and a common precision κ. A criterion sometimes used is that if cones of confidence do not intersect then the samples do not share a common mean direction. This is often phrased as saying that the mean directions are discernibly different. Although this criterion is certainly correct, it is extremely conservative; cones of confidence may overlap even though the mean directions are discernibly different. An appropriate test was devised by Watson (1956a) and re-examined by McFadden and Lowes (1981).

Suppose there are m samples, each having N_i observations giving resultant vectors of length R_i. If the samples were all drawn from a common population (which carries with it the assumption of a common κ), then

$$\left(\frac{N-m}{m-1}\right)\frac{\Sigma R_i - R^2/\Sigma R_i}{2(N-\Sigma R_i)} = f \sim F[2(m-1),2(N-m)] , \qquad (3.2.24)$$

where the summations are for i from 1 to m and $N = \Sigma N_i$.

The term $(N-\Sigma R_i) = \Sigma(N_i-R_i)$ is the sum of the within-sample dispersions and does not vary as the sample mean directions change. However, if the sample mean directions are more dispersed than is to be expected from the within-sample dispersion, then the algebraic sum of the sample resultants ΣR_i will be very much greater than their vector sum R and so f will be large compared with the relevant F statistic. This would suggest that the hypothesis of a common true mean direction is false.

For $m = 2$ (two samples) f is distributed as $F[2,2(N-2)]$, which has a particularly simple representation (McFadden and Lowes, 1981). The null hypothesis that two samples share a common mean direction may be rejected at the level of significance P if

$$\frac{(R_1 + R_2 - R^2 /(R_1 + R_2))}{2(N - R_1 - R_2)} > \left(\frac{1}{P}\right)^{1/(N-2)} - 1 \ . \tag{3.2.25}$$

With $P = 0.05$, the test is performed at the 95% level of confidence.

Test for Uniform Randomness

In some cases the directions of magnetization may be widely scattered, and the question then arises as to whether these directions could have been obtained by sampling from a uniform random population. For a truly uniform random (i.e., isotropic) population $\kappa = 0$ so that the population has no preferred, or true mean, direction. In practice, however, the observed R, and therefore the observed k (the best estimate of κ), is never zero. Watson (1956b) has devised the following test. For a sample of size N, the length of the resultant vector R will be large if a preferred direction exists, or small if it does not. Assuming that no preferred direction exists, a value R_0 may be calculated that will be exceeded by R with any stated probability. Watson (1956b) calculated R_0 for various sample sizes for probabilities of 0.05 and 0.01. To carry out the test one merely enters Watson's table at the row corresponding to the sample size N in order to find the value of R_0 which would be exceeded with given probability. For N greater than about 10, $3R^2/N$ is approximately distributed as a chi-square variate with 3 degrees of freedom. Consequently, $R_0^2 \cong \frac{1}{3} N \chi_3^2 (0.05)$ at the 95% confidence level for large N.

Testing for Conformance with a Fisher Distribution

Watson and Irving (1957) show how to use the standard chi-square test to determine whether a set of directions (or VGPs) conforms with Fisher's distribution. Unfortunately the test cannot be made in a single step; the azimuthal angle has to be tested for uniformity and the polar angle ψ has to be tested separately for conformity with the probability density $P_{\delta\psi}$ of (3.2.9). The probability of an observation making an angle greater than ψ_0 with the true mean direction is given from (3.2.9) by

$$P(\psi \geq \psi_0) = \frac{e^{\kappa \cos \psi_0} - e^{-\kappa}}{2 \sinh \kappa} \ . \tag{3.2.26}$$

Typically, neither the true mean direction nor the true precision κ is known. Thus, in order to perform the test these parameters are replaced in the equations by their maximum likelihood estimates. The maximum likelihood estimate k for κ is given in (3.2.24) as $N/(N-R)$, and that for the true mean direction is just the direction of the vector resultant of the individual unit vectors. It is therefore necessary to compare the data with the probability density of (3.2.9), the probabilities being given in (3.2.26), with k replacing κ and ψ' (the angle between an observed direction and the maximum likelihood estimate of the true mean direction) replacing ψ. If the true mean direction is known, then ψ is known directly. It is also necessary to test whether the azimuthal angle ϕ is uniformly distributed between 0 and 360°. Again, the true mean direction or its maximum likelihood estimate is needed to calculate the values of ϕ.

From (3.2.26), for N observations the expected frequency f_e of ψ' between the limits ψ'_1 and ψ'_2 is given by

$$f_e = \frac{N}{2\sinh\hat{k}}\left[e^{\hat{k}\cos\psi'_1} - e^{\hat{k}\cos\psi'_2}\right]. \qquad (3.2.27)$$

Note that previous texts (e.g., Irving, 1964) have used the approximation for large $\hat{k}$ that $2\sinh\hat{k} \approx \exp(\hat{k})$. With modern calculators and computers it is just as easy to use the correct equation (3.2.27).

To test both the azimuthal and radial distributions the observations are separated into m classes or ranges and the statistic X^2 is calculated from

$$X^2 = \sum_{i=1}^{m}\frac{(f_0 - f_e)^2}{f_e} = \sum_{i=1}^{m}\frac{f_0^2}{f_e} - N \qquad (3.2.28)$$

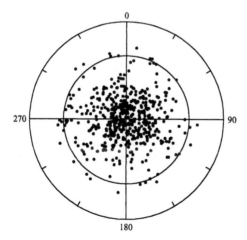

Fig. 3.3. North pole stereographic projection of VGPs observed in 508 lava flows that lie between 35° and 40° north or south latitude and have ages ≤5 Ma. The expected mean VGP is the North Geographic Pole. From McElhinny and McFadden (1997).

where f_o and f_e are the observed and expected frequencies within the range. If the observations were drawn from a Fisher distribution then X^2 will be chi-square distributed with ν degrees of freedom, χ^2_ν, where

$$\nu = m - 1 - \Pi \qquad (3.2.29)$$

and Π is the number of distribution parameters that had to be replaced by their maximum likelihood estimates in order to calculate the observed frequencies.

An example of this test is given in Fig. 3.3 for the observed VGPs from lava flows that lie within the latitude band $|35–40|°$ on the Earth's surface and have ages ≤ 5 Ma. The details of the test for a Fisher distribution are given in Table 3.1. For both the azimuthal (longitude) and radial (colatitude) variation the 508 observations were divided into 10 cells of equal expectation, so $m = 10$. Typically it would be necessary to estimate the true mean direction (i.e., two parameters) and the precision κ so that Π would be 3 for testing the radial (ψ) distribution and Π would be 2 for testing the azimuthal (ϕ) distribution (the precision is not needed to test the azimuthal distribution). In this case the expected mean VGP is known because it is the North Geographic Pole. Hence, only the precision needs to be estimated and so the two values of Π are 1 and 0. The calculated (observed) value of X^2 in each case is less than the critical value at the 5% level. Therefore, there is no reason to reject the hypothesis that the observations conform with a Fisher distribution.

Graphical methods may also be used to test the goodness-of-fit of the Fisher distribution to a set of data (see e.g., Lewis and Fisher, 1982; Fisher *et al.*, 1987).

TABLE 3.1

Test for Conformance of Observations to a Fisher Distribution for 508 Lava Flows in the Latitude Range $|35–40|°$ of Age ≤ 5 Ma[a]

	Azimuthal range (longitude)									
	0-36	36-72	72-108	108-144	144-180	180-216	216-252	252-288	288-324	324-360
Observed	52	48	49·	53	41	37	59	59	50	60
Expected	50.8	50.8	50.8	50.8	50.8	50.8	50.8	50.8	50.8	50.8

$\nu = 9$ $\quad$ X^2(observed) = 10.31 $\quad$ χ^2_9 (critical) = 16.92

	Radial range (colatitude)									
	0-5.1	5.1-7.4	7.4-9.4	9.4-11.4	11.4-13.4	13.4-15.4	15.4-17.7	17.7-20.7	20.7-25.8	>25.8
Observed	60	63	54	51	45	45	46	43	40	61
Expected	50.8	50.8	50.8	50.8	50.8	50.8	50.8	50.8	50.8	50.8

$\nu = 8$ $\quad$ X^2(observed) = 12.12 $\quad$ χ^2_8 (critical) = 15.51

[a] From McElhinny and McFadden (1997)

Analysis of Inclination-Only Data

There are many situations in which declination information is not available. The common case is that of deep-sea cores, but bore cores drilled on land are often not oriented or may be available only as a series of fragmentary pieces.

If I_0 is the inclination of the true mean direction then a simple average of the observed sample inclinations will give an estimate of I_0 that is too shallow (except for $I_0 = 0$). Suppose that the true mean inclination $I_0 = 90°$. If the directions are symmetrically distributed about this mean (as expected for a Fisher distribution), virtually all of the observed inclinations will be <90°. Their arithmetic mean will therefore always be less than the true value. If the true mean has ($D_0 = 0°$, $I_0 = 80°$), an observation at (180°, 85°) is actually 15° from the true mean, but without the declination information it is "folded back" to being only 5° from the mean. If the foldback problem is small, it is possible to obtain an approximately unbiased estimate of both I_0 and the precision κ. This has been examined by Briden and Ward (1966), Kono (1980b), McFadden and Reid (1982), and Cox and Gordon (1984).

Following the most general analysis given by McFadden and Reid (1982), consider a set of observations I_1, I_2, ..., I_N drawn from a Fisher distribution whose true mean direction has an inclination I_0. Define θ_0 as the complement of I_0 (i.e., $\theta_0 = 90° - I_0$), θ_i as the complement of I_i, and $\hat{\theta}_0$ as the maximum likelihood estimate (which in this instance is approximately unbiased) for θ_0. Let

$$C = \Sigma\cos(\hat{\theta}_0 - \theta_i) = \cos\hat{\theta}_0\Sigma\cos\theta_i + \sin\hat{\theta}_0\Sigma\sin\theta_i$$
$$S = \Sigma\sin(\hat{\theta}_0 - \theta_i) = \sin\hat{\theta}_0\Sigma\cos\theta_i - \cos\hat{\theta}_0\Sigma\sin\theta_i \ ,$$

(3.2.30)

where the summations are for i from 1 to N. The value of $\hat{\theta}_0$ is then a solution of

$$N\cos\hat{\theta}_0 + (\sin^2\hat{\theta}_0 - \cos^2\hat{\theta}_0)\Sigma\cos\theta_i - 2\sin\hat{\theta}_0\cos\hat{\theta}_0\Sigma\sin\theta_i = 0 , \quad (3.2.31)$$

which may be obtained by any simple iterative method. The correct solution is obvious as it is close to the average of the θ_i. The best estimate $\hat{I}_0$ for I_0 is then

$$\hat{I}_0 = (90° - \hat{\theta}_0)$$

(3.2.32)

with confidence limits (in degrees)

$$(\hat{I}_0 + \frac{180S}{\pi C} - \alpha_{(1-g)}) \le I_0 \le (\hat{I}_0 + \frac{180S}{\pi C} + \alpha_{(1-g)}),$$

(3.2.33)

where

$$\cos\alpha_{(1-g)} = 1 - \frac{1}{2}\left(\frac{S}{C}\right)^2 - \frac{g(N-C)}{C(N-1)}$$

(3.2.34)

and g is the critical value of $F[1,(N-1)]$ at the appropriate level of confidence. If $g = 0.05$ then the level of confidence is 95%. Note that the confidence limits are not symmetric about $\hat{I}_0$.

The best estimate k of the precision parameter κ is given by

$$k = \frac{N-1}{2(N-C)} \qquad (3.2.35)$$

and is the equivalent of (3.2.15).

3.2.4 Calculating Paleomagnetic Poles and Their Errors

The paleomagnetic pole may be calculated from the formation mean direction of magnetization (D_m, I_m) according to (1.2.6–1.2.8). The mean direction has its associated circle of confidence α_{95} and if δD_m and δI_m are corresponding errors in declination and inclination, then

$$\alpha_{95} = \delta I_m = \delta D_m \cos I_m . \qquad (3.2.36)$$

From (1.2.5) the error δI_m in the inclination corresponds to an error dp in the ancient colatitude p given by

$$dp = \tfrac{1}{2}\alpha_{95}(1 + 3\cos^2 p) . \qquad (3.2.37)$$

The error dp lies along the great circle passing through the sampling site S and the paleomagnetic pole P, and is the error in determining the distance from S to P (see Fig. 1.12). The error in the declination corresponds to a displacement dm from P in the direction perpendicular to the great circle SP, where

$$dm = \alpha_{95} \frac{\sin p}{\cos I_m} . \qquad (3.2.38)$$

The error (dp, dm) is termed the *oval of 95% confidence* about the mean pole.

Alternatively, each site direction may be specified by a virtual geomagnetic pole (VGP). The VGPs themselves may then be analyzed as Fisher-distributed observations since they correspond to points on a unit sphere. The declination is replaced by the VGP longitude east and the inclination is replaced by the VGP latitude in each case. The mean pole position will have its associated precision parameter estimate (K) and circle of 95% confidence (A_{95}). To distinguish between the analysis of VGPs and that of directions, capital symbols are always used for the analysis of poles. Khramov (1987) suggests that A_{95} and (dp, dm) can be related approximately by

$$A_{95} \approx \sqrt{dp \cdot dm} . \qquad (3.2.39)$$

It has been traditional to calculate paleomagnetic poles by first calculating the mean of the site mean directions of magnetization and then calculating the corresponding paleomagnetic pole and oval of 95% confidence given by (dp, dm). However this traditional approach has been challenged on several points. Because the mapping between the directions of magnetization and poles is nonlinear, it is not possible for both distributions to be Fisher. Creer *et al.* (1959) first noted, from studies of the paleomagnetism of the Whin Sill, that the VGPs were more likely to have a circular distribution. As a consequence it is usually assumed that VGPs have a Fisher distribution in studies of paleosecular variation (see §1.2.6), following Cox (1962). This is supported by the analysis of lava flow data for the past 5 Myr (McElhinny and McFadden, 1997) as analyzed in Table 3.1 and illustrated in Fig. 3.3. Kono (1997) used a model for the geomagnetic field in which the Gauss coefficients (§1.1.3) are random normal variates (based on the model of Constable and Parker, 1988) and showed that for such a model the distribution of the VGPs is nearly circular while that of the field directions is not. Arason and Levi (1997) show that the transformation of isotropically distributed geomagnetic poles to local site directions introduces slight apparent inclination shallowing when the directional average is compared with the geocentric axial dipole field direction. Overall it appears that the optimum procedure is first to calculate the VGPs corresponding to each site mean direction and then to average those VGPs to determine the paleomagnetic pole and its corresponding circular error A_{95}.

3.2.5 Other Statistical Distributions

It is sometimes observed that a set of directions or VGPs has an elliptical distribution rather than the circular distribution required for the Fisher distribution. A statistical distribution that allows for the treatment of such data is the *Bingham distribution* (Bingham, 1964, 1974; Onstott, 1980). The distribution is bimodal so that, unlike the unimodal Fisher distribution, one does not have to invert one of the two polarities usually observed in paleomagnetic data (which implicitly assumes that there is an axially symmetric dipole field). The distribution requires two concentration parameters κ_1 and κ_2 whose relative magnitudes describe the azimuthal dependence of the dispersion. For an isotropic distribution dispersion $\kappa_1 = \kappa_2 = \kappa$, where κ is similar to the precision for the Fisher distribution.

However, the Fisher distribution provides simple techniques for determining confidence limits and for two-sample hypothesis tests such as the comparison of mean directions. This is not the case for the Bingham distribution, so it is not used much in the analysis of paleomagnetic data. A further, and philosophically more serious, problem is that the origin of the elliptical distributions sometimes observed is not fully understood and may well relate to analytical artifacts, such as nonuniform overprinting or the presence of apparent polar wander during the time span covered by the measurements. Thus the ellipticity may be a consequence of systematic components, which is inconsistent with a model of random variations.

Of course, many paleomagnetic data sets depart noticeably from a Fisher distribution. However, in most instances, the impact of such departures upon conclusions is relatively small. Despite this, several workers have chosen to use nonparametric statistical procedures such as bootstrap methods (see e.g., Fisher *et al.*, 1987; Constable and Tauxe, 1990; Tauxe *et al.*, 1991; Tauxe, 1998) to avoid the inaccuracies of using an ill-matching distribution. The disadvantage of such nonparametric techniques is that all of the observations have to be available for hypothesis tests. In contrast, with the assumption of a Fisher distribution, a data set can be characterized by four parameters, which is all that is then needed for hypothesis tests.

3.3 Field Tests for Stability

3.3.1 Constraining the Age of Magnetization

In favorable circumstances it is possible to arrange paleomagnetic sampling so that field tests can be performed to constrain the age of magnetization. Graham (1949) proposed two such tests referred to as the *fold test* and the *conglomerate test*. Later, Everitt and Clegg (1962) introduced the *baked contact test*. The circumstances for each of these tests are illustrated in Fig. 3.4. More recently Kirschvink (1978) proposed the *unconformity test*, which is only applicable in special situations.

In the classical fold test, samples are taken from separate limbs of a fold. If the *in situ* directions of magnetization on the separate limbs differ, but agree after "unfolding" the limbs to the horizontal, then the magnetization must predate the folding and must have been stable since that time. In some cases it is observed that the maximum agreement of the directions occurs not on restoring the folds completely to the horizontal but at some intermediate folding correction. In this situation it is believed that the magnetization was imprinted during the course of the folding as determined by McClelland Brown (1983) and is referred to as a *synfolding magnetization*. The classical fold test assumes rigid-body rotation of the folds; however, the development of folds in rock formations relieves compressive tectonic stress and can involve large internal strains. In certain situations it is possible to "unstrain" the natural remanence in deformed rocks

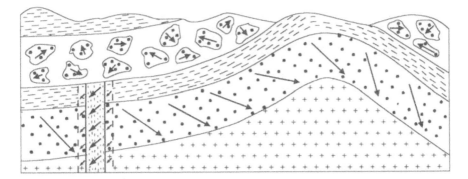

Fig. 3.4. Field relationships for the fold, conglomerate, and baked contact tests for paleomagnetic directions. From Merrill and McElhinny (1983).

and then perform a fold test. This is referred to as a *fold test with strain removal*. Details of each of these aspects of the fold test and related statistical methods are discussed in §3.3.2.

In the conglomerate test, if randomly oriented cobbles in a conglomerate were derived from the rock formation under investigation and the magnetization of these cobbles has no preferred direction, then this suggests the magnetization of the parent rock formation has been stable since the formation of the conglomerate (see §3.3.3).

In the baked contact test, if the directions of magnetization in the baked zone surrounding an intrusion are parallel to those observed in the intrusion, but differ from that of the unbaked country rock, then the magnetization of the intrusion has been stable since formation (see §3.3.4). The inverse is also true; the stable magnetization observed in the country rock, if both consistent and different from that of the intrusion, has obviously survived since the age of the intrusion. This is referred to as the *inverse contact test*.

3.3.2 The Fold Test

When applying the fold test of Graham (1949) it is generally assumed that the rocks have undergone rigid-body rotation. Rocks are assumed restored to their original untilted position by rotation of the bedding around the strike line back to the horizontal. This treatment should strictly apply only in the case of concentric folds with a horizontal axis without any strain. Often more elaborate unfolding methods are required where the axis is not horizontal, and there is then a discrepancy between net and apparent tectonic rotation due to plunging fold axes as discussed by MacDonald (1980). When there is internal deformation or strain, more sophisticated methods are required (Cogné and Perroud, 1987), which are discussed later in this section.

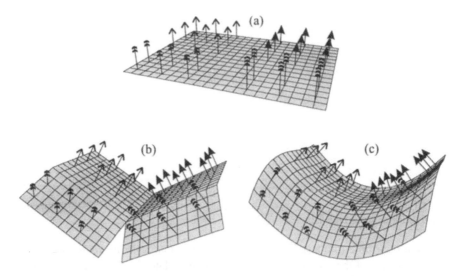

Fig. 3.5. The fold test. (a) Original geometry of rock stratum with arrows drawn perpendicular to the surface. (b) Idealized folding geometry as required for the McFadden and Jones (1981) test. (c) More typical actual folding geometry allowed for in the extended test of McFadden (1998). After McFadden (1998).

Statistical Evaluation of the Fold Test

The first statistical method for analyzing the fold test was suggested by McElhinny (1964). He used the method for the comparison of precisions given by (3.2.16) as a means of deciding whether the estimates of the precision k after restoring beds to the horizontal was significantly improved from the value obtained when the beds were *in situ*. The test was easy to use and became the standard practice for many years. Later McFadden and Jones (1981) pointed out that the criterion used by McElhinny (1964) was actually invalid but in practice was typically conservative.

The basic geological geometry required for the McFadden and Jones (1981) test is outlined in Fig. 3.5 in which the arrows are perpendicular to the surface. Figure 3.5a shows the original geometry of a rock stratum, and Fig. 3.5b shows four "limbs", the type of idealized folding geometry assumed in the test. With such an idealized geometry it is possible to use a sampling scheme that ensures there are several sites on each limb. The sites on each limb may be considered as a separate group with each group having its own common tilt correction. It is thus possible to get an estimate of the original scatter in site mean directions from each group. The requirement for a common original population demands that the scatter observed in different groups be consistent with a common precision κ.

Given the above as an essential starting point, the McFadden and Jones (1981) test determines, with the rocks in a given orientation, whether the scatter observed between groups (on different fold limbs) is consistent with the observed scatter within groups, that is, whether the different groups could have been drawn from a common distribution. If each of the groups has n observations and an estimated precision parameter of about k, then the group means can be expected to have a precision of about nk. If the precision of the groups were to differ too much from this, one would conclude that the scatter observed between groups is not consistent with the observed scatter within groups and therefore it is unlikely that the magnetization was acquired with the rocks in that particular orientation. Thus, the test is one of deciding whether the different groups share a common mean direction.

Suppose that there are m limbs that have been sampled and that R_i is the length of the resultant vector from the ith limb. If N is the total number of site mean directions and R the length of their resultant vector, then if the observations are all drawn from a common distribution the statistic f, given by

$$f = \left(\frac{N-m}{m-1} \right) \frac{\Sigma R_i - R^2 / \Sigma R_i}{2(N - \Sigma R_i)} \sim F[2(m-1), 2(N-m)], \qquad (3.3.1)$$

has an approximate F distribution with $2(m-1)$ and $2(N-m)$ degrees of freedom, where the summations are for $i=1$ to m. This is just the test of (3.2.24). Given the tilt corrections for each group and the particular set of observations, ΣR_i is constant, but R varies as the group mean directions change with respect to each other according to the tilt correction. The maximum value of R is ΣR_i, so the term $(\Sigma R_i - R^2/\Sigma R_i)$ in the numerator of (3.3.1) gives the dispersion of the group mean directions independent of the dispersion within groups. Similarly, the term $(N-\Sigma R_i) = \Sigma(N_i - R_i)$, where N_i is the number of observations in each group, in the denominator of (3.3.1) is the sum of the dispersions within the groups. Thus f gives the ratio (with division by the appropriate degrees of freedom) of the dispersion of group means to the within-group dispersion.

The expectation of f is given by $\langle f \rangle = (N-m)/(N-m-1)$, so f can be expected to be close to unity if the rocks are in the orientation in which the magnetization was acquired. If the observed value of f is much greater than unity, then this means that the dispersion of group mean directions is too large compared with the within-group dispersion. Conversely, if the observed value of f is much smaller than unity, then this means that the dispersion of group mean directions is too small compared with the within-group dispersion.

The critical values of f for the two situations (f too large or f too small) are determined using different ends of the F distribution. For f too large, if the observed value of f exceeds the critical value of the F distribution at the 5% level of significance, then the null hypothesis of a common true mean direction may be rejected (the group means are too widely dispersed). For f too small, if the observed value of f is smaller than the critical value of the F distribution at the 95% level of significance, then the group means are artificially too close to each other and the null hypothesis may again be rejected. Many F distribution tables only provide 5% significance points. To overcome this, it is sufficient to recognize that the 95% significance point for f is the same as the 5% significance point for $1/f$ with the relevant degrees of freedom swapped (i.e., the 5% significance point from the table for $F[2(N-m),2(m-1)]$ should be used). When rejection of the null hypothesis occurs, the values of f are then sometimes calculated at successively increasing values of the tilt correction from 0 to 100% to investigate whether a synfolding magnetization exists.

In practice it is difficult to apply the McFadden and Jones (1981) test because the ideal situation shown in Fig. 3.5b is rarely applicable, and often in practice it is not possible to sample more than one site on a limb. Figure 3.5c shows a more realistic situation that is more commonly observed in the field. No two of the arrows perpendicular to the surface are parallel and it is not possible to define a limb containing more than one site across which there is a common tilt correction. McFadden (1990) therefore developed a new test based on correlation between the distribution of magnetic directions and the tectonic information to cater for most of the more complicated situations. Although this is an effective test it is not apparently physically intuitive and has not gained widespread use. Therefore, McFadden (1998) then extended the McFadden and Jones (1981) test to the general case illustrated in Fig. 3.5c to make it more flexible and make few demands on the sampling scheme.

The extended test proposed by McFadden (1998) is based on the observation that if the rocks are in the orientation in which the magnetization was acquired then it matters little how the sites are grouped together to perform the McFadden and Jones test. Thus, with some loss of power, it is acceptable to group the sites together into clusters based on similar tilt corrections. The sites are initially clustered into groups with similar tilt corrections using the clustering algorithm of Shanley and Mahtab (1976). No clustering algorithm always gives the best set of clusters, so the software provided to perform the test is written so that it is a simple matter to alter the clusters using the mouse.

An example of the application of the McFadden (1998) extended McFadden and Jones test is illustrated in Fig. 3.6 from the data of Gilder *et al.* (1993). In Fig. 3.6a all the site mean directions are illustrated at 0, 70, and 100% unfolding. The site mean directions become most tightly grouped at 70% unfolding. However, the observed values of f depart significantly from unity at 0% ($f = 11.6$), at 70% ($f = 3.1$) and at 100% ($f = 4.2$) unfolding. Therefore, the position of maximum concentration (at 70% unfolding) does not necessarily identify the position at which the magnetization was acquired. This suggests that there may be some problem with the data. Gilder *et al.* (1993) noted that there were

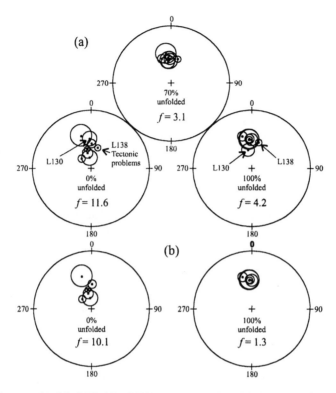

Fig. 3.6. An example of the McFadden (1998) extended version of the McFadden and Jones (1981) fold test showing that the maximum concentration of directions is not a sufficient criterion for evaluating the significance of the fold test. Stereographic projection showing site mean directions of magnetization (all on the lower hemisphere) with their circles of 95% confidence as observed in some Chinese sediments by Gilder *et al.* (1993). (a) All site mean directions are illustrated at 0, 70, and 100% unfolding. (b) The site mean directions at 0 and 100% unfolding with the two suspect sites L130 and L138 removed.. After McFadden (1998).

tectonic problems associated with site L138, and site L130 seems to have quite a different dispersion from that of the other sites. Figure 3.6b shows the data with these sites removed. At 0% unfolding $f = 10.1$ and at 100% unfolding $f = 1.3$, showing that the group mean and within-group dispersions are now statistically indistinguishable at 100% unfolded. Therefore, the magnetization was most likely acquired at 100% unfolding (probability of exceeding the observed value of $f = 1.3$ is 26.7%), giving a positive fold test.

McCabe *et al.* (1983) noted that a significantly more concentrated distribution does not necessarily imply the correct result, yet it has become quite common to use the criterion that the maximum concentration of site mean directions identifies the position in which the rocks acquired their magnetization (Watson and Enkin, 1993; Tauxe and Watson, 1994). McFadden (1998) showed that this underlying assumption is flawed because such concentration can also occur through specific relationships between the magnetization directions and the fold axes. The example given above illustrates well the fact that the position of maximum concentration of site mean directions does not necessarily identify the position at which the magnetization was acquired.

Synfolding Magnetizations

In a paleomagnetic study of fold development and propagation in the Devonian Old Red Sandstone, McClelland Brown (1983) showed that the magnetizations observed on different parts of a fold converged when the fold was partially unfolded (40% unfolded). The resulting direction of magnetization was the same as that observed in Late Carboniferous rocks, the time at which the development of the fold took place. Therefore it was proposed that the magnetization was acquired during the process of folding at the time when the fold was about 40% complete. However, some caution is required for the reliable identification of synfolding magnetization because what appears to be a synfolding magnetization may be the consequence of a complex fold with multiple rotations (see e.g., Tauxe and Watson, 1994). Sometimes it may also be the result of internal strain of a pre-existing magnetization (van der Pluijm, 1987; Cogné and Perroud, 1987). This is discussed further when considering the fold test with strain removal.

Shipunov (1997) addressed some aspects of detecting a synfolding magnetization. For folding to occur there must be brittle fracture or plastic flow at an elevated temperature. It is the internal strains and elevated temperatures that can lead to the acquisition of a synfolding magnetization. An example for a simple situation is shown in Fig. 3.7. In the unstrained case, when the classical rigid-body tilt correction is applied, the vectors cluster about their initial direction. When strain occurs, the strain induces a rotation of the vectors toward the cleavage plane so that on applying classical rigid-body unfolding the vectors bypass each other at 100% unfolding but appear to converge at some intermediate percentage unfolding. A false synfolding magnetization might then be deduced.

However, it is extremely unlikely that a synfolding magnetization would be acquired contemporaneously and at any one point in time at different locations. The process that leads to the acquisition of a synfolding magnetization suggests that the magnetization will be acquired dynamically as the folding proceeds. Thus, one would expect a synfolding magnetization to be a composite of several magnetizations whose directions would lie on a small circle with pole parallel to the folding axis for that local site. This would imply that the only way to identify a synfolding magnetization with certainty would be to identify, during demagnetization, components with these characteristics. Having identified such a

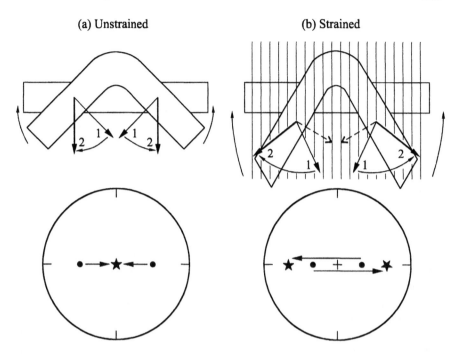

(a) Unstrained   (b) Strained

Fig. 3.7. A false synfolding magnetization can be deduced in the case of strained beds. The initial magnetization is assumed vertical to the beds. (a) When the beds are unstrained, applying the tilt correction on the assumption of rigid-body rotation (vector 1 to vector 2) makes the magnetization vectors converge as shown in the stereoplot below (star). (b) When the beds are strained, the strain induces a rotation of the vector (dashed vector to vector 1) toward the cleavage plane (shown as light vertical lines). When rigid-body rotation is assumed (100% unfolding, vector 1 to vector 2), the vectors bypass each other (stars) and are just as scattered as *in situ*. However, they would converge at some intermediate percentage unfolding and a false synfolding magnetization might be deduced. After Cogné and Perroud (1987).

synfolding magnetization, it is clear that one cannot then uniquely identify the actual direction of that magnetization from the statistics (McFadden, 1998).

Fold Test with Strain Removal
Most fold test analyses have assumed that the rocks have undergone rigid-body rotation during folding. However, if there is a change in the relative position of the particles within the rock, this can change the shape of the body, causing internal deformation or strain. Cogné and Perroud (1987) show that in the general case there is a deflection of the magnetization vector toward the flattening plane (as shown in Fig. 3.7) and probably toward the elongation axis. In the general case, therefore, strain alters the angular relationship between the tilt plane and the magnetization, leading to inconclusive results when the classical rigid-body fold test is applied (Fig. 3.7).

When a rock unit contains visibly deformed objects, such as fossils, ooliths, and reduction spots, it is easy to use these strain markers to estimate the strain state by applying the well-known tectonic and microtectonic techniques of strain analysis (e.g., Ramsay, 1967). Where such markers are rare, a strain-induced modification to some physical property can be used. The measurement of the anisotropy of magnetic susceptibility (AMS) provides a rapid method of estimating the preferred orientation of magnetic minerals. The magnetic fabric of a rock is expressed mathematically as a second-order tensor, analogous to the strain tensor. These two tensors have been shown to be related both in the direction of their axes and in the magnitude of their principal values. However, the quantitative relationship between the magnitudes of strain and AMS principal axes must be established for every rock type, and for each rock type there is a minimum number of classical strain markers generally needed.

Cogné and Perroud (1985) and Cogné *et al.* (1986) suggested a method for correcting or unstraining the paleomagnetic vectors. The treatment is based on the passive deformation theory of March (1932). The correction consists of applying the inverse strain tensor first to the magnetic vector and then to the bedding plane. The initial direction of magnetization is then recovered by a tilt correction using the unstrained bedding plane. Cogné and Perroud (1985) and Cogné *et al.* (1986) successfully applied the method to some deformed Permian and Ordovician redbeds. However, Lowrie *et al.* (1986), although acknowledging the need for correcting the bedding plane for strain before any rigid-body rotation, criticized the validity of applying inverse strain to paleomagnetic vectors. Cogné and Perroud (1987) dispute this and provide a useful summary of the present state of the debate regarding the unstraining of paleomagnetic vectors.

3.3.3 Conglomerate Test

To use this test it is necessary to identify conglomerate cobbles in either the same formation or another formation that have been derived from the beds whose stability of magnetization is being investigated, as illustrated in Fig. 3.3. If the directions of magnetization of the conglomerate cobbles are random then this suggests that the magnetization of the parent formation has been stable since the deposition of the conglomerate (Graham, 1949). The cobbles need to be large enough so that the mechanical forces far exceed the magnetic aligning forces during deposition. The test for randomness of directions may then be used (§3.2.3). This is a general test and Fisher *et al.* (1987) and Shipunov *et al.* (1998) provide more powerful tests against specific alternative hypotheses. Breccias by their nature imply minimum transport of material and in this case the randomizing effect of the mechanical forces involved can be insufficient for them to be useful for the application of the conglomerate test.

The test can be enhanced if the cobbles can be shown to have the same demagnetization characteristics as the parent beds (e.g. Buchan and Hodych, 1989). If the cobbles have still retained indications of their original stratification, it might be possible to restore the magnetic directions to the paleohorizontal. In this case only the resulting inclinations would have significance and, when corrected to the horizontal, they should converge. However, the tops or bottoms of the bedding in the cobbles are likely to be unknown so that the inclinations after correction may be positive or negative without necessarily indicating reversals (Van der Voo, 1993). A special case of the conglomerate test occurs when the conglomerate itself lies within the formation and was deposited not long after the parent beds were deposited. This is referred to as an *intra-formational conglomerate test* and, if the test is positive, can provide an important constraint that the age of the magnetization of the parent beds dates from the time of formation.

3.3.4 Baked Contact Test

When an igneous rock intrudes an older rock formation the intrusion heats a region of the surrounding country rock. The baked region of the country rock will then cool in the same magnetic field as the intrusion and so will acquire a direction of magnetization the same as that of the intrusion. In addition, the direction of the magnetization in the unbaked country rock will differ from that of the intrusion. Since the country rock and the igneous intrusion are generally very different rock types, agreement between the direction of magnetization of the intrusion and that of the baked region of the country rock provides strong evidence for the stability of the magnetization of the intrusion. This situation also applies to the baked rock underlying an extruded lava flow. To complete the test it is necessary to show that the direction of magnetization of the unbaked country rock differs from that of the intrusion because this shows that neither the intrusion nor the unbaked country rock has been subjected to any regional remagnetization. However, in certain situations the test can be shown to be positive if it can be demonstrated that there are changes in properties with distance from the baked contact that correspond with the diminishing effects of the intrusion (Everitt and Clegg, 1962).

It is useful to define three gradational zones outside the igneous body as illustrated in Fig. 3.8. In the *metamorphic zone* there will be extensive changes in the magnetic minerals, whereas in the *heated zone* these changes will be comparatively small, although in both cases the rock will have acquired a total TRM from the reheating episode. In the *warmed zone*, the temperature never rises to the Curie temperature of the magnetic minerals and only a partial TRM is induced. This then combines vectorially with any component in the country rock that is stable to the temperatures involved in the warmed zone. Outside the

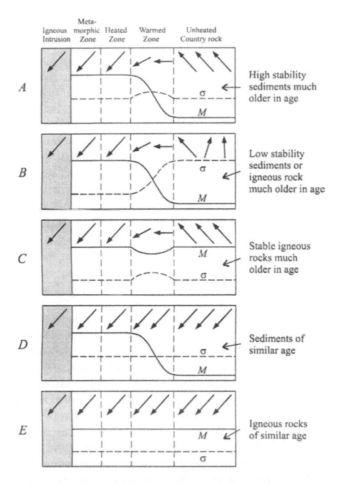

Fig. 3.8. The baked contact test. The variation in direction of magnetization (arrows), magnetization (*M*), and scatter (σ) with distance from an igneous intrusion in five possible situations is illustrated schematically. After Irving (1964), with permission from John Wiley & Sons.

warmed zone is the unheated country rock. Five possible situations are illustrated in Fig. 3.8.

In the first case (Fig. 3.8A), the country rock is much older than the intrusion and consists of sedimentary material of high stability. The occurrence of high magnetization (*M*) at the contact falling to low values at a distance, coupled with agreement between the directions of magnetization of the intrusion and baked contact rock, indicates that both are stable and that the magnetization was acquired at the time the igneous intrusion cooled. The increase in the intensity of *M* may be up to a factor of 10, since the magnetization of the sediment is usually due to CRM, whereas the baked rock now has a TRM. The scatter ($\sigma = 1/\kappa$) of the directions may or may not change, except in the warmed zone due to the addition of two differing vector directions. Consistent directions in the unheated sedimentary rock, which differ from both that of the present Earth's field and that of the intrusion, provide further evidence for the stability of the unheated sedimentary rock (other than evidence from consistency or laboratory tests, etc.). The baked contact test now provides evidence that the magnetization of the unheated sediment has been stable at least since the time of the intrusion. This is referred to as the inverse contact test as defined in §3.3.1.

In Fig. 3.8B, the country rock is still much older than the intrusion but consists of sedimentary or igneous material of low magnetic stability. M decreases from the contact into the unheated zone, whereas σ increases and scattered directions are observed. In Fig. 3.8C, the country rock is again much older than the intrusion but consists of stable igneous material. This situation is similar to that in Fig. 3.8A but in this case no change in the intensity of M need be observed between the baked rock and the unheated zone. The magnetization of the unheated country igneous rock is thus shown to have been stable since the time of the intrusion representing another example of an inverse contact test.

When the country rock is only a little older or of comparable age to that of the intrusion, the situation is less favorable. If the country rock is sedimentary material (Fig. 3.8D) the main variation will be a decrease in the intensity of M between the baked zone and the unheated zone, a change that could be up to a factor of 10 if the original magnetization in the sedimentary rocks was a CRM. The most unfavorable situation (Fig. 3.8E) is when the country rock is stable igneous rock of the same age as the intrusion. No variations will be seen from the baked zone to the unheated zone, a situation that could also occur from general heating due to a period of regional metamorphism.

It should be noted that the inverse is also true. If stable magnetizations are observed in unbaked rocks that provide positive evidence for a baked contact test such as illustrated in Figs 3.8A and Fig. 3.8C, then this also provides evidence that the unbaked sediments have retained their magnetization at least since the time of the baking. This is referred to as the *inverse contact test*.

3.3.5 Unconformity Test

The unconformity test can be applied in the special case when successive zones of normal and reverse magnetization are truncated by an unconformity in the sequence, as proposed by Kirschvink (1978). In the example shown in Fig. 3.9, the lower part of a rock sequence has recorded zones of normal and reverse magnetization. Sedimentation ceased and erosion took place to produce the unconformity prior to deposition of the upper part of the sequence. If the polarity zones in the upper younger sequence do not match those in the lower older

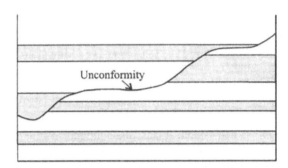

Fig. 3.9. The unconformity test. Zones of normal (shaded) and reverse (white) magnetizations in the lower sequence are truncated by an unconformity. If the polarity zonation in the upper sequence differs from that of the lower sequence, then the magnetization of the lower beds predates the unconformity. After Kirschvink (1978), with permission from Elsevier Science.

sequence, then the magnetization in the lower beds is older than the episode of erosion that created the unconformity. On the other hand, if the polarity zones show continuity across the unconformity, then the sediments must carry a CRM that postdates the younger sedimentation.

3.3.6 Consistency and Reversals Tests

If a single geological unit or formation can be sampled over a wide area and through a considerable thickness, in which are represented a variety of rock types with differing mineralogy, and if consistent directions of magnetization are observed, then there is good reason to believe that the magnetization has been stable since the time of formation. This is generally referred to as the *consistency test*. However, it is important that the directions observed should also differ from that of the present Earth's field or of any other field direction that has been observed in rocks of younger age from the same craton or block. Agreement with known directions of a younger age can be a strong indication that remagnetization has taken place.

The presence of reversals of magnetization represented by two groups of directions that are 180° apart is a much stronger consistency test than simple consistency of directions without reversals. If, subsequent to formation, both groups acquire a secondary component of magnetization, they will both change in the same direction towards the secondary magnetic field direction. The two resultant groups of directions will then not be 180° apart (Fig. 3.10). It is thus necessary to be able to test whether these opposing directions of magnetization differ discernibly from being 180° apart. Such a test is called the *reversals test*. Originally the simple procedure was to invert one of the directions by 180° and then test if the resulting two directions of magnetization were discernibly

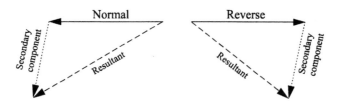

Fig. 3.10. Normal and reverse magnetizations originally 180° apart each acquire a secondary component causing the resultant directions of magnetization to deviate from being antiparallel.

different using the comparison method discussed in §3.2.3. However, McFadden and McElhinny (1990) showed that this method is flawed because the larger the circles of confidence the harder it would be to show that the two directions were discernibly different from one another. Thus the worst data sets were the easiest ones with which to achieve a positive reversals test!

McFadden and McElhinny (1990) therefore formalized the computation of the reversals test by using a classification system based on the amount of information that is available for the test. First, one of the two directions and its circle of confidence is rotated through 180° for comparison with the other direction. The angle γ_o between the two means can then be calculated and it is also possible to calculate the angle γ_c at which the two directions become significantly different at the 95% confidence level. The reversals test obviously fails if $\gamma_o > \gamma_c$. Otherwise the reversals test is classified as "A" if $\gamma_c \leq 5°$, as "B" if $\gamma_c \leq 10°$, and "C" if $\gamma_c \leq 20°$ and as "indeterminate" if $\gamma_c > 20°$. An example is given in Fig. 3.11. The computation of the critical angle γ_c will depend on the

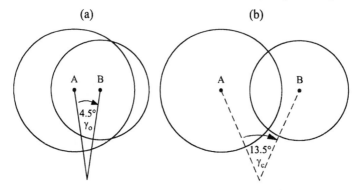

Fig. 3.11. Example of the classification of the reversals test (McFadden and McElhinny, 1990). One of the two opposing mean directions, together with its circle of 95% confidence, is rotated by 180°. (a) The two mean directions, A and B, are separated by $\gamma_o = 4.5°$. (b) These must be separated by an angle $\gamma_c = 13.5°$ to differ significantly at the 95% confidence limit. The reversals test is then classified as "C". Note that in (b) the circles of 95% confidence still overlap slightly following (3.2.18) in §3.2.3.

statistical properties of the normal and reverse populations and on the number of observations. If the number of samples in each population is small or if the two populations do not share a common precision, calculation of γ_c should be performed using a simulation or bootstrap method. McFadden and McElhinny (1990) show how to do this using a suggestion made by Watson (1983).

In the common case where there are several observations N_1 and N_2 in the two populations having resultants R_1 and R_2 and they share a common precision, the critical angle γ_c may be calculated from

$$\cos\gamma_c = 1 - \frac{(N - R_1 - R_2)(R_1 + R_2)}{R_1 R_2}\left[\left(\frac{1}{P}\right)^{\frac{1}{N-2}} - 1\right]. \qquad (3.3.2)$$

where $N = N_1 + N_2$ and $P = 0.05$ (McFadden and McElhinny, 1990).

3.4 Laboratory Methods and Applications

3.4.1 Progressive Stepwise Demagnetization

After collection and measurement of their NRM, rock samples are subjected to various methods of progressive demagnetization in the laboratory in an attempt to elucidate the magnetic history of the rock since formation. The most common methods used are *alternating field* (AF) *demagnetization, thermal demagnetization* and *chemical demagnetization*. These methods are often referred to as *magnetic cleaning, thermal cleaning,* and *chemical cleaning,* respectively. During demagnetization experiments samples are subjected to stepwise increasing values of alternating field, temperature, or time (in the case of chemical demagnetization) in the presence of zero steady magnetic field (field-free space). The magnetization is remeasured after each such treatment and the resultant changes in both direction and intensity are displayed and analyzed as discussed below in §3.4.2. A general discussion and description of instruments in use for demagnetization is given in Collinson (1983).

Alternating Field Demagnetization
If a rock specimen is placed in an alternating magnetic field with peak value B, then all grains with microscopic coercive force $B_c' < B\cos\theta$ (where θ is the angle between B and the grain microscopic coercive force) will follow the field as it alternates. If the alternating field can be applied to all possible grain orientations, then as the alternating field is slowly decreased to zero, grains with

progressively lower coercive force become fixed in different orientations and hence ultimately the magnetization of grains with coercive forces less than B will have their directions randomized. The procedure for a single demagnetization step involves taking the specimen around successively larger hysteresis loops (see Fig. 2.5 and §2.1.4) until the desired peak field is reached and then around successively smaller ones.

The technique of AF demagnetization was first carried out on a routine basis by As and Zijderveld (1958). In their apparatus they demagnetized the specimen three times, once along each of three mutually perpendicular axes at progressively higher peak alternating magnetic fields produced by a solenoid. The main objection to this technique is that not all directions in the specimen are exposed to the same peak field (for a theorectical analysis of the consequences, see McFadden, 1981). Creer (1959) introduced the technique of tumbling the specimen about two axes at right angles and perpendicular to the axis of the demagnetizing coil. Doell and Cox (1967b) employed a more elaborate three-axis tumbler. In the demagnetizing procedure it is important that the production of ARM (see §2.3.5) be avoided, so the procedure must be carried out in field-free space. Unfortunately a direct effect of tumbling the sample is to induce a *rotational remanent magnetization* (RRM) antiparallel to the rotation vector of the sample (Doell and Cox, 1967b; Wilson and Lomax, 1972).

Smith and Merrill (1980) and Stephenson (1980) provided a theoretical explanation of rotational remanent magnetization, which is now known to be a form of *gyroremanent magnetization* (GRM). GRM can also be produced during static demagnetization in rocks that are even slightly anisotropic and which contain small magnetically hard particles of magnetite, titanomagnetite, or maghemite (but not hematite). The effect is then seen when high alternating fields are used to reveal the hardest magnetization component (Stephenson, 1993). When tumbling is used the effect can be eliminated by arranging for the sense of rotation to be reversed so as to give equal time for the forward and reverse rotations while increasing and decreasing the field (Wilson and Lomax, 1972). Alternatively the sample can be demagnetized twice at each step with the rotation of the tumbler being reversed the second time. The average of the magnetizations then eliminates the effect of GRM (Hillhouse, 1977). For static demagnetization along three axes, the effect can be eliminated by measuring the x, y, and z components after each demagnetization along the x, y, and z axes and then averaging the three measurements in each direction (Stephenson, 1993). The AF demagnetization equipment available in most laboratories can produce peak alternating fields in the 100–200 mT range. This is adequate for dealing with magnetite bearing rocks but not sufficient for hematite-bearing rocks, which respond more effectively to thermal or chemical demagnetization.

For SD grains the relaxation time τ is directly related to the microscopic coercive force B_c' as shown by (2.3.15) so that progressive AF demagnetization of a rock specimen randomizes the magnetization of those grains with low coercive force, which for any given grain volume are also those with the shortest relaxation time. If the NRM of a rock specimen is capable of withstanding high alternating fields then the grains in which the NRM resides have long relaxation times, and are capable of retaining their magnetization for considerably longer than the age of the rock.

Apart from an interest in the stability of rocks as judged by their ability to withstand alternating magnetic fields, it is important to be able to judge the type of the magnetization being studied and thereby the possible origin. Studies of the demagnetization of TRM and CRM produced in weak magnetic fields show that they have similar demagnetization characteristics that are quite different from that of IRM (Kobayashi, 1959), as illustrated in Fig. 3.12. The IRM acquired by magnetic fields of 3 mT is effectively destroyed in a peak alternating field of similar value, but the TRM acquired in 50 µT and the CRM acquired in 1 mT decrease only slightly at 10 mT with a measurable part remaining at 50 mT. Thus, it is relatively simple to distinguish IRM from TRM or CRM using alternating field demagnetization but not to distinguish CRM from TRM. The discrimination of TRM from CRM, however, can be made through special thermal demagnetization procedures (McClelland, 1996).

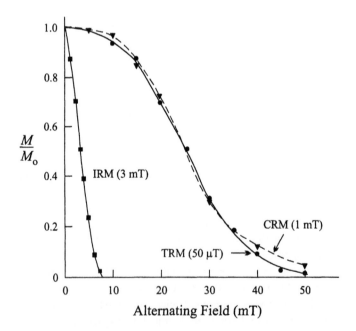

Fig. 3.12. Alternating field demagnetization of IRM, CRM, and TRM in magnetite samples. The field used in each process is indicated in brackets. After Kobayashi (1959).

Thermal Demagnetization

When a rock specimen is heated to temperature T and then cooled in field-free space, all grains with blocking temperature $T_B \leq T$ will be reset to zero magnetization. For SD grains, T_B for a grain of volume v is defined as that temperature at which the relaxation time τ becomes small (e.g., <100 s) as given by (2.3.14). A high blocking temperature indicates that the grains will have a long relaxation time at room temperature. The blocking temperature spectrum can be investigated through thermal demagnetization studies. Two methods have been used. In the *continuous method*, the magnetization of the sample is measured while still hot by using a small furnace close to a suitable magnetometer (Wilson, 1962c). In the *progressive method*, the sample is heated to successively higher temperatures, and cooled in field-free space after each heating, the magnetization being remeasured at each step (Irving *et al.*, 1961).

Because short relaxation time components also have lower blocking temperatures, thermal demagnetization progressively eliminates the short relaxation time components. The technique is most effective with hematite-bearing rocks whose coercivities are much greater than the peak fields available in AF demagnetization equipment. Unlike AF demagnetization, thermal demagnetization appears to have little use in removing large IRM components, such as those produced by lightning strikes. Irving and Opdyke (1965) describe the blocking temperature spectrum as consisting of two basic types as illustrated in Fig. 3.13a. *Thermally discrete components* are those of great stability that remain unchanged up to temperatures near the Curie temperature, whereas *thermally distributed components* are those that possess a range of blocking temperatures. Thermally distributed components therefore include the less stable components that are more capable of acquiring secondary magnetizations.

The blocking temperature spectrum for an identical set of magnetite grains will be different for TRM and CRM (McClelland, 1996). Therefore, unlike AF demagnetization, it is possible to distinguish between CRM and TRM through suitable thermal demagnetization procedures. Although CRM and TRM will unblock at a common temperature in magnetite and at a common temperature in hematite, the TRM and CRM magnitudes will be different because the net fractional alignment for CRM is controlled by the blocking volume and the reaction temperature (§2.3.6), whereas TRM is controlled by the final volume and the blocking temperature (§2.3.5).

The procedure outlined by McClelland (1996) to discriminate between TRM and CRM is to apply thermal demagnetization to a sample at successively higher temperatures up to at least 450°C. At each step the sample is given a PTRM (§2.3.5) so that the PTRM acquisition properties can be measured for each temperature interval. There should be a linear relationship between the PTRM lost and the PTRM gained for each temperature interval for both CRM and TRM up to about 350°C, but this will start to change significantly for CRM above this

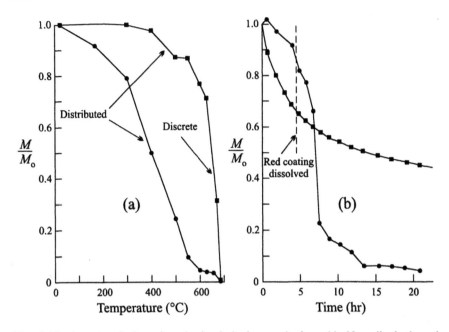

Fig. 3.13. Aspects of thermal and chemical demagnetization. (a) Normalized thermal demagnetization curves of two samples from the Waterberg redbeds of South Africa, one dominated by a thermally discrete component near the Curie point and the other composed of thermally distributed components up to the Curie point (after Jones and McElhinny, 1967). (b) Normalized decay of magnetization (circles) and decrease in iron content (squares) during chemical demagnetization of a sample from the Chugwater Formation of Wyoming (after Collinson, 1965).

temperature. If the PTRM checks fail below about 400°C, then it is not possible to distinguish CRM from TRM.

Chemical Demagnetization

The technique of chemical demagnetization was developed by Collinson (1965) for investigating the magnetization of red sediments. In this method specimens are immersed in hydrochloric acid for suitable lengths of time, after which they are washed and dried and their magnetization is remeasured. Collinson (1965, 1966) showed that the red pigment is extracted preferentially by dissolving in cold concentrated hydrochloric acid and that the leeching causes a demagnetization, as illustrated in Fig. 3.13b (see also Park, 1970; Roy and Park, 1974; Henry, 1979).

Chemical demagnetization experiments depend critically on the porosity of the samples under investigation and some of the experiments undertaken by Collinson (1965) involved chemical treatment of up to 24 hr. In other cases chemical treatment for more than 4500 hr has been used (Roy and Park, 1974). Henry (1979) describes experiments to determine the best methods of sample preparation to optimize the procedures for chemical demagnetization. Because of the high surface area to volume ratio for small grains, chemical treatment preferentially removes the smaller grains. Such grains have smaller relaxation times since, from (2.3.14), the relaxation time is directly related to the volume of a grain. The selective removal of fine-grained hematite means that the secondary components more likely to be carried by these grains can be eliminated in the process.

The classical experiments of Roy and Park (1974) showed that the appropriate combination of thermal demagnetization with chemical demagnetization can be valuable in understanding the process and timing of the magnetization acquired by certain redbeds. In their study Roy and Park detected three components of magnetization and the results suggest that the magnetization was acquired at different stages in a constant field direction that repeatedly reversed its polarity through 180°. It was concluded that the first magnetization was due to the alignment of small specular hematite particles (observed in specimens leached for more than 3000 hr) during or shortly after deposition to form a DRM. The next stage was the acquisition of a CRM_A by hematite growth around the deposited particles, with the final stage being the acquisition of CRM_B during and/or filling and cementing of all grains during final lithification.

3.4.2 Presentation of Demagnetization Data

Changes in the vector magnetization of a rock sample during demagnetization involve both its direction and its intensity. One can illustrate changes in direction on a stereonet and changes in intensity by plotting the intensity as a function of the treatment being used. This requires two diagrams to display the two components of the magnetization vector. Alternatively, an orthogonal vector diagram may be used that displays both parts of the vector at the same time.

Such a diagram is termed a *Zijderveld diagram* following Zijderveld (1967) and is the standard method for presentation of the results of stepwise demagnetization. Some examples of Zijderveld diagrams for a two-component system are illustrated in Fig. 3.14.

The method of presentation in the Zijderveld diagrams of Fig. 3.14 is as follows. Imagine the direction and intensity at each successive treatment to be represented by the direction and length of the vector in three-dimensional space starting at the origin. The endpoint of the vector at each demagnetization step is then projected onto two planes at right angles to one another. The solid points are these endpoints when projected onto the horizontal plane containing axes NS and EW, whereas the open points are these endpoints when projected onto the vertical plane containing axes NS and Up-Down. The NS axis is common to both planes. Note that the EW axis is apparently inverted so that the direction N-E-S-W is in an anticlockwise direction rather than the normally expected clockwise direction. This is a matter of convenience for the presentation. Inverting the EW axis means that the solid circles plot in the upper part of the diagram rather than overlapping with the open circles in the bottom part.

The effectiveness of the resulting diagrams depends on the amount of overlap

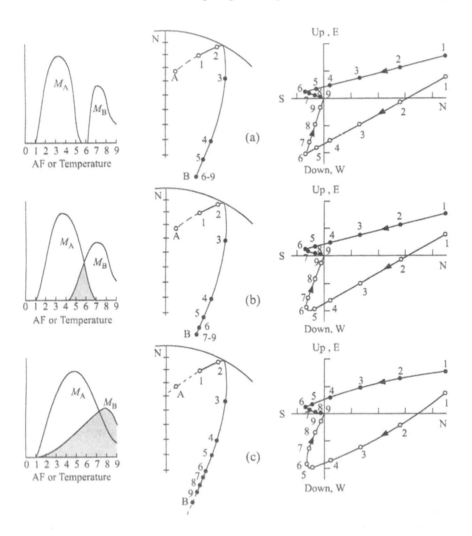

Fig. 3.14. Component coercivity or blocking temperature spectra (left) for two components of magnetization together with the resulting stereoplots (middle) and Zijderveld diagrams (right) for (a) nonoverlapping spectra, (b) partially overlapping spectra but wide windows of nonoverlap, and (c) total overlap of the spectra. On the stereoplots solid circles are on the lower hemisphere and open circles are on the upper hemisphere. On the Zijderveld diagrams solid circles lie on the horizontal plane and open circles lie on the vertical plane. After Dunlop (1979), with permission from Elsevier Science.

of the coercivity or blocking temperature spectra of the two components. Figure 3.14a shows the case where the two components have nonoverlapping coercivity or blocking temperature spectra and therefore exhibit ideal behavior on the Zijderveld diagram. The first component (*A*), removed between steps 1 and 6 in the demagnetization procedure, is directed NNE (as shown by the solid circles in the horizontal plane) and has an upward inclination (as shown by the open circles in the vertical plane). The second component (*B*), removed between steps 6 and 9, decays to the origin and is directed SE with steep downward inclination. This final component that decays to the origin is termed the *characteristic remanent magnetization* (ChRM).

In Fig. 3.14b the spectra overlap only partially but there are wide windows of nonoverlap. The resulting Zijderveld diagram easily identifies the two components, with the straight line segments that identify them merging from one to the other through a small curve that corresponds to the overlap in the spectra. In Fig. 3.14c the spectra of the *A* and *B* components now overlap completely so that the segments become curves and neither component can be accurately determined.

The declination (*D*) of any component represented by a straight line segment on a Zijderveld diagram is readily determined from the direction of the line in the horizontal plane imagined to commence from the origin in each case. For example, the declination of component *A* in Fig. 3.14a is determined by transposing the line 1–6 so that point 6 is at the origin and then measuring the angle the line makes with true north. In the vertical plane the angle between the corresponding line and the horizontal axis (transposed in the same way) is the *apparent inclination* (I_a), which is related to the true inclination (*I*) by

$$\tan I = \tan I_a \, |\cos \vartheta|, \tag{3.4.1}$$

where ϑ is the angle between the line in the vertical plane and the common axis that lies in the horizontal plane. In Fig. 3.14 the common axis is NS, in which case $\vartheta = D$. However, if the common axis is EW (as might be chosen when *D* is closer to 90° or 270° than 0° or 180°) then $\vartheta = D\text{-}90°$.

The principles shown in Fig. 3.14 can be extended to more than two components with an obvious further degree of complexity depending on the relative overlap of the spectra of the three or more components. However, if the blocking temperature spectra overlap, this does not necessarily mean that the coercivity spectra will overlap in the same way. Therefore the choice of demagnetization technique can be critical for resolving the different magnetization components. It is often not clear which of the various demagnetization methods should be used when treating a set of samples collected from a rock unit. In general, it has been found that AF demagnetization is appropriate in the case of igneous rocks and thermal demagnetization in the

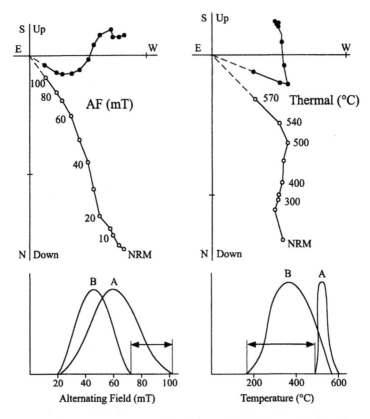

Fig. 3.15. Zijderveld diagrams from AF (left) and thermal (right) demagnetization of a sample of some Ordovician volcanics. Solid (open) circles are for projections onto the horizontal (vertical) plane. The corresponding coercivity and blocking temperature spectra are shown at the bottom for components *A* and *B*. The arrows indicate the nonoverlapping part of the spectra in each case. After Perroud (1983).

case of sediments. However, it is always best to experiment initially using at least two methods to determine which method is the most appropriate. In many rocks both magnetite and hematite are present and the components derived from these minerals can easily be detected by thermal demagnetization where only the magnetization due to hematite survives heating above 580°C, the Curie point of magnetite.

An example is given in Fig. 3.15 where both AF and thermal demagnetization are needed to separate out two magnetization components *A* and *B* in some Ordovician volcanics (Perroud, 1983). AF demagnetization is only capable of isolating the high-coercivity component *A*, corresponding also to the high blocking temperature component during thermal demagnetization. The lower coercivity, lower blocking temperature component *B* is not isolated by AF

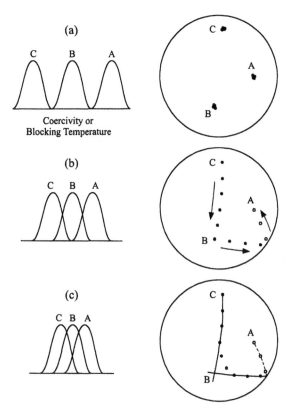

Fig. 3.16. Difference vector method of analysis for three components A, B, C with varying overlap in their coercivity or blocking temperature spectra (left) and corresponding stereoplots of the difference vector paths (right). After Hoffman and Day (1978), with permission from Elsevier Science.

demagnetization because of almost complete overlap of its coercivity spectrum with that of A. However, the blocking temperature spectrum of B hardly overlaps with that of A and component B is therefore easily isolated during thermal demagnetization.

Hoffman and Day (1978) propose an alternative method, illustrated in Fig. 3.16, that is simple and effective for analyzing demagnetization data. In this method the data are displayed stereographically as the vectors that are removed during successive demagnetization steps, referred to as *difference vectors* or *subtracted vectors*. Figure 3.16 shows the case in which there are three components A, B and C, where A is the most stable and C is the least stable component. In Fig. 3.16a, the coercivity or blocking temperature spectra do not overlap so that the difference vectors are represented by three points on the stereoplot corresponding to the three components A, B and C.

In Fig. 3.16b there is overlap of *A* with *B* and *B* with *C* but no overlap between *A* and *C*. On the stereoplot the path of the difference vector consists of two great circles, one from *C* to *B* and the other from *B* to *A*. The intersection, or kink, indicates the direction of the *B* component.

In Fig. 3.16c there is overlap of *A* with *B* and *B* with *C*, with appreciable overlap of *A* with *C* as well. The path of the difference vector is now more complicated with a smearing of the two great circle paths near the direction corresponding to component *B*. However, although the Zijderveld diagram method in this case fails to determine the direction of component *B*, extrapolation of the two great circle segments enables the direction of component *B* to be estimated.

3.4.3 Principal Component Analysis

Kirschvink (1980) applied the classic multivariate technique of *principal component analysis* (PCA) to estimate the directions of lines and planes of best least-squares fit along the demagnetization paths on a Zijderveld plot. From a set of successive data points PCA determines the best fitting line, whose precision is estimated by the *maximum angular deviation* (MAD). Although there is no general rule for acceptable values of the MAD value, line fits from PCA that yield MAD $\geq 15^\circ$ are often considered ill defined and questionable, whereas those with MAD $\leq 10^\circ$ would be considered to be reasonably good. Of particular importance when dealing with the ChRM is to decide whether the origin should be included as a data point because the ChRM is determined from the trend of the data points towards the origin. There are three situations that can be considered as follows:

(i)   force the line to pass through the origin (anchored line fit);
(ii) use the origin as a separate data point (origin line fit);
(iii) do not use the origin at all (free line fit).

For determination of the ChRM either the anchored line fit or origin line fit would obviously be used, whereas the free line fit would be used for other components.

Demagnetization planes found with PCA can be used in place of the difference vector paths of the Hoffman and Day (1978) method shown in Fig. 3.16, thus avoiding noise amplification caused by vector subtraction. Kirschvink (1980) gives two methods that can be used for jointly estimating an average remanence direction from demagnetization lines and planes. Schmidt (1982) introduced a refinement to the line fitting procedure used by Kirschvink (1980) that has been termed *linearity spectrum analysis* (LSA). LSA attempts to overcome the problem of deciding which points to incorporate in determination of the best fitting line because a covert bias can be introduced if the criterion for rejecting or accepting a point at the beginning or end of a line is made too coarse. Kent *et al.*

(1983) produced a more elaborate procedure for determining linear and planar structure in demagnetization data using a sophisticated procedure encapsulated in a process that they called LINEFIND. A fundamental aspect of the LINEFIND process is that the statistical model used assigns measurement errors to each of the remanence vectors. The errors involved in estimating a particular direction in a multicomponent system are fundamental to PCA, LSA, and LINEFIND. McFadden and Schmidt (1986) have shown how these errors can be used in estimating overall mean directions and cones of confidence.

3.4.4 Analysis of Remagnetization Circles

When rocks that contain more than one component of magnetization are progressively demagnetized, it is usually observed that, during the preferential demagnetization of one component, the resultant direction of magnetization moves along a great circle (see Fig. 3.14). In most cases the individual components can be identified by linear segments on the Zijderveld diagram with the final linear segment progressing toward the origin giving the ChRM. However, in some cases, either because of overlap in the stability spectra (Fig. 3.14c) of two components and/or because the intensity decreases below the sensitivity of the magnetometer, the ChRM is not obtained. The only available information regarding the ChRM resides in the great circle. A single great circle on its own provides insufficient information to estimate the ChRM, but if more than one great circle is available (e.g., from different specimens) and these great circles converge, then an estimate of the ChRM may be obtained.

In practice it is possible that some specimens may give an estimate of the ChRM from the Zijderveld diagram (a direct observation) and some may only have great circle information. Jones *et al.* (1975) first formulated a method, but gave no mathematical details, that combined the information from direct observations (set points) with those from great circles to give an estimate of the ChRM that made use of all the available information. Halls (1976) and Bailey and Halls (1984) presented an alternative method of analysis, but Schmidt (1985) showed that the method could lead to bias in the estimate of the ChRM and hence systematic errors. McFadden and McElhinny (1988) formalized the method proposed by Jones *et al.* (1975) but introduced the concept of *sector constraints*, which assists in eliminating the bias observed by Schmidt (1985).

An example of the application of the great circle analysis of McFadden and McElhinny (1988) is given in Fig. 3.17. In Fig. 3.17a the effect of applying the sector constraints becomes obvious. For each of the great circles the motion of the resultant vector is from the open circles to the closed circles. Without any sector constraints the solution chosen would be the intersection of the two great circles. However, this intersection occurs more than 180° along each of the circles from their starting point and thus cannot be correct. By defining

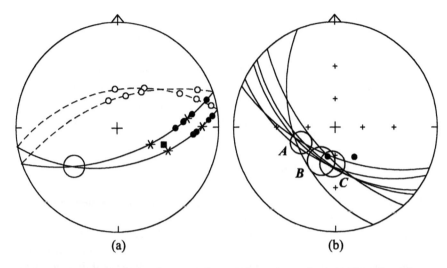

(a)   (b)

Fig. 3.17. Combined analysis of great circles and direct observations. (a) The effect of sector constraints. The motion of the resultant vector in each great circle is from the open circles to the closed circles. The point of intersection of the two circles (indicated by the large open circle) is more than 180° along each circle from the starting point. Sector constraints in each case are defined by the asterisks and then give the result indicated by the solid square. (b) Analysis of six great circles and two set points. *A*; no sector constraints used and no set points included. *B*; sector constraints used but no set points included. *C*; sector constraints and set points all included. Solid circles are the two set points and the open circles are circles of 95% confidence for the results *A*, *B* and *C*. From McFadden and McElhinny (1988), with permission from Elsevier Science.

acceptable arcs of the great circles (shown with asterisks) outside of which the true direction cannot lie, the result indicated by the square is obtained. When there are set points and great circles the analytical procedure developed by McFadden and McElhinny (1988) is an iterative one.

Figure 3.17b shows an example from the application of the method to the data from a site in some red sandstones from the Tarim Block, China (McFadden *et al.*, 1988b). There are six great circles and two set points. The variation in the best estimate of the mean direction is shown for three data combinations. Result *A* is for the great circles only without any sector constraints, *B* incorporates the information provided by the sector constraints but not from set points and the final result *C* incorporates the information from sector constraints and the set points. Details of the statistical formulation and iterative procedure used are provided in McFadden and McElhinny (1988).

3.5 Identification of Magnetic Minerals and Grain Sizes

3.5.1 Curie Temperatures

Curie temperatures in rocks should, in principle, provide a simple diagnosis of the magnetic minerals that are present. Table 2.2 gives the Curie temperatures for the most common magnetic minerals, and it is immediately apparent that some of these either have similar values or may overlap with others depending on the composition. *Thermomagnetic curves* (or M_s–T curves) are produced by heating samples in the presence of a steady magnetic field of suitable strength and then measuring the resulting magnetization as a function of temperature. The

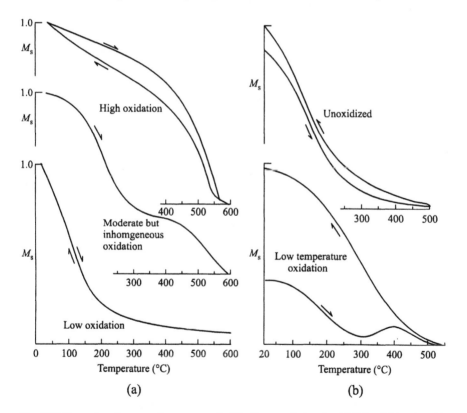

Fig. 3.18. Examples of vacuum thermomagnetic (M_s–T) curves in basalts. (a) Subaerial basalts with high, moderate but inhomogeneous, and low oxidation states. (b) Deep-sea basalts containing unoxidized titanomagnetite (reversible curve) and titanomaghemite from low-temperature oxidation (irreversible curve due to the inversion of titanomaghemite during heating). After Dunlop and Özdemir (1997).

temperature is generally cycled from room temperature to about 700°C and back again. However, the instruments used for this purpose are often not sensitive enough or the concentration of magnetic minerals not high enough to yield data of sufficiently high quality.

To ensure that true Curie temperatures are measured, the thermomagnetic analysis must be carried out in fields of sufficient strength to saturate minerals like hematite and goethite. This is virtually impossible using the electromagnets that are available in a typical paleomagnetic laboratory. Therefore the use of thermomagnetic curves (Fig. 3.18) is generally restricted to rocks containing strong magnetic minerals such as magnetite. Curie temperatures are determined as a matter of course during the normal thermal demagnetization procedures used on rocks (§3.4.1), but this does not necessarily reveal the Curie temperature of those minerals (goethite, pyrrhotite, etc.) that are minor constituents and which revert to other minerals at intermediate temperatures (see §2.2.4).

Figure 3.18 gives examples of some vacuum thermomagnetic curves from subaerial and deep-sea basalts. Fresh subaerial basalts, when heated in vacuum (Fig. 3.18a), may exhibit a high Curie point (500–580°C) due to high-temperature deuteric oxidation, a low Curie point due to homogeneous titanomagnetite with $x \approx 0.6$-0.7 in a low oxidation state, or rarely both Curie points for moderate but inhomogeneous oxidation states (see §2.2.2). The thermomagnetic curves are practically reversible. In the case of deep-sea basalts (Fig. 3.18b), unoxidized basalts have reversible curves due to titanomagnetite, but basalts containing titanomaghemite due to low-temperature oxidation have irreversible curves due to the inversion of the titanomaghemite during the measuring process.

3.5.2 Isothermal Remanent Magnetization

After a rock specimen has been demagnetized, it is common practice to measure the IRM acquired in increasing steady fields (§2.1.4) as a means of measuring the intrinsic coercivity spectrum. A plot of this progressive acquisition of IRM is referred to as an *IRM acquisition curve* (Fig. 3.19). When saturation (SIRM) is reached the sample is then exposed to reverse fields of increasing strengths until the remanence reduces to zero at the coercivity of remanence, B_{cr}. Figure 3.19 shows two examples of IRM curves, one for a hematite-rich and one for a magnetite-rich basalt sample. The magnetite-rich sample shows a steep rise in IRM acquisition up to 0.3 T (300 mT) at which point single domain magnetite grains are all aligned in the field direction. The IRM curve still continues to rise because of the presence of hematite and at 2 T is near to saturation. For this sample $B_{cr} = 61$ mT. The hematite-rich sample has no initial sharp rise in IRM and again is near to saturation at about 2 T, with a value of $B_{cr} = 615$ mT (10 times that of the magnetite-rich sample).

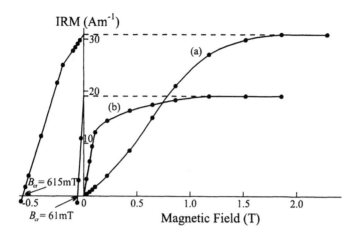

Fig. 3.19. IRM acquisition curves for two samples from the Antrim Plateau Volcanics of northern Australia. (a) Hematite-rich sample; (b) magnetite-rich sample. From McElhinny (1973a).

Dunlop (1972) suggested that the coercivity spectrum be estimated from the IRM acquisition curve by calculating the fraction of the IRM acquired over successive equal incremental values of the applied field. Figure 3.20 shows examples of such coercivity spectra in sediments in which coercivities <0.1 T are mostly due to magnetite, those between 0.1 and 0.3 T are due to coarse detrital hematite, and those >0.3 T due to authigenic hematite and goethite. There is considerable overlap in the coercivity ranges of these phases. Some uses of coercivity spectrum analysis are given by Lowrie and Heller (1982).

The interpretation of the magnetic mineralogy can be considerably improved by stepwise thermal demagnetization of the acquired IRM (Heller, 1978). As

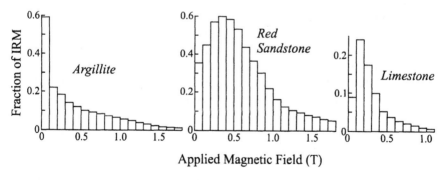

Fig. 3.20. Coercivity spectra determined from IRM acquisition curves for three types of sediment. The dominant magnetic minerals in the limestone and red sandstone samples are detrital and pigmentary hematites. Magnetites with coercivities <0.1 T are important in the argillite. After Dunlop (1972).

TABLE 3.2

Maximum Coercivities and Unblocking Temperatures for Some Common Magnetic Minerals

Magnetic mineral	Max coercivity (T)	Max unblocking temp.(°C)
Magnetite	0.3	580
Maghemite	0.3	≈350
Titanomagnetite $x = 0.3$	0.2	350
$x = 0.6$	0.1	150
Pyrrhotite	0.5-1	325
Hematite	1.5-5	675
Goethite	>5	80-120

shown in Table 3.2, minerals with similar maximum coercivities generally have different unblocking temperatures. A modification of the method of coercivity spectral analysis was therefore proposed by Hirt and Lowrie (1988) and further refined by Lowrie (1990). It appears to work well in sediments and is especially useful in weakly magnetized limestones.

In this refined method Lowrie (1990) first observed the IRM acquisition up to the maximum available magnetizing field along the *z* axis of the sample. A maximum field of at least 1 T is desirable, but 5 T is preferable (Fig. 3.21) because the contribution of goethite to IRM is often pronounced above 1.5 T. Next, a field of 0.4 T is applied perpendicular to the first field along the *y* axis, thus remagnetizing the coercivity fraction softer than 0.4 T along the *y* axis and leaving the high coercivity minerals magnetized along the *z* axis. Finally, a field of 0.12 T is applied normal to each of the other fields along the *x* axis. The sample is then thermally demagnetized in field-free space in the usual way (see §3.4.1).

Two examples of the Lowrie (1990) method are illustrated in Fig. 3.21 for some Swiss sediments. The method of analysis consists of plotting and evaluating the thermal decay of each of the three components separately. In the first case (Fig. 3.21a), the smooth IRM acquisition curve rises steeply at first, but does not reach saturation even in 5 T. Thermal demagnetization of the soft (< 0.12 T), medium (0.12–0.4 T), and hard (0.4–5 T) fractions shows distinct unblocking temperatures. The hard and medium coercivity fractions show strong evidence for the unblocking of pyrrhotite below 330°C and hematite at 640°C, whereas the soft fraction demagnetizes completely by 330°C. There is no indication of goethite, maghemite, or magnetite.

In the second example (Fig. 3.21b), the IRM acquisition curve climbs steeply in low fields but again still does not reach saturation in 5 T. Thermal demagnetization shows an abrupt drop of the hard fraction at only 80°C, indicating the presence of a sizeable goethite component. Above 80°C up to 640°C, the hard fraction shows a monotonic decay due to hematite. The soft and medium coercivity fractions show no significant discontinuities, indicating

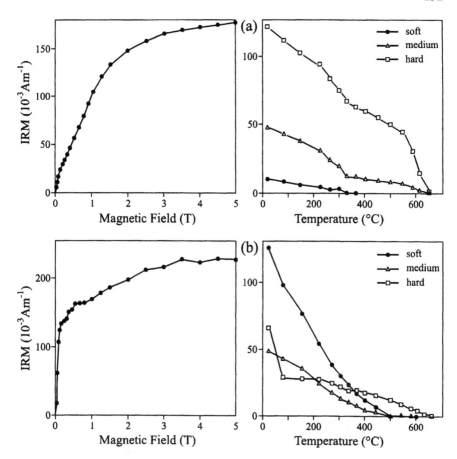

Fig. 3.21. IRM acquisition curves and thermal demagnetization of a three-component IRM produced by magnetizing the sample in 5 T along the *z* axis, followed by 0.4 T along the *y* axis and then 0.12 T along the *x* axis for two Swiss limestones. The soft (<0.12 T), medium (0.12–0.4 T), and hard (0.4–5 T) components are shown during thermal demagnetization. (a) Sample dominated by pyrrhotite (unblocking at 330°C) and hematite. (b) Sample dominated by magnetite and hematite with an appreciable fraction of goethite. After Lowrie (1990).

maghemite or pyrrhotite, and demagnetize smoothly to zero at 540°C, indicative of magnetite. The sample is therefore dominated by magnetite and hematite with an appreciable fraction of goethite.

3.5.3 The Lowrie–Fuller Test

Lowrie and Fuller (1971) proposed a test that would enable the simple identification of those igneous rocks that were dominated either by SD grains or

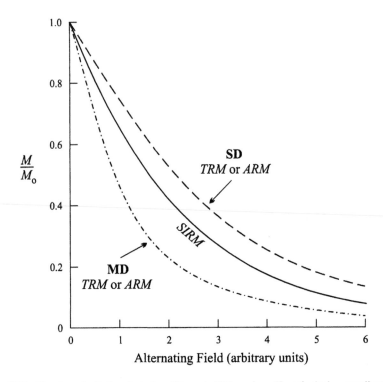

Fig. 3.22. The Lowrie–Fuller test for SD and MD grains. Hypothetical normalized AF demagnetization curves for weak-field TRM or ARM and SIRM. For SD grains the TRM (ARM) curve always lies above the corresponding SIRM curve, whereas for MD grains the TRM (ARM) curve always lies below the corresponding SIRM curve.

by MD grains of magnetite as carriers of the remanence. This would enable the rapid selection of those igneous rocks in which SD grains dominated the remanence. The test is based on the experimental observation that normalized AF demagnetization curves (see §3.4.1) of weak-field TRM and SIRM have different relationships for SD and large MD grains of magnetite. In large MD grains SIRM is more resistant to AF demagnetization than weak-field TRM, whereas for SD grains the opposite is true (Fig. 3.22). Interestingly, theory suggests that noninteracting SD assemblages should exhibit Lowrie–Fuller trends opposite in sense to those obtained experimentally (Schmidt, 1976; Halgedahl, 1998). A modification to the test (Johnson *et al.*, 1975) uses weak-field anhysteretic magnetization (ARM) instead of weak-field TRM since they exhibit similar AF demagnetization characteristics and the sample does not have to be heated for the test.

As shown in Fig. 3.22, for MD grains the normalized AF demagnetization curve of ARM (or weak-field TRM) always lies below that of SIRM, whereas for SD grains the ARM (or weak-field TRM) curve always lies above that of SIRM. It appears that PSD grains of magnetite up to around 10–15 μm display the same characteristics as SD grains (Bailey and Dunlop, 1983) so that SD or PSD grains may be identified in this way. However, in rare situations, such as in some low-stress hydrothermal magnetites (Heider *et al.*, 1992), the Lowrie–Fuller test appears to fail. In practice, however, it is still valuable as a grain-size discriminator for magnetite in igneous rocks whose magnetization is of thermal origin.

3.5.4 Hysteresis and Magnetic Grain Sizes

The measurement of hysteresis curves or loops for weakly magnetized rocks has now become quick and reliable through the introduction of automated instrumentation for this purpose (e.g., Flanders, 1988). When rocks are saturated through the application of an applied magnetic field during hysteresis (§2.1.4 and Fig. 2.5), the nonmagnetic matrix has a diamagnetic or paramagnetic signal (§2.1.2 and Fig. 2.2) that is still present above the saturation field. This effect is corrected for in modern automated instrumentation. The common parameters that are measured are M_s, M_{rs}, B_c and B_{cr} or the ratios M_{rs}/M_s and B_{cr}/B_c.

The remanence ratio M_{rs}/M_s and the coercivity ratio B_{cr}/B_c have diagnostic values for SD, PSD, and MD grains (Table 3.3). For a random assemblage of uniaxial SD grains that have isotropic random orientations of their easy axes (i.e., they are dominated by shape anisotropy), $M_{rs}/M_s = 0.5$ (§2.3.4 and Fig. 2.13). In practice the magnetization of each grain rotates back to the nearest easy direction on removal of the applied field. When the population of SD grains is dominated by magnetocrystalline anisotropy ([111] easy axis for magnetite) $M_{rs}/M_s = 0.87$. In the case of hematite, if saturation is not reached, then M_{rs}/M_s can be as high as 0.95 because M_s may still remain pinned in the easy basal plane. For ideal MD grains $M_{rs}/M_s \leq 0.1$. The coercivity ratio B_{cr}/B_c for SD grains cannot be less than 1 and a value of 1.5 is often used as an upper SD limit (Day *et al.*, 1977), but this is a purely arbitrary value. For ideal MD grains a value of $B_{cr}/B_c \geq 4$ is often used. PSD grains would be expected to have values intermediate between SD and ideal MD values (Table 3.3).

Following Day *et al.* (1977) it is now common practice to plot M_{rs}/M_s versus B_{cr}/B_c. Such a plot for some young (100 ka) basalts and some Jurassic basalts is

TABLE 3.3

Limiting Values for the Remanence Ratio M_{rs}/M_s and Coercivity Ratio B_{cr}/B_c for Single-Domain (SD), Pseudo-Single-Domain (PSD), and Multidomain (MD) Grains

Ratio	SD	PSD	MD
M_{rs}/M_s	≥ 0.5	0.5-0.1	≤ 0.1
B_{cr}/B_c	1.0-1.5	1.5-4.0	≥ 4.0

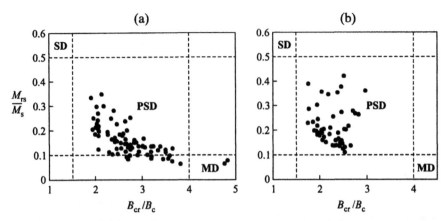

Fig. 3.23. M_{rs}/M_s versus B_{cr}/B_c plot for (a) the young Vulcano Island basalts from the Aeolian Islands, Sicily (from Laj *et al.*, 1997, with permission from Elsevier Science) and (b) the Jurassic Lesotho basalts of southern Africa (Kosterov *et al.*, 1997). The magnetic grain size boundaries are those listed in Table 3.3.

given in Fig. 3.23. Most grain sizes fall within the PSD grain size range as might be expected. For sediments these plots have been found useful in distinguishing between remagnetized and unremagnetized limestones (§2.3.6). Channell and McCabe (1994) found that the unremagnetized Maiolica limestones from Italy followed the typical SD–PSD trends on these plots (Fig. 3.24). However, the remagnetized limestones from both North America and Europe followed an unusual trend with high SD–PSD values of M_{rs}/M_s but PSD–MD values of B_{cr}/B_c from about 3 to 10. These high values of B_{cr}/B_c could result from a mixture of SP and stable SD grains as the scanning electron microscope images of Fig. 2.18 suggest since high values of B_{cr}/B_c are a characteristic of SP as well as MD grains. The intercept of the trend of the remagnetized limestones at $B_{cr}/B_c = 1$ is $M_{rs}/M_s \approx 0.87$ implying that magnetocrystalline anisotropy dominates, as would be expected for spherical grains. If this is so, then the observed M_{rs}/M_s values have been considerably reduced by the addition of SP grains.

The classical hysteresis loops from SD, PSD, and MD grains are frequently found in rocks. However, rocks rarely contain just one domain state. A sample containing a mixture of grain sizes of a single magnetic mineral or a mixture of different magnetic minerals can sometimes give rise to hysteresis loops that are similar to those of a population of grains of uniform composition and identical size. Further complication can occur because the loops are frequently distorted, having constricted middle sections (*wasp-waisted loops*) or spreading middle sections and slouching shoulders (*potbellied loops*). It seems that the requirement for such distortions is the presence of at least two magnetic components with strongly contrasting coercivities. Numerical simulation of hysteresis loops shows that both wasp-waisted and potbellied loops can easily be

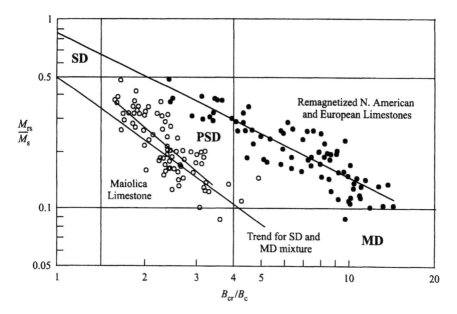

Fig. 3.24. M_{rs}/M_s versus B_{cr}/B_c plot for unremagnetized Maiolica limestones from Italy and for remagnetized North American and European limestones. The SD and MD mixture line corresponds to that determined from synthetic and crushed natural magnetites from Parry (1982). The magnetic grain size boundaries are those listed in Table 3.3. After Channell and McCabe (1994).

generated from populations of SD and SP grains (Tauxe *et al.*, 1996). Wasp-waisted loops require a contribution from SP grains that saturate quickly having a steep initial slope to the hysteresis curve. Potbellied loops require low initial slopes with the SP contribution approaching saturation at higher fields.

3.5.5 Low-Temperature Measurements

Both magnetite and hematite exhibit transitions in their magnetic properties when cooled below 0°C. When hematite is cooled through the Morin transition (−15°C, see §2.2.3), the spin-canted remanence vanishes and only the isotropic defect moment remains. On reheating in zero magnetic field a fraction (*memory*) of the spin-canted remanence is recovered. It is likely that the defect moment renucleates the spin-canted moment during reheating. Fuller and Kobayashi (1964) therefore proposed that cycling across the Morin transition is a non-destructive alternative to thermomagnetic analysis for the detection of hematite.

In magnetite the magnetic properties that depend on magnetocrystalline anisotropy in MD grains change abruptly around the Verwey transition temperature ($\approx$ -150°C, see §2.2.2). The magnetocrystalline anisotropy constant K_1 momentarily becomes zero (changing from positive to negative) at the *isotropic point* (≈ -135°C) just above the Verwey transition. Therefore cycling through this transition in zero field is also a useful means of detecting magnetite. However, the anisotropy of elongated SD grains of magnetite depends only on shape, so these grains exhibit a *low-temperature memory* on reheating. Thus, cycling through the isotropic point of magnetite in zero field can be used as a means not only of detecting and recovering the remanence of SD grains that depend on shape anisotropy but also of demagnetizing MD magnetite grains.

Chapter Four

Magnetic Field Reversals

4.1 Evidence for Field Reversal

4.1.1 Background and Definition

Synoptic observation long ago revealed much about the behavior of the geomagnetic field (§1.1.1). Direct measurements made by a range of different people, including mariners, have given us a detailed global picture of the modern geomagnetic field and a good record of how the field has changed over the past century. The record is reasonable on a larger scale back to about 1840, but beyond that it fades away (Bloxham and Jackson, 1992). Recently, satellites and a sophisticated observatory network have provided detailed information on the very short-term behavior of the field. Reversals, however, are different – there are no direct observations. Furthermore, the reversal process, despite being almost instantaneous on the geological time scale, is so slow on the human time scale that no single human will ever observe this remarkable phenomenon in its entirety. Thus, with regard to reversals, the evidence relies totally upon the indirect observations afforded by paleomagnetism. An excellent review of magnetic field reversals is given by Jacobs (1994).

The mathematical equations governing the geodynamo are notoriously difficult, so much so that there is as yet no analytical solution for the full set of equations. However, the symmetry of the field equations demands that for any solution there is another solution in which the field is everywhere reversed.

(Merrill *et al.*, 1979; see also Cox, 1981). Thus, even without a solution to the equations, it is evident that there must be another possibility for the geodynamo in which the field is everywhere reversed relative to today. Furthermore, such a "reverse" state should have the same statistical properties as today's "normal" field. In contrast, it is not immediately evident from the equations whether a mechanism exists in the Earth to produce a transition from one (quasi-) stable solution to the other. Certainly there is evidence that such a mechanism exists elsewhere, since the Sun, for example, has a general magnetic field that appears to reverse its polarity frequently. Also, the magnetic field of the Milky Way (the galaxy in which the solar system resides) tends to be perpendicular to the plane of the galaxy but the sign of this field appears to vary on a large regional scale, suggesting reversals of that field.

The now incontrovertible documentation that the Earth's magnetic field has reversed its polarity many hundreds of times in the past must be recognized as one of the great triumphs of paleomagnetism. Glen (1982) gives an excellent history of the development of this field of endeavor and Opdyke and Channell (1996) provide an outstanding technical review of magnetic stratigraphy. Knowledge of reversals and of their ages has been used in diverse applications, such as the dating of beach ridges (Idnurm and Cook, 1980), paleoanthropological sites (McFadden *et al.*, 1979), vertebrate fossil sites (MacFadden *et al.*, 1987; Whitelaw, 1991a,b, 1992), and the onset of aridity and dune building (Cheng and Barton, 1991). The knowledge has also been used to gain an improved understanding of the dynamo process and its boundary conditions (Laj *et al.*, 1991; McFadden and Merrill, 1984, 1997; Courtillot and Besse, 1987; Hoffman, 1992a, 1996; see also Merrill *et al.*, 1996). The Geomagnetic Polarity Time Scale (GPTS) has become a central tool in the construction of Late Cretaceous–Cenozoic geological time scales and facilitates correlation between other measures of geologic time, such as isotope stratigraphy and biostratigraphy.

The field at the surface of the Earth is dominated by a geocentric axial dipole field (the g_1^0 term in a spherical harmonic analysis; §1.2.3) so an apparently acceptable definition of a field reversal is that a reversal occurs when the geocentric axial dipole field changes its polarity. This might be an acceptable definition from a geomagnetic perspective given that it is possible to obtain many simultaneous observations and perform a spherical harmonic analysis to determine the value and sign of g_1^0. However, a paleomagnetic observation at one locality from rocks of a given age with field direction approximately opposite to the known mean field of that age is not sufficient evidence for a reversal of the axial dipole. This is because the nondipole field can produce large local deviations of the field, even to the extent of appearing locally to be a reversal. For example, the International Geomagnetic Reference Field for 1995 shows that the magnetic declination can be as much as –40° at a latitude of 40°S and there are locations at high latitudes where the declination can be 180°. Therefore, it is necessary to obtain effectively simultaneous observations of inverted field directions well distributed over the Earth's surface before it can be concluded that the axial dipole field has reversed. The change in sign of the field must also exhibit some stability before it can be considered to be a reversal. A reversal of the Earth's magnetic field is thus usefully defined as a globally observed 180° change in the axial dipole field averaged over a few thousand years.

4.1.2 Self-Reversal in Rocks

The earliest observations of lava-flow magnetic directions roughly opposed to that of the present Earth's magnetic field were obtained by David (1904) and Brunhes (1906). Using the magnetic directions from more than 100 Japanese and Manchurian lava flows, Matuyama (1929) produced the first (crude) reversal chronology. However, these studies did not by themselves provide convincing evidence for reversals of the geomagnetic field. Because the evidence relied totally on the indirect paleomagnetic observation, it was necessary to prove that the magnetization was a valid record of an inverted Earth's magnetic field and not a false record created by some self-reversal mechanism in the rock. Indeed, Néel (1955) produced theoretical models for self-reversal, highlighting the need for convincing evidence of genuine reversal of the field. Since then other possible self-reversing mechanisms have been proposed by several authors (e.g., Verhoogen, 1956; Uyeda, 1958; Ishikawa and Syono, 1963; O'Reilly and Banerjee, 1966; Hoffman, 1975, 1992b; Tucker and O'Reilly, 1980; McClelland and Goss, 1993), and Dunlop and Özdemir (1997) have recently provided an excellent detailed review.

All current models for self-reversal share a conceptual framework that requires at least two magnetic phases in the rock. One phase becomes magnetized first, parallel to the external magnetic field, and then through some interaction of the first phase with the second, the second phase subsequently becomes magnetized antiparallel to the first. This occurs either because there is a negative exchange interaction acting between the two phases or simply because the magnetic field of the first phase swamps that of the external magnetic field (*magnetostatic interaction*). An example of the latter is illustrated in the model of Fig. 4.1 for a thermoremanent self-reversal. In this model the field of interest is that *inside* the material so the magnetic field $\mathbf{H}$ is used rather than the magnetic induction $\mathbf{B}$ (see §2.1.1). Phase A has the higher Curie temperature so on cooling it becomes magnetized parallel to the external field $\mathbf{H}$ while phase B is still above its Curie temperature. On further cooling, phase B becomes magnetized in the total field $\mathbf{H}+\mathbf{H}_A$ (Fig. 4.1a), where $\mathbf{H}_A$ is the magnetic field due to phase A. In this model $\mathbf{H}_A$ within phase B is opposite to the external field $\mathbf{H}$. If $|\mathbf{H}_A| > |\mathbf{H}|$ then phase B

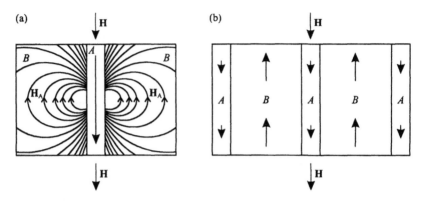

Fig. 4.1. Model for a self-reversal mechanism in rocks with two magnetic phases A and B. A has the higher Curie temperature but B has the higher room temperature value of saturation magnetization (M_s). (a) On cooling, A becomes magnetized first in a direction parallel to the external field $\mathbf{H}$, and on further cooling phase B becomes magnetized in the total field $\mathbf{H}+\mathbf{H}_A$, where $\mathbf{H}_A$ is the field in phase B caused by the magnetization of A and is antiparallel to $\mathbf{H}$. The relative magnitudes of $\mathbf{H}$ and $\mathbf{H}_A$ then determine the direction of the total field in which B becomes magnetized. (b) If $\mathbf{H}_A$ is larger than $\mathbf{H}$ then phase B will acquire a magnetization antiparallel to $\mathbf{H}$ on further cooling. Because B has the higher room temperature M_s, the sample will then have a self-reversed magnetization.

will be magnetized opposite to $\mathbf{H}$, and if at room temperature the total magnetization of B exceeds that of A then the sample will have *self-reversed* (Fig. 4.1b). After formation of the rock, one of the phases may be altered by chemical changes (e.g., oxidation) in such a way that the self-reversal may not be detected in a subsequent laboratory experiment.

Unfortunately, self-reversal is not restricted to the realm of theory, for Nagata (1952), Nagata *et al.* (1957), and Uyeda (1958) demonstrated self-reversal in samples from the Haruna dacite, a hyperthermic dacite pumice. One of the minerals in this dacite, a titanohematite containing roughly 50 mole percent ilmenite and hematite, is experimentally the best understood and most studied self-reversal mineral (e.g. Uyeda, 1958; Ishikawa and Syono, 1963; Hoffman, 1975, 1992b). The self-reversal mechanism is complicated, involving Fe–Ti ordering and possibly exsolution. However, the essence of the mechanism is that a negative exchange interaction occurs between a weakly magnetic phase that has a high Curie temperature and a much stronger magnetic phase with a lower Curie temperature.

McClelland and Goss (1993) and Merrill and McFadden (1994) suggest that it is also possible for self-reversal to occur in sediments if two anticoupled phases coexist during chemical alteration. For example, McClelland and Goss show that under some circumstances the conversion of maghemite to hematite would result in self-reversal. Documentation of this phenomenon is effectively impossible because a sediment self-reversal that occurred at ambient temperatures would typically take so long that it could not be reproduced in a laboratory experiment.

The very existence of self-reversing minerals/rocks demanded convincing field tests before there could be acceptance that the Earth's magnetic field had actually reversed. Many instances of such tests exist (§4.1.3) and provide strong evidence in favor of field reversal. Fortunately, thermoremanent self-reversal is rare and sediment self-reversal is probably even more rare. Furthermore, great care is taken in documenting a reversal and there is overwhelming evidence (§4.1.3) that the vast majority of reverse magnetizations observed in rocks arise from changes in the polarity of the Earth's magnetic field. Thus, it is extremely unlikely that self-reversal could have caused any errors in the commonly used reversal chronologies.

Because of the strong evidence in favor of field reversal, it was somewhat surprising that an apparent correlation between oxidation state and polarity was purported to have been found in the late 1960s (Wilson and Watkins, 1967; see also Smith, 1970b). Although sometimes still referred to, this apparent correlation has been shown to be due to a data artifact (Merrill, 1985).

4.1.3 Evidence for Field Reversal

Three field tests provide compelling evidence with regard to reversals. If the Earth's magnetic field has indeed undergone reversals, then:
(i)   studies of baked contacts adjacent to intrusive igneous rocks or underlying lava flows should show agreement in the polarity of the magnetization of the igneous rock and the baked rock;
(ii) rocks worldwide in which the magnetization was acquired simultaneously should exhibit a common polarity; and
(iii) there should be rock sequences that show the polarity of the magnetic field changing continuously from one state to the other.
The first two tests rely on observations in different rock types (and therefore with different magnetic mineralogy) of magnetizations acquired simultaneously. This makes it difficult for the rare circumstance of self-reversal to compromise a conclusion that the Earth's magnetic field has indeed reversed.

The evidence from the studies of baked contacts was first compiled by Wilson (1962b). This was later updated by Irving (1964) and then by McElhinny (1973a). The successive increase, up until 1973, in the overwhelming evidence showing the correspondence in polarity between the baked and baking rocks is given in Table 4.1. Conventionally, magnetizations in the same sense as the present magnetic field are termed *normal* (N) and those in the opposite sense are termed *reverse* (R). Directions intermediate between these (defined by the virtual geomagnetic pole being further than 45° from the axis of rotation) are termed *intermediate* (I). Comparisons of polarities for the case of reverse magnetizations

TABLE 4.1
Comparison of Polarities Observed in Igneous Rocks and Their Baked Contacts[a]

Igneous	Baked contact	Number of observations		
		Wilson (1962b)	Irving (1964)	McElhinny (1973a)
N	N	14	34	47
R	R	34	49	104
I	I	1	2	3
N	R	3	3	3
R	N	0	0	0

[a]N, normal; R, reverse; I, intermediate.

were of course central to confirmation of the reality of the field reversals and this is seen by the much greater emphasis given to this situation in Table 4.1. The three cases of disagreement reported in the table refer to measurements made before the advent of magnetic cleaning techniques in paleomagnetic studies and are probably in error. Today the hypothesis of field reversal is so well accepted that baked contact studies are now used not so much as a check on the reality of field reversals but as a check on the age of the magnetization of igneous rocks (see §3.3.4). This is especially true in studies of old Precambrian rocks in which the importance of evidence relating to the age of the magnetization being measured is paramount.

Wilson (1962a) undertook a study into the magnetic record in a doubly-baked rock that was particularly convincing in the field-reversal versus self reversal debate. The situation is illustrated in Fig. 4.2. A lava flow, reversely magnetized, had heated an underlying laterite. The baked zone of the laterite had a direction of magnetization the same as that of the lava – reverse. Subsequently, both lava and baked laterite were intruded by a dike whose magnetization was also reverse

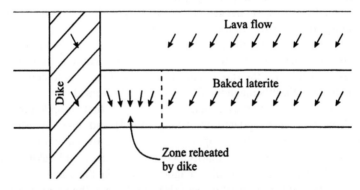

Fig. 4.2. Evidence for field reversal from the magnetic record in a doubly baked rock. A lava flow originally heated the laterite horizon. Subsequently a dike intruded both, reheating both lava flow and laterite in its adjacent region. In this region the laterite was thus first reheated by the lava and then reheated a second time by the dike. From Merrill and McElhinny (1983).

but whose direction differed by about 25° from that of the lava and baked laterite. A study of the zone of second heating of the laterite caused by the dike showed a gradual change from the dike direction, adjacent to the dike, to the lava direction at a distance where the second heating had produced no effect. Thus, in the same region both superimposed magnetizations were of reverse polarity. It is difficult to explain this observation by any reasonable mechanism other than that the Earth's magnetic field had changed polarity and was reverse during both heating episodes.

4.2 The Geomagnetic Polarity Time Scale

4.2.1 Polarity Dating of Lava Flows 0-6 Ma

Mercanton (1926) first realized that if rocks containing reverse magnetizations were due to reversals of the Earth's magnetic field, then this should be registered in rocks worldwide and so he obtained samples from Spitsbergen, Greenland, Iceland, the Faroe Islands, Mull, Jan Mayen Land and Australia as a test. He found that some were magnetized in the same sense as the present Earth's field and others were roughly reversed from it. Matuyama (1929) observed similar effects in lavas covering the past 2 million years from Japan and Manchuria. He noticed that the reverse lavas were always older than the lavas magnetized in the same sense as the present field, which was the first suggestion of a time sequence associated with reversely magnetized rocks. Based on observations of volcanic rocks from the Chaîne des Puys in France, Roche (1951, 1956) concluded that the most recent reversal of the Earth's magnetic field took place in the middle of the Early Pleistocene. Similar observations were made by Hospers (1953, 1954) in lava sequences in Iceland, by Opdyke and Runcorn (1956) in rocks from the United States, and by Khramov (1955, 1957, 1958) in sedimentary sequences in western Turkmenia. These observations suggested that an ordered sequence of polarity inversions might exist in the geological record. This conclusion was further emphasized by the evidence that almost all rocks of Permian age are reversely magnetized (Irving and Parry, 1963).

In the above studies the assessment of the ages of magnetizations relied on the relatively imprecise methods used for dating rocks on the basis of fossil occurrences. It was only in the early 1960s that developments (by Evernden, McDougall and Dalrymple) in the K–Ar isotopic dating method enabled the dating of quite young volcanic rocks with some precision. Rutten (1959) was the first to use K–Ar dating to assess the age of magnetic polarities. He concluded that the present normal polarity had existed since at least 0.47 million years ago and that an earlier period of normal polarity existed about 2.4 million years ago.

In an attempt to define a polarity time scale, systematic studies using joint magnetic polarity and K–Ar age determinations on young lava flows were undertaken in both the United States and Australia. The first time scale put forward by Cox *et al.* (1963a) appeared to be consistent with a periodicity of magnetic reversals at about 1 million year intervals. However, as new data appeared in the literature (Cox *et al.*, 1963b, 1964a; McDougall and Tarling, 1963, 1964) it rapidly became apparent that there was no simple periodicity; the lengths of successive polarity intervals varied haphazardly, some being long (~1 Myr) and others short (~0.1 Myr). Cox *et al.* (1964b) proposed that within intervals of predominantly one polarity lasting of the order of 1 Myr, there were short intervals of opposite polarity of the order of 0.1 Myr. The longer intervals were termed magnetic polarity epochs and the shorter intervals were called events. The epochs were named after pioneering scientists in geomagnetism (Brunhes, Matuyama, Gauss, and Gilbert) whereas the events were labeled from the location of their discovery (e.g., Jaramillo, Olduvai, Kaena, and Sidufjall). The terms *chron* and *subchron* have subsequently been officially adopted to replace the terms epochs and events (see Table 4.3). However, the terms "epoch" and "event" are still used but with somewhat different meaning. For example, the term *reversal event* is often used to describe the phenomenon of two reversals that occur close to each other in time, whereas the term subchron refers to the time interval of the stratigraphic record of an event (see §4.3.1).

A few of the earliest compilations, covering the years 1959–1966, of the GPTS for the past 4 million years are shown in Fig. 4.3. These early studies combined conventional K–Ar dating with measurements of magnetic polarity from widely spaced localities worldwide; they were not carried out on continuous sequences so the ordering relies on the accuracy of the K–Ar dates. In general, the evolution of this time scale has been marked by the inclusion of an increasing number of reversals. Glen (1982), in an excellent history of this evolution, notes that the classic work by Khramov (1955, 1957, 1958) on sedimentary sequences in what is now western Turkmenistan clearly influenced the early igneous rock work on polarity time scales of Cox, Doell, and Dalrymple in the United States and of McDougall and Tarling in Australia in the early 1960s.

The development of high-precision $^{40}Ar/^{39}Ar$ dating techniques has shown that many of the ages determined by the conventional K–Ar method were too young, so those ages have had to be corrected. Furthermore, as increasingly more measurements have been made, questions have arisen regarding the identification of some of the short subchrons. The problem that arises with some of these is to decide whether they represent true reversals or some other geomagnetic field behavior such as changes in intensity.

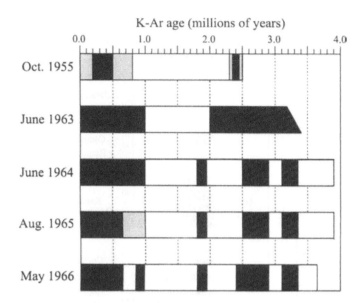

Fig. 4.3. Early evolution of the geomagnetic polarity time scale 1959–1966. Black represents normal polarity, white represents reverse polarity, and gray indicates uncertain polarity. Abstracted with permission from Cox (1969). © American Association for the Advancement of Science.

Figure 4.4 shows the present state of the GPTS for the past 6 Myr following Cande and Kent (1995). This represents the time interval during which all the polarity chrons and subchrons have been named. Cande and Kent (1995) also identified various cryptochrons (see Table 4.3), which may be due to geomagnetic behavior such as geomagnetic intensity changes that occur at 0.49, 1.20, (Cobb Mountain) and 2.42 Ma.

Although natural with the evolution of the GPTS, the development of the terminology of polarity chrons (such as the reverse Matuyama chron) containing polarity subchrons (such as the normal Jaramillo, Olduvai, and Réunion subchrons) was perhaps unfortunate. The problem is that the nomenclature suggests a reasonable interval of time (the chron) during which the polarity was biased toward one value while containing short, aberrant intervals (subchrons) of the opposite polarity. This concept of polarity bias persists in the literature today (e.g., Johnson *et al.*, 1995; Algeo, 1996). As discussed in §4.5.4, the individual normal and reverse intervals (whether they be intervals of the chron polarity or subchrons of the opposite polarity) are equivalent, independent intervals drawn from a random process.

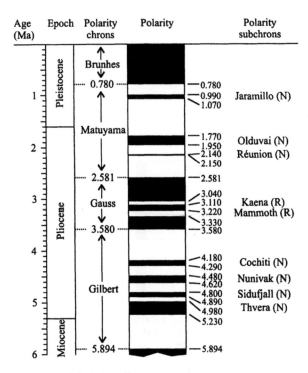

Fig. 4.4. Geomagnetic polarity time scale for the past 6 Myr based mainly on ^{39}Ar/^{40}Ar and paleomagnetic data on igneous rocks. Black represents normal polarity, and white represents reverse polarity. From Merrill *et al.* (1996).

4.2.2 Geochronometry of Ocean Sediment Cores

Volcanic activity is intermittent so the study of lava successions does not produce a continuous sequence of polarity information. In contrast, continuous sequences can be obtained using ocean-bottom cores from deep-sea sediments, providing an independent method of determining the polarity time scale. The cores are not usually oriented but they are taken nearly vertically into the ocean bottom so that changes in sign of the magnetic inclination measured in the cores or changes of 180° in declination can easily be identified as records of polarity change. Deposition rates are relatively low in oceanic sediments, typically being in the range 1–10 mm per 1000 years. This means that the Matuyama–Brunhes boundary at 0.78 Ma will typically be found at a depth of between 0.78 and 7.8 m and the Gilbert–Gauss boundary at 3.58 Ma will typically be found at a depth of between 3.58 and 35.8 m. This slow sedimentation rate smooths the magnetic signal but does allow one to go further back in time for a given length of core. Oceanic sedimentation rates elsewhere (e.g., continental margins) can be much higher and provide a more detailed record of magnetic changes but over a shorter interval of time.

The earliest investigations of marine sediment cores rapidly confirmed the reality of the land-based polarity time scale (Ninkovich *et al.*, 1966; Opdyke and Glass, 1969). Two examples from the North Pacific are given in Fig. 4.5, with the sedimentation rates differing by about 50%. In both cases there is positive (downward pointing) inclination (normal polarity) in the upper part of the core corresponding to the Brunhes chron. An abrupt change to negative (upward

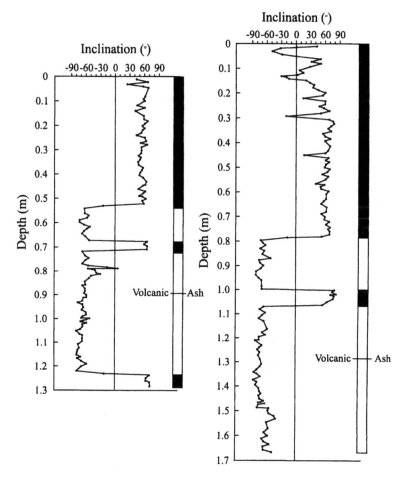

Fig. 4.5. Variation of magnetic inclination with depth in two deep-sea sediment cores from the North Pacific. (Left) V20-107, (right) V20-108. On the right of each log, the black strip indicates normal polarity, and the white strip indicates reverse polarity. From Ninkovich *et al.* (1966), reprinted with permission from Elsevier Science.

pointing) inclination (reverse polarity) occurs some distance down each core, corresponding to the Matuyama–Brunhes boundary. Further down each core there is a brief return to normal polarity corresponding to the Jaramillo subchron, and so on. The correlation of the polarity changes down the cores with the GPTS enables various horizons to be dated precisely so that the rate of sedimentation can be determined. This enables the lengths of the various polarity chrons and subchrons to be determined with some precision.

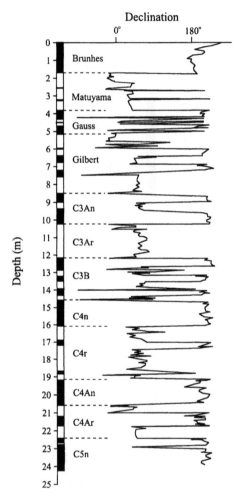

Fig. 4.6. Magnetic stratigraphy in a long (24-m) deep-sea sediment core (Conrad 12-65) from the equatorial Pacific. Changes in declination with respect to a scribe mark on the core are plotted against depth. The black (normal) and white (reverse) bar diagram on the left is a proposed extension of the GPTS. From Foster and Opdyke (1970).

4.2.3 Extending the GPTS to 160 Ma

Early on it was recognized that the GPTS could not be extended beyond a few million years using only absolute dating methods without an independent means of ordering the older reversals. Beyond about 5 Ma the dating errors in the methods of §4.2.1 would become too large to delineate reversals and order them correctly. A 3% error in radiometric dating of a lava flow with age near 10 Ma would misplace it by more than the mean polarity interval for the past few million years. Therefore, it becomes impossible to date a lava flow accurately enough to conclude that it erupted in the same polarity chron as another lava flow in a different locality. However, in continuously deposited sediments the ordering is automatically retained and so studies of long deep-sea sediment cores enabled the polarity sequence to be determined back to about 10 Ma, as illustrated in Fig. 4.6. Foster and Opdyke (1970) examined two cores from the equatorial Pacific that were 24 and 28 m in length. For these cores it was more reliable to measure changes in declination because the magnetic inclination is near zero in the equatorial region. The results from the 24-m Conrad 12-65 core, shown in Fig. 4.6, were the first reported measurements of magnetic changes extending down through 11 polarity chrons. McDougall *et al.* (1976a,b, 1977) showed that by studying long sequences of lava flows it was also possible to extend the GPTS beyond 5 Ma using radiometric dating of lava flows on land.

The most valuable extension to the GPTS has, however, come from sea-floor magnetic anomalies (see chapter 5). Vine and Matthews (1963) proposed that the lineated magnetic anomalies observed on either side of oceanic ridges were a consequence of sea-floor spreading (as put forward by Hess, 1962) and the alternating polarity of the geomagnetic field. This was essentially confirmed by Vine and Wilson (1965). Subsequently, Vine (1966) and Pitman and Heirtzler (1966) extended the magnetic reversal pattern back to 10 Ma. This extended pattern was dated by assuming an age for the base of the Gauss chron and by assuming a constant spreading rate.

The beauty of the sea-floor spreading pattern is that the ordering of the reversal pattern is automatically maintained. It rapidly became apparent that the sea-floor magnetic anomalies, extending for thousands of kilometres on either side of the ridge crests of the South Atlantic, South Pacific, and North Pacific oceans, were in principle correlatable back to the age of the oldest sea floor. Heirtzler *et al.* (1968) examined anomaly patterns from all the oceans in an attempt to construct a GPTS. In order to resolve several problems they took the bold step of assuming, on the basis of the evidence before them, that the South Atlantic ocean had spread at an almost continuous rate since the Late Cretaceous. They produced a GPTS for the Late Cretaceous and for the whole of the Cenozoic and put the age of the K/T boundary at 60 Ma. Since then, there have been several geomagnetic polarity time scales published, most using the

Heirtzler *et al.* (1968) interpretation of the anomaly sequences (LaBrecque *et al.*, 1977; Berggren *et al.*, 1985a,b; Harland *et al.*, 1982, 1990) but with some important modifications for some of the anomalies by, for example, Blakely (1974), Klitgord *et al.*, (1975), and Cande and Kristoffersen (1977). Ages within these time scales are generally determined by interpolation between several absolute age calibration points.

The Harland *et al.* (1982) chronology of reversals for the past 170 Myr combined data from lavas, deep-sea sediments, and marine magnetic anomalies. However, this chronology did not incorporate the more recent changes based on

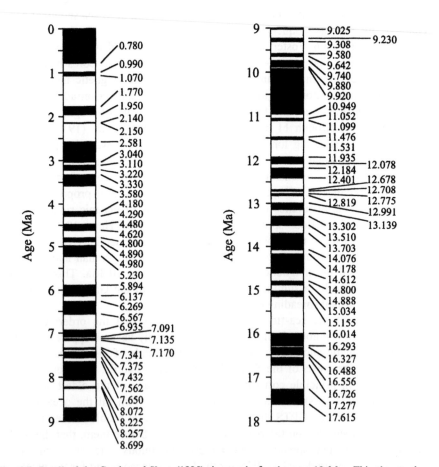

Fig. 4.7. Detail of the Cande and Kent (1995) time scale for the past 18 Myr. This time scale includes only those subchrons that are recorded both in marine magnetic anomaly records and in magnetostratigraphic sections.

^{40}Ar/^{39}Ar dating. Kent and Gradstein (1986) presented a geomagnetic and geologic time scale for the past 160 Myr based on the integration of marine magnetic anomaly data with biostratigraphic, magnetostratigraphic, and radiometric data.

Despite having a rather slow rate of spreading, the South Atlantic has not suffered major plate reorganization and both flanks of the ridge crest are well preserved. Thus, in the first major analysis of magnetic anomaly profiles from the world's ocean basins since that of Heirtzler *et al.* (1968), Cande and Kent (1992a) concluded that the South Atlantic magnetic anomalies remained the most appropriate as the basis for a GPTS. However, their revised GPTS for the Late Cretaceous and Cenozoic had significant changes from the earlier time scales. This time scale was later refined by Cande and Kent (1995), who still used the South Atlantic as a reference ocean but assumed that its spreading rate had smooth variations, which they approximated by splining between calibration points. Additionally, the Cande and Kent (1995) time scale (Table 4.2) adopted astrochronological estimates for the ages of the Plio–Pleistocene reversals. Only those subchrons that are recorded both in marine magnetic anomaly records (see Chapter 5) and in magnetostratigraphic section (see §4.3) are included in the Cande and Kent (1995) time scale. Because of the relatively rapid reversal rate, Fig. 4.7 shows details of this time scale back to 18 Ma.

In general, the GPTS for the past 10 or 20 Myr has stabilized significantly, with minor improvements continuing to occur. For example, the time scales of Baksi (1993) for the past 18 Myr and of Cande and Kent (1995) differ only in minor detail. However, there are two issues of possible contention; calibration points and assumptions about the South Atlantic spreading rate (§5.3.1). Wei (1995) disputed five of the nine calibration points used by Cande and Kent (1992a, 1995) and used other calibrations instead. This has led to a time scale that differs significantly from the Cande and Kent time scales, with the greatest differences being in the Miocene, as may be seen in Table 4.2.

In Fig. 4.8 the Late Jurassic and Early Cretaceous part of the of Kent and Gradstein (1986) reversal chronology has been combined with the Cande and Kent (1995) chronology to give a complete reversal time scale for the past 160 Myr. The presence of two suggested zones of mixed polarity between 100 and 107 Ma (Gradstein *et al.*, 1994) has been omitted, since these are not seen in marine magnetic anomalies, and their existence in magnetostratigraphic data can be disputed (Opdyke and Channell, 1996). Extending the time scale to ages older than 160 Ma where the marine magnetic anomaly record ends relies on the integration of magnetostratigraphic sections sampled around the world. This is discussed further in §4.3.

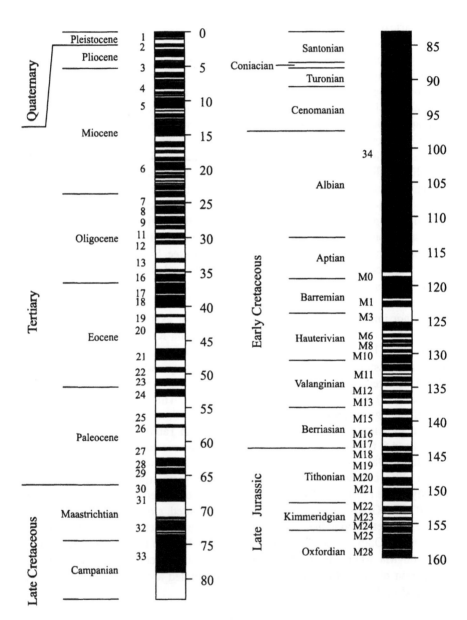

Fig. 4.8. Geomagnetic polarity time scale with magnetic anomaly numbers (see §5.3) for the past 160 Myr using the combined scales of Kent and Gradstein (1986) and Cande and Kent (1995).

Table 4.2

Late Cretaceous–Cenozoic Normal Polarity Intervals (Ma) in Recent Time Scales

Chron	C & K[a] (1995)	Wei (1995)	Chron	C & K[a] (1995)	Wei (1995)
C1n	0.000-0.780	0.000-0.780	C6Aar.2n	22.459-22.493	23.284-23.312
C1r.1n	0.990-1.070	0.990-1.070	C6Bn.1n	22.588-22.750	23.392-23.527
C2n	1.770-1.950	1.770-1.950	C6Bn.2n	22.804-23.069	23.572-23.794
C2r.1n	2.140-2.150	2.140-2.150	C6Cn.1n	23.353-23.535	24.031-24.183
C2An.1n	2.581-3.040	2.580-3.040	C6Cn.2n	23.677-23.800	24.302-24.405
C2An.2n	3.110-3.220	3.110-3.220	C6Cn.3n	23.999-24.118	24.573-24.673
C2An.3n	3.330-3.580	3.330-3.580	C7n.1n	24.730-24.781	25.191-25.235
C3n.1n	4.180-4.290	4.180-4.290	C7n.2n	24.835-25.183	25.281-25.579
C3n.2n	4.480-4.620	4.480-4.620	C7An	25.496-25.648	25.850-25.982
C3n.3n	4.800-4.890	4.800-4.890	C8n.1n	25.823-25.951	26.136-26.247
C3n.4n	4.980-5.230	4.980-5.230	C8n.2n	25.992-26.554	26.284-26.784
C3An.1n	5.894-6.137	5.829-6.051	C9n	27.027-27.972	27.214-28.100
C3An.2n	6.269-6.567	6.173-6.450	C10n.1n	28.283-28.512	28.400-28.625
C3Bn	6.935-7.091	6.795-6.943	C10n.2n	28.578-28.745	28.690-28.854
C3Br.1n	7.135-7.170	6.986-7.019	C11n.1n	29.401-29.662	29.514-29.779
C3Br.2n	7.341-7.375	7.183-7.216	C11n.2n	29.765-30.098	29.886-30.228
C4n.1n	7.432-7.562	7.271-7.398	C12n	30.479-30.939	30.633-31.103[b]
C4n.2n	7.650-8.072	7.483-7.902	C13n	33.058-33.545	33.313-33.812
C4r.1n	8.225-8.257	8.055-8.088	C15n	34.655-34.940	34.922-35.200
C4An	8.699-9.025	8.543-8.887	C16n.1n	35.343-35.526	35.586-35.760
C4Ar.1n	9.230-9.308	9.106-9.191	C16n.2n	35.685-36.341	35.909-36.518
C4Ar.2n	9.580-9.642	9.490-9.560	C17n.1n	36.618-37.473	36.771-37.543
C5n.1n	9.740-9.880	9.670-9.827	C17n.2n	37.604-37.848	37.660-37.877
C5n.2n	9.920-10.949	9.874-11.089	C17n.3n	37.920-38.113	37.941-38.112
C5r.1n	11.052-11.099	11.214-11.273	C18n.1n	38.426-39.552	38.389-39.382
C5r.2n	11.476-11.531	11.738-11.806	C18n.2n	39.631-40.130	39.451-39.892
C5An.1n	11.935-12.078	12.314-12.495	C19n	41.257-41.521	40.898-41.135
C5An.2n	12.184-12.401	12.628-12.904	C20n	42.536-43.789	42.064-43.245
C5Ar.1n	12.678-12.708	13.256-13.294	C21n	46.264-47.906	45.731-47.511
C5Ar.2n	12.775-12.819	13.379-13.436	C22n	49.037-49.714	48.778-49.540
C5Aan	12.991-13.139	13.654-13.813	C23n.1n	50.778-50.946	50.734-50.921
C5Abn	13.302-13.510	14.050-14.312	C23n.2n	51.047-51.743	51.034-51.802
C5can	13.703-14.076	14.555-15.021	C24n.1n	52.364-52.663	52.478-52.800
C5and	14.178-14.612	15.147-15.677	C24n.2n	52.757-52.801	52.897-52.947
C5Bn.1n	14.800-14.888	15.901-16.007	C24n.3n	52.903-53.347	53.057-53.527
C5Bn.2n	15.034-15.155	16.178-16.320	C25n	55.904-56.391	56.113-56.584
C5Cn.1n	16.014-16.293	17.289-17.592	C26n	57.554-57.911	57.691-58.027
C5Cn.2n	16.327-16.488	17.628-17.800	C27n	60.920-61.276	60.850-61.188
C5Cn.3n	16.556-16.726	17.872-18.051	C28n	62.499-63.634	62.359-63.479
C5Dn	17.277-17.615	18.617-18.955	C29n	63.976-64.745	63.821-64.613
C5En	18.281-18.781	19.603-20.074	C30n	65.578-67.610	
C6n	19.048-20.131	20.321-21.295	C31n	67.735-68.737	
C6An.1n	20.518-20.725	21.633-21.814	C32n.1n	71.071-71.338	
C6An.2n	20.996-21.320	22.047-22.324	C32n.2n	71.587-73.004	
C6Aan	21.768-21.859	22.703-22.780	C32r.1n	73.291-73.374	
C6Aar.1n	22.151-22.248	23.025-23.107	C33n	73.619-79.075	
			C34n	83.000-	

[a]Cande and Kent (1995). [b]A typographic error in Wei (1995) has been corrected.

4.3 Magnetostratigraphy

4.3.1 Terminology in Magnetostratigraphy

Hospers (1955) appears to have been the first to suggest the use of reversals as a means of stratigraphic correlation and Khramov (1955, 1957) undertook the first application. In his book Khramov (1958, English translation 1960) suggested the possibility of a strict worldwide correlation of volcanic and sedimentary rocks and the creation of a single geochronological paleomagnetic time scale valid for the whole Earth.

The application of the well-known principles of stratigraphy to the observed reversal sequences or magnetic properties of the strata in sedimentary records is referred to as *magnetostratigraphy* or *magnetic stratigraphy* (see review by Opdyke and Channell, 1996). Here the interest is specifically in the magnetic polarity and the term magnetostratigraphy is used to refer to the stratigraphy of observed reversal sequences. The terminology has been formalized by the IUGS Sub-Commission on the Magnetic Polarity Time Scale (Anonymous, 1979). The basic unit in magnetostratigraphy is the magnetostratigraphic polarity zone, which may be referred to simply as a *magnetozone*. Polarity zones may consist of strata with a single polarity throughout, they may be composed of an intricate alternation of normal and reverse units, or they may be dominantly either normal or reverse magnetozones containing minor subdivisions of the opposite polarity. Because of changes in sedimentation rate, polarity zones of the same thickness may not represent intervals of the same time, so the term polarity chron is used to represent a time interval. The recommended hierarchy in magnetostratigraphic units and polarity chron (time) units is given in Table 4.3 following Opdyke and Channell (1996).

Unlike most branches of stratigraphy, the chronologic system in use in magnetostratigraphy was established as it was being developed. Type sections

TABLE 4.3

Hierarchy in Magnetostratigraphic Units and Polarity Chron (Time) Units as Recommended by the IUGS Sub-Commission on the Magnetic Polarity Time Scale[a]

Magnetostratigraphic polarity units	Geochronologic (time) equivalent	Chronostratigraphic equivalent	Approximate duration (years)
Polarity megazone	Megachron	Megachronozone	10^8-10^9
Polarity superzone	Superchron	Superchronozone	10^7-10^8
Polarity zone	Chron	Chronozone	10^6-10^7
Polarity subzone	Subchron	Subchronozone	10^5-10^6
Polarity microzone	Microchron	Microchronozone	$<10^5$
Polarity cryptozone	Cryptochron	Cryptochronozone	Existence uncertain

[a]After Opdyke and Channell (1996).

were never established for the Brunhes, Matuyama, Gauss, and Gilbert chrons during the original development of the GPTS through polarity determinations on dated lava flows (§4.2.1). However, type localities are known for many subchrons, such as the Jaramillo Creek in New Mexico for the Jaramillo subchron. Because of the historical development of the subject, the four named chrons are used for the Pliocene and Pleistocene, with the remainder of the time scale, as shown in Fig. 4.8, being subdivided into polarity chrons designated by numbers correlated to marine magnetic anomalies (see §5.3.2 for an explanation of the nomenclature). The geomagnetic polarity history as preserved in the sea floor has become the template (a sort of type section) for the GPTS for the past 160 Myr. Thus the designation of type sections, in the classical stratigraphic sense, for the geomagnetic polarity pattern since 160 Ma is unnecessary (Opdyke and Channell, 1996). On the other hand, type sections for magnetobiochronology (the correlation of the biological record with the magnetic polarity sequence) might be desirable. For example, Opdyke and Channell suggest that a potential type section for the Paleogene might be the Contessa section at Gubbio in Italy studied by Lowrie *et al.* (1982).

4.3.2 Methods in Magnetostratigraphy

For times older than the present sea-floor record, the use of type sections for geomagnetic polarity history is the only way in which the record of polarity changes can be established. The establishment of polarity sequences must then be carried out using classical stratigraphic principles involving type sections. Opdyke and Channell (1996) point out that a good example of a viable magnetostratigraphic type section is that recording the Kiaman Superchron (see §4.3.6) in Australia, where the superchron is thought to be tied directly to the rock record (Irving and Parry, 1963; Opdyke *et al.*, 1999).

Polarity subchrons as short as 20 kyr in duration are present in the polarity record, so any sampling scheme should ideally attempt to sample intervals of this duration. It is common practice when sampling marine sedimentary cores to remove samples at 10 cm spacings. Such sampling density in pelagic sediments with sedimentation rates of 10 mm kyr^{-1} (10 m Myr^{-1}) would resolve polarity chrons with duration greater than ~10 kyr. However, terrestrial sediments rarely have the same homogeneity as deep-sea sediments with respect to lithology and sedimentation rates. Terrestrial sections are thus usually selected for sampling for some particular reason, such as the presence of important vertebrate faunas, radiometrically dated horizons, or climatically indicative levels in loess, tills, or pollen-rich deposits.

The basic problem in magnetostratigraphic studies is that the observed series of normal and reverse polarity zones usually has a pattern that could correlate with one or more segments of the GPTS. At first a constant sedimentation rate, or rate of extrusion of lavas, is assumed and trial correlations with the GPTS are made. Such correlations are aided by additional information such as radiometric ages or biostratigraphic events in the section that have been correlated with the GPTS elsewhere. An excellent example of this is illustrated in Fig. 4.9 for the Haritalyangar section in India (Johnson *et al.*, 1983). Here the correlation of polarity zones with the GPTS is aided by magnetostratigraphic data from Pakistan where some of the same Miocene vertebrate fossils are correlated with radiometric ages on volcanic ashes. On this basis the reversal sequence can be correlated with the GPTS quite unequivocally on the basis of pattern fit. The resulting regression is significant at the 99% level.

Using the available magnetostratigraphic data, Gradstein *et al.* (1994) proposed an integrated geomagnetic polarity and stratigraphic time scale for the whole Mesozoic, the framework of which involves ties between radiometric dates, biozones, and stage boundaries and between biozones and magnetic

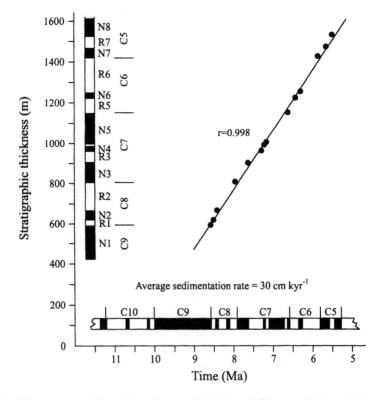

Fig. 4.9. Magnetostratigraphy of the Haritalyangar section in India and correlation with the GPTS of Mankinen and Dalrymple (1979). The high linear regression value (*r* = 0.998) indicates excellent agreement and suggests roughly constant sedimentation rate. After Johnson *et al.* (1983), with permission from Elsevier Science.

reversals observed from marine magnetic anomalies and in sediments. This integrated time scale differs slightly from that shown in Fig. 4.8 in that magnetic anomaly M28 is estimated to be about 3 Myr younger. A period by period review of the current state of global studies in magnetostratigraphy is given by Opdyke and Channell (1996).

4.3.3 Quality Criteria for Magnetostratigraphy

Opdyke and Channell (1996) proposed a reliability index with 10 criteria, listed in Table 4.4, for studies in magnetostratigraphy. These criteria are based on the principles established by Van der Voo (1990a) for paleomagnetic studies relating to determinations of pole positions (see §6.2.2 and Table 6.1).

Obviously there will be few studies able to meet all 10 of the criteria listed. Radiometric dating is often not possible and field tests usually cannot be made in flat-lying beds or deep-sea cores. However, ratings of at least 5 out of the possible 10 should be achievable in modern magnetostratigraphic studies. The global database of magnetostratigraphy established by McElhinny *et al.* (1998) indicates that only two studies at the present time have managed a perfect score of 10. These are the studies of the Barstow Formation of Colorado (MacFadden *et al.*, 1990) and the Newark Supergroup of New Jersey (Kent *et al.*, 1995). Both of these studies therefore represent classic examples of the application of magnetostratigraphy in geology.

TABLE 4.4

Reliability Index Criteria for Studies in Magnetostratigraphy[a]

Criterion	Brief description
1	Stratigraphic age known at the stage level and associated paleontology presented adequately
2	Sampling localities placed in a measured stratigraphic section
3	Complete thermal or AF demagnetization undertaken and illustrated using Zijderveld diagrams
4	Directions determined from principal component analysis (Kirschvink, 1980)
5	Data published completely as VGP latitudes and/or declination and inclination versus stratigraphic distance plots, together with fully documented statistical parameters
6	Magnetic mineralogy determined
7	Field tests (fold, conglomerate, etc. tests) for the age of magnetization undertaken where possible
8	A positive reversals test determined (McFadden and McElhinny, 1990)
9	Radiometric ages, especially $^{40}Ar/^{39}Ar$ or U–Pb ages from volcanic ashes or bentonites, available in the stratigraphic section
10	Multiple sections studied

[a]After Opdyke and Channell (1996)

4.3.4 Late Cretaceous–Eocene: The Gubbio Section

The position of the Cretaceous–Tertiary boundary within the GPTS has been of great interest because of the associated mass extinctions. In most of the documented Late Cretaceous–Early Tertiary sections of the world this critical boundary falls within a hiatus. However, the marine sediments at Gubbio, Italy, contain an unbroken sequence across this boundary (Luterbacher and Premoli

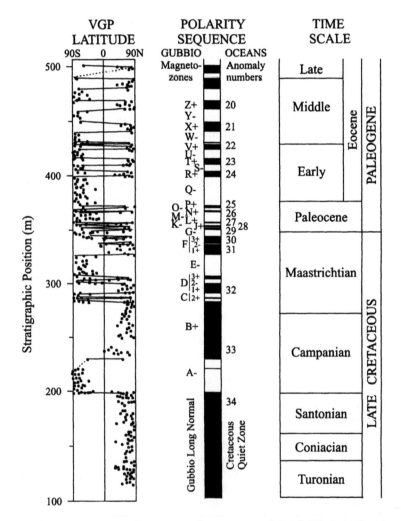

Fig. 4.10. The Late Cretaceous to Eocene reversal sequence recorded at Gubbio, Italy. The observed polarity zones are compared with the GPTS. Normal polarity zones are in black. After Lowrie and Heller (1982).

Silva, 1964). The sections contains a long, continuous sequence of magnetic polarity zones intercalibrated with planktonic foraminiferal zones (Alvarez *et al.*, 1977; Arthur and Fischer, 1977; Lowrie and Alvarez, 1977; Premoli Silva, 1977; Roggenthen and Napoleone, 1977). The sequence has become a classic (see Fig. 4.10) and is particularly valuable because of its unambiguous correlation to magnetic anomaly profiles and to oceanic cores (Tauxe *et al.*, 1983b).

4.3.5 Late Triassic GPTS

Extension of the GPTS back beyond about 175 Ma is currently difficult because there is no extant modern ocean floor that allows the construction of a complete reference sequence of polarity reversals. In general the development of a GPTS for earlier times has relied on the piecing together of separate magnetostratigraphic records of variable length, variable reliability, and usually with insufficient absolute or relative chronological control. However, studies in the Newark rift basin of eastern North America show the potential for GPTS extension in thick, complete continental sedimentary sections.

The initial magnetostratigraphic results (McIntosh *et al.*, 1985) from the Newark basin were from discontinuous outcrop and industry boreholes. An improved documentation of the polarity sequence was provided in later studies by Witte and Kent (1989) and Witte *et al.* (1991). These later studies were based on more dense sampling and better demagnetization procedures, and they incorporated field tests to constrain the age of magnetization. These studies showed the potential for detailed correlation of cyclostratigraphy and magnetostratigraphy in the Newark basin and stimulated a comprehensive drilling program. Under this program continuous coring with near-complete recovery at seven drill sites produced a total of 6770 m of core. This represented almost the whole of the Upper Triassic continental lacustrine sediments together with some of the lowermost Jurassic interbedded continental sediments and lavas of the Newark igneous extrusive zone. Stratigraphic overlapping of the cored sections provided about 30% redundancy and it was possible to assemble a (normalized) 4660-m-thick composite section (Kent *et al.*, 1995). The remaining part of the Jurassic section was studied using test borings by the Army Corps of Engineers (Fedosh and Smoot, 1988; Witte and Kent, 1990; Witte *et al.*, 1991; Olsen *et al.*, 1996b).

The lacustrine deposits display a pronounced cyclic variation in lithofacies linked to the McLoughlin cycle (a Milankovitch cycle). Furthermore, the Watchung basalts at the top of the sedimentary sequence are known to be contemporaneous with the Palisades Sill, which is well dated at 201±2.7 Ma (^{40}Ar/^{39}Ar; Sutter, 1988) and 202±1 Ma (U-Pb zircon; Dunning and Hodych, 1990). Hence, it is possible to date events within the drill cores by using the cyclostratigraphy to count down from the Watchung basalts.

Paleomagnetism: Continents and Oceans

Combining the magnetostratigraphic analysis of the 4660-m-thick composite section with the cyclostratigraphic analysis, it was possible to provide a GPTS for almost the whole of the Late Triassic (Kent *et al.*, 1995; Olsen and Kent, 1996; Olsen *et al.*, 1996a). Recently, by undertaking additional sampling to confirm short polarity intervals and to determine the thickness and duration of polarity transition zones, and by extending the cyclostratigraphy to older strata, Kent and Olsen (1999) presented a refinement of the initial magnetostratigraphic results and were able to calculate an astronomically tuned GPTS for almost 25 million years of the Late Triassic. They further extended the record by 6 Myr by downward extrapolation of the sedimentation rates determined by cyclostratigraphy. The lithostratigraphy, magnetostratigraphy, and cyclostratigraphy of the composite section are shown in Fig. 4.11.

The details of the resulting GPTS are provided in Table 4.5. It is impressive, and encouraging for the future potential for extending the GPTS through magnetostratigraphy, that this GPTS is good enough to perform a statistical analysis of the reversal sequence. Kent and Olsen (1999) note that the interval lengths have a mean duration of about 0.54 Myr (corresponding to an average reversal rate of about 1.8 Myr^{-1}) and that there is no discernible polarity bias.

TABLE 4.5

Late Triassic GPTS[a]

Chron	Age (Ma)	Chron	Age (Ma)	Chron	Age (Ma)
E24n	202.021	E18n	208.624	E10n	223.219
E23r	202.048	E17r	210.006	E9r	224.487
E23n	203.026	E17n	210.466	E9n	225.016
E22r	203.330	E16r	210.762	E8r	226.634
E22n.2n	203.565	E16n	212.559	E8n	227.245
E22n.1r	203.576	E15r.2r	212.871	E7r	227.619
E22n.1n	203.941	E15r.1n	212.903	E7n	228.924
E21r.3r	204.256	E15r.1r	213.110	E6r	229.284
E21r.2n	204.277	E15n	213.942	E6n	229.990
E21r.2r	204.410	E14r	215.427	E5r	230.384
E21r.1n	204.458	E14n	216.656	E5n.2n	230.531
E21r.1r	204.647	E13r	217.469	E5n.1r	230.645
E21n	205.177	E13n.2n	218.390	E5n.1n	231.047
E20r.2r	206.554	E13n.1r	218.425	E4r	231.226
E20r.1n	206.590	E13n.1n	218.953	E4n	231.649
E20r.1r	206.765	E12r	219.784	E3r	231.817
E20n	206.884	E12n	219.953	E3n	232.054
E19r	207.567	E11r	221.956	E2r	232.412
E19n	207.717	E11n	222.232	E2n	232.818
E18r	208.098	E10r	222.728	E1r(part)	233.153

[a]After Kent and Olsen (1999).

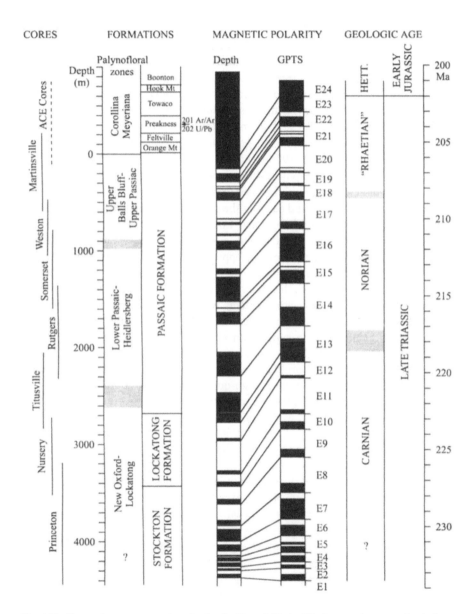

Fig. 4.11. Composite magnetostratigraphy for the Newark Basin. Black represents normal polarity and white reverse polarity. The observed magnetozones (left column) are converted into a Late Triassic GPTS (right column). After Kent and Olsen (1999).

4.3.6 Superchrons

A superchron is an interval of time in the range 10^7–10^8 years during which the geomagnetic polarity remained constant (Table 4.3). The existence of two such superchrons was suggested before the reversal chronology was well developed. These are the Permo-Carboniferous (*Kiaman*) Reverse Superchron (Irving and Parry, 1963) and the *Cretaceous* Normal Superchron (Helsley and Steiner, 1969). The Cretaceous Superchron, which extends from approximately 118 to 83 Ma (Cande and Kent, 1995), is the best documented, since it is evident both in marine magnetic anomalies and in magnetostratigraphic data (Fig. 4.8). The Kiaman Superchron appears to be longer, extending from about 316 to about 262 Ma (Opdyke and Channell, 1996; Opdyke *et al.*, 1999); a substantial body of worldwide magnetostratigraphic data supports its existence, as summarized in Fig. 4.12 for the Late Carboniferous and Permian.

Johnson *et al.* (1995) used data from the global paleomagnetic database (Lock and McElhinny, 1991; McElhinny and Lock, 1993) to interpret, from indirect statistical arguments, the existence of another superchron during the Early Ordovician. However, there does not appear to be any direct evidence at this stage to support the interpretation. There is some suggestion from Siberian sections (Gallet and Pavlov, 1996) of a reverse polarity zone lasting 15 Myr during the Arenig, but there are problems with time resolution and correlations with other parts of the world, where the best published magnetostratigraphic data do not show evidence for a superchron at that time (Ogg, 1995; Opdyke and Channell, 1996). It would be surprising if the two known superchrons were unique, and so despite the problems of the Johnson *et al.* (1995) interpretation their work shows the need for more high-quality magnetostratigraphy.

Figure 4.13 shows a histogram of the lengths of all known polarity intervals from 330 Ma to the present as compiled by Opdyke and Channell (1996). The durations of the vast majority of polarity intervals lie in the range 0.1 to 1.0 Myr and there are few with duration longer than 2 Myr. It has been suggested that each of the two superchrons contains a short interval of the opposite polarity (see Opdyke and Channell, 1996), which would mean that each of the superchrons would in reality be made up of two exceptionally long intervals of the same polarity separated by a very short interval. The evidence for these short intervals is not convincing in either case and, as discussed in §4.5.4, their existence would be surprising. The lengths of the Cretaceous and Kiaman Superchrons shown in Fig. 4.13 are their total lengths; they are clearly outliers of the general distribution with or without the short interval of opposite polarity. As shown by McFadden and Merrill (1995), and discussed in §4.5.4, the Cretaceous Superchron is so long that it cannot be part of the general reversal process; the suggestion is that the reversal process ceased to operate during the superchron. Presumably this would also be the case with the Kiaman Superchron.

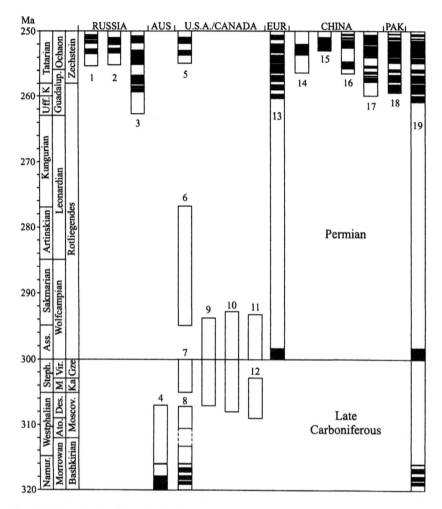

Fig. 4.12. Compilation of Late Carboniferous–Permian magnetostratigraphy data. Black represents normal polarity and white reverse polarity. AUS – Australia, EUR – Europe, PAK – Pakistan. 1, Eastern Russia (Molostovsky, 1992); 2, Gusinaya Zemlya Peninsula (Gurevich and Slautsitais, 1985); 3, Urals (Khramov *et al.*, 1974; Burov *et al.*, 1996); 4, New South Wales (Opdyke *et al.*, 1999); 5, Dewey Lake Fm, Texas (Molina Garza *et al.*, 1989); 6, Cutler Fm, Utah (Gose and Helsley, 1972); 7, Supai Grp, Arizona (Steiner, 1988); Cumberland Grp, Nova Scotia/New Brunswick (DiVenere and Opdyke, 1990, 1991); 8, Cumberland Grp, Nova Scotia/New Brunswick (DiVenere and Opdyke, 1990, 1991; Opdyke *et al.*, 1999); 9, Maroon-Minturn Fm, Colorado (Miller and Opdyke, 1985); 10, Casper Fm, Wyoming (Diehl and Shive, 1981); 11, Ingleside Fm, Wyoming (Diehl and Shive, 1979); 12, Minturn Fm, Colorado (Magnus and Opdyke, 1991); 13, Rotliegendes, Germany (Menning *et al.*, 1988); 14, Sth Tien Shan, Northern Tarim (McFadden *et al.*, 1988b); 15, Dalong Fm, Sichuan (Heller *et al.*, 1988); 16, Wujiaping, Sichuan (Steiner *et al.*, 1989); 17, Shihezi and Shiqiangfen Fms, Shanxi (Embleton *et al.*, 1996; Menning and Jin, 1998); 18, Wargal and Chidru Fms (Haag and Heller, 1991); 19, Composite magnetostratigraphic column. Compiled from Opdyke (1995) with additions.

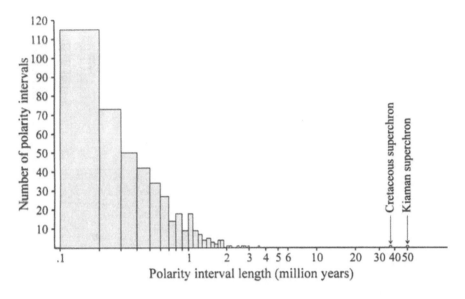

Fig. 4.13. Histogram of polarity intervals for the past 330 Myr. After Opdyke and Channell (1996).

4.4 Polarity Transitions

4.4.1 Recording Polarity Transitions

The final compelling piece of evidence that the Earth's magnetic field changes polarity comes from observations of the magnetization in rock sequences in which the polarity changes continuously from the one state to the other. A comprehensive review of geomagnetic polarity transitions has recently been provided by Merrill and McFadden (1999).

For the sake of clarity the term *reverse* is always taken to indicate either the state of magnetization of a rock or the polarity of the Earth's magnetic field in the sense when they are opposed to the present-day situation. The actual period of change from one polarity to the other is termed a *polarity change* or, more often, a *polarity transition*. Directions of magnetization observed in rocks during such transitional times are termed *transitional*, which is distinct from the term *intermediate* (I) used in §4.1.3.

The duration of a typical geomagnetic polarity transition is not well known but probably lies between about 1000 and 8000 years (§4.4.4). Although this is slow on the human time scale, it is but a fleeting moment on the geological time scale. Furthermore, it represents only a brief interval in the formation time of a typical rock unit. Consequently, there are several data acquisition problems in attempting to characterize polarity transitions.

First, it is a difficult task to identify rocks that have recorded a given polarity transition. There are only 127 transitional records currently available in the polarity transition database initially described by McElhinny and Lock (1996). These records represent a total of 46 different transitions, the best documented transition being the Matuyama–Brunhes with 32 records. The Athanasopoulos *et al.* (1996) compilation shows 23 sedimentary records for the Matuyama–Brunhes transition. Love and Mazaud (1997) have undertaken a comprehensive review of data for the Matuyama–Brunhes reversal transition field. They present a data set comprising sedimentary and lava data; they considered 62 different sites, of which only 11 satisfy their quality criteria.

Second, there is the potential for polarity transitions to be recorded in sediments, intrusives, and lava flows. The characteristic of the recording, the information available, and the reliability of the recording process itself in each of these different rock types are, however, quite different. This means that at times it can be difficult to reconcile the records from different rock types. In sedimentary rocks it is possible to obtain a good continuous record of a transition if the sedimentation has been rapid. In marine sediments the magnetic record is locked in over a 10 to 30-cm range so that the magnetic signal recorded by them is a filtered one. However, sediment records have proved to be one of the most valuable sources of information on the transitional behavior of the geomagnetic field (e.g. Opdyke *et al.*, 1973; Valet and Laj, 1981; Valet *et al.*, 1986, 1989; Clement and Kent, 1991; van Hoof and Langereis, 1992a, b). Terrestrial lava flows extruded during times of polarity transitions provide excellent records of the transitional field but only at discrete, instantaneous points in time. The first such detailed record of a polarity transition from a lava sequence was obtained by van Zijl *et al.* (1962a, b) from the Jurassic Stormberg lavas of Lesotho in southern Africa, now studied in more detail by Kosterov and Perrin (1996).

One of the most studied records in recent times, largely because of the available detail and because of some of the peculiarities in the record, is that from the Steens Mountain in Oregon (Watkins, 1969; Prévot *et al.*, 1985; Coe and Prévot, 1989; Coe *et al.*, 1995). Intrusive rocks formed during a polarity transition may also provide a record of the transitional geomagnetic field if the cooling is sufficiently slow. However, there are significant problems in identifying such transition records, in obtaining reasonable estimates of the passage of time, and in proving that chemical remagnetization has not occurred during the subsequent slow cooling. Consequently, there are few successful studies of such transitional records (Dunn *et al.*, 1971; Dodson *et al.*, 1978).

This means that it has been extraordinarily difficult to obtain the large amounts of spatially and temporally well distributed and well-correlated data required to provide robust characterization of the transitional process (Merrill and McFadden, 1999). In order to obtain a spherical harmonic description of the magnetic field out to just a degree 4 it is necessary to have a minimum of 24 independent but simultaneous measurements well-distributed over the Earth's surface (§1.1.3). There is some evidence to suggest that during a reversal there may have been significant changes in direction on time scales less than 100 years (e.g., Laj *et al.*, 1988; Okada and Niitsuma, 1989; Coe *et al.*, 1995; Worm, 1997). Thus, to be useful for a spherical harmonic analysis, the time of acquisition of the magnetization must be constrained to better than 100 years for each observation. This is not currently possible and the prospect for such accuracy seems remote. Thus it is effectively impossible to determine the structure of the transitional field at any point in time let alone determine the structure at multiple points in time so as to be able to obtain the morphology of a reversal transition.

In the absence of appropriate data to undertake spherical harmonic analyses, transitional directions have commonly been characterized by their VGPs (§1.2.3) and transitional intensities by their VDMs (§1.2.4). Both the VGP and VDM concepts assume that the field is dominantly that of a geocentric dipole, which seems unlikely for the transitional field. The portrayal of transitional directions and intensities as VGPs and VDMs must therefore be seen as nothing more than a convenient mapping and not as a meaningful characterization. The paucity of genuine information about field structure in the available data has meant that there are few real constraints on the morphology of transitional fields. Thus the question of transitional field behavior has been fertile ground for innovative models that have succumbed in the light of subsequent data.

4.4.2 Directional Changes

There have been at least seven phenomenological reversal models suggested to date. These have been summarized by Merrill and McFadden (1999) (see also McFadden and Merrill, 1995), and the basics of the models are as follows.

(i)   The dipole field either decays and then builds up in the opposite direction or rotates without changing intensity (Creer and Ispir, 1970). See Fig. 4.14.

(ii) The transitional field is predominantly nondipolar (Hillhouse and Cox, 1976). This model was proposed when it became apparent that VGP paths for the same reversal, but obtained from different locations, were not coincident.

(iii) The transitional field is dominated by nondipole zonal harmonics (Hoffman, 1979; Fuller *et al.*, 1979). It is now accepted that the transitional field is not axially symmetric, but the Hoffman–Fuller hypothesis

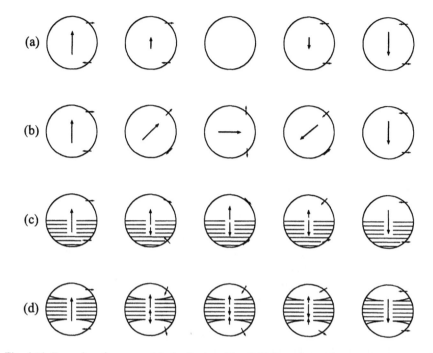

Fig. 4.14. Examples of some models for the transition field for a change from reverse to normal polarity. (a) Decrease in dipole moment which remains axisymmetric during the process. (b) Rotation of the main dipole without changes in moment. (c) A quadrupole transition model in which the polarity change is initiated in the southern hemisphere of the core. (d) An octupole transition model in which the polarity change is initiated in the low latitude zone of the core. From Merrill and McElhinny (1983), compiled from Fuller *et al.* (1979).

represented a conceptual leap and ushered in the era of "modern" polarity transition studies. See Fig. 4.14.

(iv) The transitional field is nondipolar but the transitional VGPs are longitudinally confined and are biased to fall within two (preferred) antipodal longitude bands of width about 60° (Clement, 1991). This model was based on sedimentary data.

(v) The transitional field is predominantly dipolar (although this aspect was later dropped) and the VGPs are not only longitudinally confined and biased toward two preferred bands but also, importantly, the same two bands are preferred for each of the most recent reversals (Laj *et al.*, 1991, Tric *et al.*, 1991). This hypothesis is particularly exciting for, if true, it may imply mantle control of processes in the geodynamo, at least during transitions. The preferred paths lie near the Americas or near an antipodal path (Gubbins and Coe, 1993) through western Australia and Asia. This model was also based on sedimentary data.

(vi) The transitional VGPs from lava flows clump in two patches, one on the American path above, lying between central South America and Antarctica, with the other cluster falling in western Australia and the eastern Indian Ocean (Hoffman, 1991a). The same clusters are occupied by VGPs for different reversals.

(vii) There is no consistent pattern for transitional VGPs from different reversals and the VGPs within a reversal are not longitudinally confined (Prévot and Camps, 1993). This model emerged from analyses of 362 transitional VGPs from 121 volcanic units of Miocene age or younger and is an extension of an earlier suggestion by Valet *et al.* (1992).

If the transitional field were strongly dipolar then VGPs for the same time but for different sites around the globe would be consistent with each other (just as they are today). The transitional VGP paths would then be consistent for observation sites around the globe. Conversely, if different sites give widely divergent VGP positions for the same time then this is strong evidence that the field is not dominantly dipolar. Oddly enough though, consistency in transitional VGP paths from sites around the globe does not necessarily mean that the field was strongly dipolar (e.g., Merrill and McFadden, 1999). This exposes interpretation of transitional VGPs to several major difficulties, as is evidenced by the number of phenomenological models suggested and by their diversity (almost to the point of total contradiction).

The model probably most discussed today is model v, based on sedimentary data. In this model it is suggested that: transitional VGPs from a given site are confined in longitude; that for a given reversal there is a bias for the VGPs to fall into one of two preferred antipodal bands; and that the same bands are preferred for different reversals. Full testing of this model demands that each of its three components be properly tested. To test the hypothesis that the preferred bands are the same for different reversals it must first be confirmed that individual reversals do indeed exhibit a bias toward preferred bands for their VGPs, which itself requires confirmation of longitude confinement for individual VGP paths. Unfortunately, the paucity of data has meant that it has not been possible to separate out and resolve these individual questions. Consequently, investigators have tended to lump together all of the available data and then seek an overall clustering of some parameter (such as the longitude of the transitional VGP equator crossing). Thus, it remains difficult to interpret the results with any genuine clarity.

A major problem in assessing the reversal models is that the polarity transition records obtained from volcanics and from sediments appear to show different characteristics, as is illustrated in Fig. 4.15. The plot of 362 transitional VGPs from 121 volcanic records of reversals less than 16 Myr in age (Fig. 4.15a) does not exhibit any obvious clustering or preferred longitude sectors (Prévot and Camps, 1993). The record for reversals less than 12 Myr in age recorded in

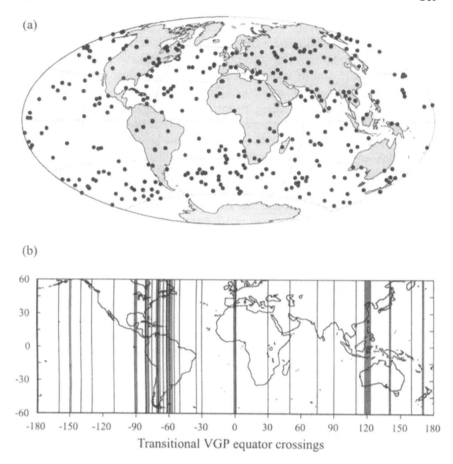

Fig. 4.15. Comparison of the records of transitional VGPs observed in volcanics and sediments. (a) 362 transitional VGPs from 121 volcanic records of reversals less than 16 Myr old. From Prévot and Camps (1993), reproduced with permission from *Nature*. (b) Plot of equator crossings for the available sedimentary VGP transition paths. From McFadden and Merrill (1995).

sediments is summarized in Fig. 4.15b, which shows a plot of the longitude at which each VGP transition path crosses the equator. The claimed clustering of the paths in two approximately antipodal positions is readily apparent. As shown by McFadden *et al.* (1993), this clustering is significantly greater than would be expected from a uniform random distribution.

However, it should be noted that both Valet *et al.* (1992) and McFadden *et al.* (1993) questioned whether model v is actually supported by the very sediment data on which it is based; they both note that the clustering might be linked to the poor distribution of the observation sites, which are themselves strongly grouped in two antipodal longitude bands with the angles between the corresponding site longitude and VGP path strongly clustering around ±90°.

During polarity transitions the intensity of the geomagnetic field decreases on average to about 25% of its value before or after the transition, although values as low as 10% have been observed (§4.4.3). The problem that arises is whether the magnetic torque on the settling magnetic particles has become too weak to overcome other effects so that there is inclination shallowing as discussed in §2.3.7. Quidelleur and Valet (1994) considered some aspects of this concept and then Quidelleur *et al.* (1995) carried out redeposition experiments to test it. Using the relation of (2.3.37), they obtained values of the inclination shallowing factor f as low as 0.23 with an ambient field of 4.6 μT, similar to the field in the middle of a transition. Barton and McFadden (1996) showed that, for this to be an important factor in the clustering of VGP transition paths, values of f below 0.3 are required and that the effect becomes pronounced when f approaches 0.1. Thus, there is the potential that clustering of transitional VGPs may have been enhanced in sedimentary records by a combination of rock magnetic problems and a poor distribution of sites.

Despite the conclusion by Prévot and Camps (1993) that the lava data do not show any obvious clustering or preference for longitudinal bands, there is some evidence to the contrary. Constable (1992) analyzed nontransitional lava data from flows spanning the past 5 Myr to determine if there is any nonuniformity in longitudinal distribution. Her analysis included more than 2000 VGPs and showed a distinct bias toward two longitude bands, which were similar to the preferred bands identified in the sediment data. Some of this bias is almost certainly a consequence of the present orientation of the equatorial dipole. Love (1998) has repeated the test suggested by Prévot and Camps (1993) but arrived at the dramatically different conclusion that the transitional VGPs do exhibit longitudinal preference for the Americas and for the antipodal Asian path (he did not, however, find evidence for latitudinal clustering). The initial data were similar (transitional VGPs younger than 20 Ma) but the selection criteria were different; Love (1998) considers VGPs from different lavas to be independent measurements, despite the counterview that lava flows typically bunch together in short intervals of time (Holcomb, 1987). The critical step is that Love weights each datum (VGP) according to its latitude and the number of transitional VGPs in that record. Histograms of the number of weighted datums versus longitude then show a distribution which, at the 99% level of confidence, is nonuniform. Despite the fact that the analysis fails to separate the critical issues noted above (Merrill and McFadden, 1999), the finding by Love (1998) that the two most occupied transitional longitudes are similar to those identified by analyses of sediment data is impressive.

Hoffman (1999) analyzed VGP data from the Love and Mazaud (1997) database for the Matuyama–Brunhes reversal transition. The data indicate the existence of four major groupings of virtual poles, leading to the suggestion of preferred locations for transitional VGPs (see also Hoffman, 1991a, 1992a).

Interpretation of the transitional field directions remains enigmatic. The problem is exacerbated by the fact that the claimed preference of transitional VGPs for certain longitude bands is small. For example, Love (1998) finds approximately one quarter of the weighted data fall into the two most populated longitudinal bands, whereas for a uniform distribution one might expect about one sixth of the data in those bands. This highlights the fact that the questions can only be resolved by observation, but the current paucity of relevant data precludes a clear resolution.

4.4.3 Intensity Changes

Reliable paleointensity information is typically much more difficult to obtain than reliable directional data. Thus, it is to be expected that even less would be known about the transitional field strength than about the transitional field directions. Nevertheless, there appears to be more agreement about the nature of intensity changes during a transition than about the directional changes, and some robust conclusions can be drawn.

Absolute paleointensities determined from lavas and relative paleointensity estimates from both sedimentary and igneous rocks recording transition directions indicate that the field intensity decreases substantially during a polarity change. These data indicate reductions sometimes to only 10% of the usual field intensity outside the transition. Figure 4.16 provides a summary of absolute paleointensities relating to polarity transitions for lavas less than about 10 Ma. Absolute paleointensities for this time interval have been analyzed by Tanaka *et al.* (1995) using only those determinations made by the Thellier (Thellier and Thellier, 1959a,b; see §1.2.4) or Shaw (1974) methods. The 323 values include many from the central part of polarity transitions that are characterized by low-latitude VGPs. In order to allow for the different site locations all intensity values have been converted to VDMs (see §1.2.4 and §1.2.5). Despite the attendant problems of using VDM during transitional times, the use of VDM (as with VGP) should be seen as a convenient mapping that attempts to take account of the effects of site location on the surface of the Earth. Average VDM for VGP latitude bands of width 20° between -90° and +90° are shown in Fig. 4.16. Note that when the VGP latitude lies between -45° and +45°, there is a dramatic drop in VDM values. These absolute intensity values indicate that *on average* the field at the central part of transitions (VGP latitude 0°) is about 25% its usual value.

The reduction in intensity values for low-latitude transitional VGPs provides robust evidence against a reversal model in which polarity reversal occurs by rotation of the dipole field with no change in magnitude (Merrill and McFadden,

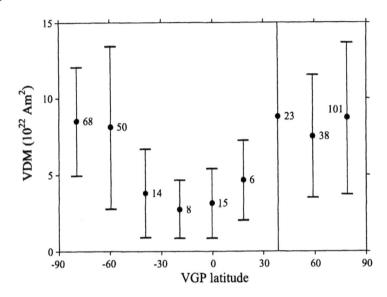

Fig. 4.16. Mean VDM versus VGP latitude for data covering the past 10 Myr averaged over 20° VGP latitude bands. The number of values averaged in each band is indicated and 95% confidence limits are shown. After Tanaka *et al.* (1995).

1999). This is because a reduced intensity for low-latitude transitional VGPs demands a reduction in the dipole field strength during a polarity transition.

4.4.4 Polarity Transition Duration

In order to understand a process it is necessary to know the order in which events occurred and how long they took to occur. Consequently it is important to determine the duration of geomagnetic polarity transitions. Unfortunately estimates of the duration can vary by more than an order of magnitude for the same transition. Merrill and McFadden (1999) conclude that the duration of an average geomagnetic polarity transition is not well known but probably lies between 1000 and 8000 years (see also Bogue and Merrill, 1992).

There are four main methods for estimating the duration of polarity transitions.

(i) Rapidly deposited sediments. This requires evidence of a constant rate of deposition, which is then estimated using absolute or relative dating methods to obtain ages before and after the transition. The duration of the transition can then be estimated from the depth interval in which the directions are transitional. A variant of this method combines relative intensity information with the directional information. The best estimates come from oceanic cores, for example, 4000 years by Harrison and

Somayajulu (1966), 4700 years by Niitsuma (1971) and 4600 years by Opdyke (1972). Other estimates also cluster around 4000–5000 years.

(ii) Statistical approach with lava flow data. This method uses the ratio of the number of transitional directions to the number of non-transitional directions in some time window and (under the assumption of a uniform random sample in time) apportions the times linearly to estimate the average time the field has spent in a transitional state. This was first undertaken by Cox and Dalrymple (1967) who obtained an average duration of 4600 years. Kristjansson (1985) obtained an average value near 6000 years. Unfortunately, magnetic field excursions (§4.4.5) also produce transitional directions. Lund *et al.* (1998) note that there is evidence for at least 14 excursions in the Brunhes alone, so they may be relatively common. This would tend to produce a bias toward an overestimate for the average duration of genuine reversal transitions.

(iii) Cooling rate in intrusive igneous rocks. If a transition occurs during the cooling of an intrusive igneous rock and the cooling is sufficiently slow, then the magnetic field changes are recorded as the cooling front sweeps through the intrusive (Dunn *et al.*, 1971; Dodson *et al.*, 1978). This method appears less reliable than the others because of uncertainties in estimating cooling rates and because the total magnetization at any point is not acquired instantaneously.

(iv) Precise absolute dating. This method is typified by Singer and Pringle (1996), who used precise $^{40}Ar/^{39}Ar$ dating on lava flows from different localities that have recorded transitional directions for the same reversal. There are significant difficulties in resolving such a short time span with radiometric dating: but this process may in the future provide our best estimates.

Theoretical considerations suggest that the reversal process proper probably takes longer than the time over which directional changes are seen. It has also been suggested that the intensity variations in a polarity transition occur over a longer time interval than the directional changes. These suggestions are based on transition records observed in plutons (Dunn *et al.*, 1971), lavas (Mankinen *et al.*, 1985; Bogue and Paul, 1993), and loess (Zhu *et al.*, 1993). Using the statistical properties of the reversal time scale, McFadden and Merrill (1993) also suggested that the reversal process is longer than that observed in paleomagnetic directional data.

Lava flows at Steens Mountain in Oregon recorded a reverse to normal polarity transition about 16 Ma. Interpretations of directional and intensity data from some of these flows suggest that extremely rapid changes sometimes occurred in the transitional magnetic field (Mankinen *et al.*, 1985; Prévot *et al.*, 1985; Coe *et al.*, 1995); the astonishingly rapid change of 6° per day averaged over several days has been suggested. Merrill and McFadden (1999) query these interpretations, concluding that processes such as remagnetization associated with chemical changes could be responsible for the observations.

4.4.5 Geomagnetic Excursions

Wide departures from the geocentric axial dipole field direction have been observed to occur at a single locality, even though the field does not appear to change polarity but returns to its previous state. Such departures have been termed *geomagnetic excursions*. Verosub and Banerjee (1977) defined an excursion to have occurred when the VGP departs more than 45° from the geographic pole (but that departure is not associated with a polarity transition). Lund *et al.* (1998) have shown that there is evidence for at least 14 such excursions during the Brunhes, so it is likely they are a common feature of geomagnetic field variations. Sometimes it is difficult to distinguish when an excursion has occurred because short subchrons ($\sim 10^5$ years in duration) are known to be present in the geomagnetic record (see §4.2.1).

Excursions are known to occur in lava successions, the best known of which is the Laschamp excursion observed in the Chaîne des Puys in France (Bonhommet and Babkine, 1967). They also occur in sedimentary sequences, but the interpretation is often equivocal because their time scale is short ($< 10^4$ years) and only a narrow band of sediment is typically involved. Also, it can be argued that the departures are due to some effect of the sedimentation itself rather than the geomagnetic field (Verosub and Banerjee, 1977). Champion *et al.* (1988) identified 10 events in the late Matuyama and Brunhes (Fig. 4.17); they claimed that these events were actually field reversals and classified them as subchrons. However, the necessary evidence to conclude that these events are genuine reversals is lacking. Merrill and McFadden (1994), Langereis *et al.* (1997), and Lund *et al.* (1998) all classify these events as excursions. Indeed, the existence of several very short subchrons in the Brunhes would be at odds with the observed structure of the GPTS since the Cretaceous (§4.5.4).

Two possible explanations for geomagnetic excursions appear likely; they reflect either large amplitude secular variation or aborted reversals of the geomagnetic field. Distinguishing between these alternatives may not be possible unless aborted reversals can be shown to have a different signature from large-amplitude secular variation. It is theoretically possible for the nondipole field to become large enough that local reversals of the field could occur and thus both polarities would appear to exist simultaneously over a wide region (Coe, 1977; Merrill and McFadden, 1994). Gubbins (1999) notes that an excursion may represent a reversal of the magnetic field in the outer core but not in the inner core, so that the field would return to its original polarity. Consequently, an excursion could be global as suggested by Langereis *et al.* (1997) for six well-dated excursions during the Brunhes.

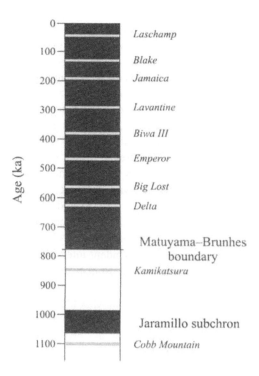

Fig. 4.17. Late Matuyama–Brunhes geomagnetic excursions (names in italics) identified by Champion *et al.* (1988). Black, normal polarity; white, reverse polarity; shaded, excursions. The duration of the excursions is $<10^4$ years.

4.5 Analysis of Reversal Sequences

4.5.1 Probability Distributions

Cox (1968) recognized the apparent randomness of interval lengths (§4.2.1) and developed a probabilistic model in which he assumed that polarity changes occur as a result of interaction between random processes and a steady oscillation in the intensity of the axial dipole field. However, the observed distribution of field intensities is inconsistent with that predicted by Cox's model (Kono, 1972; McFadden and McElhinny, 1982). Also, Laj *et al.* (1979) noted that this model suffered from mechanical problems arising from the different time constants of the random fluctuations of the nondipole field and the steady oscillations of the dipole field. Nevertheless, Cox (1968) ushered in the era of statistical characterization of the reversal process and the use of such characterization to attempt to understand processes occurring in the deep Earth.

The time taken for a polarity transition is, in geological terms, short (§4.4.4), and is short relative to the typical length of a polarity interval. Thus within the GPTS it is a reasonable first order approximation to think of a reversal as a rapid, and indeed almost instantaneous, event. Together with the apparent randomness, this leads naturally to consideration of the reversal process as a general renewal process. This would require that the lengths of individual polarity intervals be independent. Such independence was claimed early on (Cox, 1968, 1969; Nagata, 1969) but was subsequently challenged by Naidu (1974). The later work of Phillips *et al.* (1975) and Phillips and Cox (1976) shows that there is no significant correlation between the lengths of polarity intervals. Thus it is a reasonable first-order approximation to regard geomagnetic reversals as a random process with independent intervals. This is quite different from some other systems, such as for the Sun in which reversals occur about every 11 years.

Recognizing that a reversal is a rare event, Cox (1969) showed that his model led to a Poisson process and that the relevant distribution of interval lengths was therefore the exponential distribution. For a *Poisson* process the probability density p(x) of interval lengths x is given by

$$p(x)dx = \lambda e^{-\lambda x}dx , \qquad (4.5.1)$$

where λ is the rate of the process and $\mu = 1/\lambda$ is the mean interval length. Despite a reasonable fit to the observations, there were fewer very short intervals than required by the Poisson process. Naidu (1971) showed that a gamma distribution provided a good fit to the then observed intervals of the Cenozoic time scale and Phillips (1977) confirmed this in an extensive study of geomagnetic reversal sequences. For a *gamma* process the probability density p(x) of interval lengths x is given by

$$p(x)dx = \frac{1}{\Gamma(k)}(k\Lambda)^k x^{(k-1)}e^{-k\Lambda x} dx$$
$$= \frac{1}{\Gamma(k)}\lambda^k x^{(k-1)}e^{-\lambda x} dx ; \lambda = k\Lambda \qquad (4.5.2)$$

where the mean interval length is now given by $\mu = k/\lambda = 1/\Lambda$.

From (4.5.1) and (4.5.2) it is apparent that the gamma process leads to a family of distributions depending on the value of k, and that the Poisson process is simply the special case of a gamma process with $k = 1$. Appropriately scaled probability densities for this family of distributions are shown in Fig. 4.18. It is

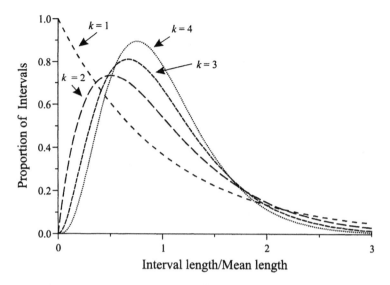

Fig. 4.18. Proportions of interval lengths (probability densities) of gamma distributions plotted against the interval length (scaled to the mean length). Plots are for the Poisson distribution ($k = 1$) and for $k = 2$, 3, and 4. Note the relatively large number of short intervals for the Poisson distribution. After Merrill *et al.* (1996).

immediately apparent that a gamma process with $k > 1$ has far fewer very short intervals than a Poisson process.

4.5.2 Filtering of the Record

Obtaining a reliable GPTS from the marine magnetic anomaly record is not a trivial matter (see chapter 5). The major problems relate to accurate dating of the individual events and reliable recognition of the shorter intervals. A small error in the dating of a reversal leads to small consequences in the analysis of the reversal sequence. However, if a short polarity interval is missed then this is a much more significant error. As shown by the examples given in Fig. 4.19, when a short interval is missed it is combined with the preceding and succeeding intervals of the opposite polarity. This means that the short interval is missed from its own polarity sequence and incorrectly produces a long interval of the opposite polarity, which is then the sum of at least three intervals.

McFadden (1984a) has shown that if n observations from a gamma process with index k have been joined together then the resulting interval appears to have come from a gamma process within index nk. Naturally this is the case for a Poisson process ($k = 1$) as well. Therefore, if a short interval is missed from a Poisson process then the resulting long interval of the opposite polarity looks like an observation from a gamma process with $k = 3$. This has two immediate

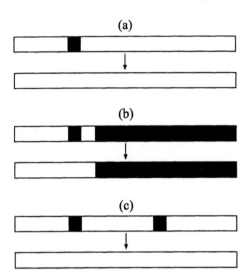

Fig. 4.19. Reduction in the number of polarity intervals caused by preferential filtering of very short intervals. (a) Three intervals reduced to one, (b) four intervals reduced to two, and (c) five intervals reduced to one. From Merrill and McElhinny (1983).

consequences since it is likely that some short intervals have been missed. First, it is convenient to analyze the reversal process as a gamma process because this automatically accounts for any filtering in the observed sequence. Second, the parameter k can be a fairly sensitive indicator of polarity intervals that have been missed (McFadden and Merrill, 1984). This means that estimates of the parameter k can vary by a large amount over small intervals of time without necessarily implying that the process itself has changed much.

McFadden and Merrill (1984) concluded that k is about 1.25 for the period from about 80 Ma to the present. This suggested that several short intervals in the actual reversal process had not been identified in the GPTS. However, subsequent polarity time scales did not change this situation, and so McFadden and Merrill (1993) discussed the alternative interpretation that the reversal process itself is actually gamma. They showed that this would imply the existence of an interval of about 50 kyr immediately following a reversal during which the probability for another reversal would be depressed. The observed sequence is similar (statistically) to that which would be expected from a Poisson process that had been filtered to join into surrounding intervals any interval less than 30 kyr in duration. This is consistent with the conclusion of Parker (1997) that the maximum resolution of the GPTS from marine magnetic anomalies is about 36 kyr. Both interpretations remain viable, with the hope being that new techniques with higher resolution (see §5.1.3) will resolve the matter.

4.5.3 Nonstationarity in Reversal Rate

If the rate at which reversals occurs is variable, then the process is referred to as nonstationary. This is of central geophysical interest because the existence of nonstationarity implies a change in the properties of the origin of the process. The variation in the reversal rate, using the reversal chronology of Cande and Kent (1995) for the interval 0–118 Ma and that of Kent and Gradstein (1986) for the interval 118–160 Ma, is shown in Fig. 4.20. The use of sliding windows is necessary because of the variation in reversal rate (nonstationarity), which precludes considering all of the intervals of the past 160 Myr as a random sample from a single distribution. Consequently Fig. 4.20 has been constructed using a sliding window containing 50 polarity intervals (using constant length windows is inappropriate, see below). Because the Cretaceous Superchron is not part of the surrounding reversal process (§4.5.4), the sliding window does not run across the superchron. The characteristic time of these changes is the same as that associated with mantle processes, leading to suggestions that nonstationarities in the reversal record are associated with changes in the core–mantle boundary conditions (e.g., Jones, 1977; McFadden and Merrill, 1984, 1993, 1995, 1997; Courtillot and Besse, 1987). Suggestions that spatial variations in the core–mantle boundary conditions affect reversals are now supported by some dynamo theory (Glatzmaier *et al.*, 1999).

Gallet and Hulot (1997) proposed an alternative nonstationarity for the reversal rate. They suggest a model in which the rate is essentially constant from

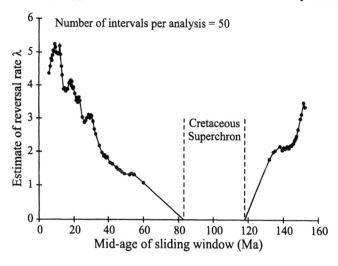

Fig. 4.20. Estimated reversal rate for the geodynamo for the past 160 Myr. Constructed from the time scales of Kent and Gradstein (1986) and Cande and Kent (1995), following the methods of McFadden (1984a). From Merrill *et al.* (1996).

158 to 130 Ma and from 25 Ma to the present, with an intermediate nonstationary segment including the Cretaceous Superchron. In an innovative approach Constable (1999) uses a method that does not rely on the details of any specific probability density. Although she is unable to reject either model, the gross structure of her estimated reversal rate closely parallels that of Fig. 4.20.

The question arises as to whether the fine structure observed in curves such as shown in Fig. 4.20 is meaningful. For example, several studies (Negi and Tiwari, 1983; Mazaud *et al.*, 1983; Mazaud and Laj, 1991; Marzocchi and Mulargia, 1990; Rampino and Caldeira, 1993; Raup, 1985; Stothers, 1986) have suggested either a 15- or a 30-million-year periodicity in the reversal chronology record. McFadden (1984b) showed that similar apparent periodicities were produced by the use of fixed-length sliding windows to analyze a Poisson process with a uniformly changing reversal rate. McFadden and Merrill (1984) did not observe these periodicities when using sliding windows with a fixed number of intervals. Lutz (1985), Stigler (1987), McFadden (1987), and Lutz and Watson (1988) all showed that the perceived periodicities are more likely an artifact of the methods of analysis than real geophysical phenomena.

4.5.4 Polarity Symmetry and Superchrons

McElhinny (1971) and Irving and Pullaiah (1976) introduced the concept of *polarity bias*, which incorporated the idea of differential stability in the normal and reverse polarity states. That is, it was perceived that during times of *normal polarity bias* the field remained normal most of the time because that polarity was substantially more stable than the reverse polarity and vice versa for times of *reverse polarity bias*. Cox (1982) formalized this into polarity bias superchrons. This concept was interesting in that it was at odds with one of the few obvious and truly robust conclusions from the dynamo equations. The equations are symmetric in the magnetic field. This means that the velocity field in the geodynamo is unable to sense the direction of the magnetic field and so the statistical properties of the normal and reverse polarity states should be identical (Merrill *et al.*, 1979).

Using the analysis tools of McFadden (1984a), McFadden and Merrill (1984) were able to show that, at least since about 165 Ma, the normal and reverse polarity states do not exhibit any differences in their relative stabilities either before or after the Cretaceous Superchron (see also McFadden *et al.*, 1987). Merrill and McFadden (1994) were able to show that the Cretaceous Superchron is so long that it cannot be a member of the typical process describing reversals (see also McFadden and Merrill, 1995; Opdyke and Channell, 1996). Thus, it would seem that the geodynamo has two basic states – a reversing state and a non-reversing state. Together with the observed nonstationarity in reversal rate this leads to a simple broad-scale interpretation consistent with theory and with the observed polarity data for the past 160 Myr. That is, the boundary conditions imposed on the core by the lowermost mantle gradually changed in such a way as to reduce the reversal rate until it eventually reached zero (sometime soon after the start of the Cretaceous Superchron). The superchron was then not a consequence of polarity bias but of the fact that the reversal process had ceased. The fact that it was of normal polarity was merely a matter of the polarity the field was in (with a 50% chance for either polarity) at the time the reversal process ceased. The boundary conditions continued to change, and at some time before the end of the superchron the reversal process started again; the reversal rate gradually increasing to the current rate of about 4½ reversals per million years.

Kent and Olsen (1999) found that the interval lengths of their Late Triassic (Newark) GPTS (Fig. 4.11) have a mean duration of about 0.54 Myr, which corresponds to an average reversal rate of about 1.8 Myr^{-1} (§4.3.5). They concluded that the distribution of interval lengths, ranging from about 0.02 to about 2 Myr, is well approximated by a Poisson distribution ($k \approx 1$). Critically, they note that there is no discernible polarity bias. Thus there is little evidence in the Newark GPTS for a prominent interval of low reversal rate (Johnson *et al.*, 1995) or strong normal polarity bias (Algeo, 1996). Kent and Olsen (1999) point out that the Hettangian was a time of widespread igneous activity and of thick accumulation of sediment during a time of predominantly normal polarity. Results from the Hettangian tend to be over-represented in paleomagnetic compilations and this may provide a more satisfactory explanation for the claims of Johnson *et al.* (1995) and Algeo (1996).

The suggestion by Champion *et al.* (1988) that there are several reverse subchrons in the Brunhes (§4.4.5) has fundamental implications for the reversal process because it would imply that the reverse polarity state has been substantially less stable during the Brunhes than the normal polarity state. Since the dynamo equations show that the two polarities must be statistically identical, such a difference in stability can only be produced by mantle-imposed boundary conditions. Because the two stabilities are the same throughout the rest of the GPTS, these boundary conditions would have had to have changed in much less than 10^5 years. However, the mantle changes on a time scale of 10^7–10^8 years, so there is a major incompatibility. Merrill and McFadden (1994) concluded that the evidence that the observed abnormal directions represent genuine field reversals is not sufficiently compelling for them to be included in the GPTS.

Short polarity subchrons during the Cretaceous or Kiaman Superchrons would have similar implications for the reversal process. Consequently, the evidence for any such subchrons during either superchron needs to be thoroughly tested. Within the Cretaceous Superchron Tarduno *et al.* (1992) observed seven reverse polarity zones in the middle of the Albian. However, the magnetization in these zones is carried by hematite and may be late diagenetic in origin. Furthermore, the reverse magnetization components are not antipodal to the normal magnetization components but are offset toward directions consistent with the Late Cretaceous or Paleogene. Opdyke and Channell (1996) conclude that it is doubtful these mid-Albian reverse polarity zones represent the geomagnetic field at the time of deposition of the sediments, and consequently they should not be incorporated into the GPTS.

Chapter Five

Oceanic Paleomagnetism

5.1 Marine Magnetic Anomalies

5.1.1 Sea-Floor Spreading and Plate Tectonics

The sea-floor spreading hypothesis was first formulated by Hess (1960, 1962). The mid-ocean ridges, which are characterized by unusually high heat flow along their crests, are the largest topographic features on the surface of the Earth. On Hess's model the mid-ocean ridges are interpreted as representing the rising limbs of mantle convection where hot magma comes right through to the surface and new oceanic crust is formed as the magma cools. It was originally thought that the intrusion of new material forced the cooling crust to move away from the ridge symmetrically on either side. It is now generally believed that the crust and part of the upper mantle are under tension at a spreading center. Thus, the oceanic crust is pulled apart, allowing magma to rise to the surface and the whole oceanic crust is part of a conveyor belt system, rising up at the mid-ocean ridges and eventually sinking down at the oceanic trenches. The spreading rate across a mid-ocean ridge is defined as the relative rate of separation of the plates on either side of the ridge, sometimes referred to as the "full rate". Consequently, the spreading rate on one side of a ridge is often referred to as the "half rate". Values are usually quoted in km Myr^{-1} (mm yr^{-1}).

The theory of plate tectonics was formulated by McKenzie and Parker (1967), Morgan (1968) and Le Pichon (1968). The plate tectonics model is now accepted

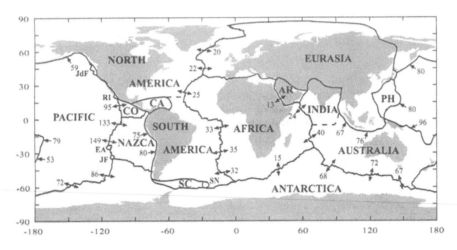

Fig. 5.1. Aseismic plates on the surface of the Earth bounded by seismically active zones. The nine larger plates are named, and eight smaller plates are labelled AR (Arabia) CA (Caribbean) CO (Cocos) JdF (Juan de Fuca) PH (Philippine) RI (Rivera) SC (Scotia) SN (Sandwich) together with two microplates indicated as EA (Easter) JF (Juan Fernandez). Directions and rates of relative motion (km Myr^{-1}) between plates are shown at selected points on their boundaries. Double arrows indicate separation at mid-ocean ridges and single arrows indicate convergence at oceanic trenches (subduction zones).

as the first-order explanation of global tectonics. To a first approximation it is possible to divide the Earth's surface into several essentially aseismic plates or blocks bounded by seismicity associated with active ridge crests, faults, trenches, and mountain systems. The plates can be composed entirely of continental crust, or of oceanic crust, or of a combination of both. Figure 5.1 shows the surface of the Earth divided into 17 plates (see DeMets *et al.*, 1990, 1994) together with two microplates. There does not appear to be any formal definition of a microplate. Originally, Le Pichon (1968) proposed there were six major aseismic plates, but since that time it has been found that relative motion exists between North and South America and between India and Australia (Wiens *et al.*, 1985) and that there are several smaller plates. Other minor plates not shown in Fig. 5.1 are Nubia (west Africa) and Somalia (east Africa).

The sea-floor spreading hypothesis maintains that these plates are in constant motion and that seismic boundaries between them delineate zones where oceanic crust is created or destroyed, continental crust extended or compressed, and crustal plates are translated laterally along faults without change in their surface area. Some of the principal points of the plate tectonics model are illustrated in Fig. 5.2. Three flat-lying layers are distinguished. The *lithosphere*, which generally includes the crust and upper mantle, has significant strength and is of the order of 100 km in thickness. The *asthenosphere*, which is a layer of

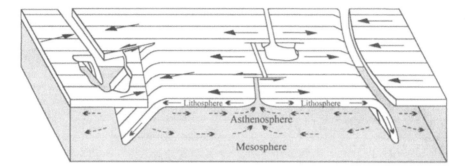

Fig. 5.2. Model illustration of plate tectonics showing the roles of the lithosphere, asthenosphere and mesosphere. Arrows on the lithosphere indicate relative motions and in the asthenosphere represent possible compensating flow in the mantle. An arc to arc transform fault appears at the left between oppositely facing island arcs, two ridge–ridge transform faults along the ocean ridge are in the center, and a simple arc structure is at the right. From Isacks *et al.* (1968).

effectively no strength on the appropriate time scale, extends from the base of the lithosphere to a depth of several hundred kilometres. The *mesosphere*, which may have strength, makes up the lower remaining portion of the mantle and is relatively passive, perhaps inert, in tectonic processes.

The lithosphere as defined above corresponds with the *seismic lithosphere*, a region of high seismic velocity at the top of the mantle overlying a low-velocity zone. The bottom of the seismic lithosphere is characterized by an abrupt decrease in shear-wave velocities at depths of 150–200 km under the continents and 10–50 km under the ocean floor depending on age (Regan and Anderson, 1984). This corresponds roughly to the 600°C isotherm below the ocean floor, which approximates the *effective elastic plate thickness* (i.e., the thickness that reacts to loads and deformation as an elastic sheet). The asthenosphere is often equated with the seismic low-velocity zone that arises from the increase of temperature with depth. It should be noted that the *thermal lithosphere* has been defined as having its lower boundary as the depth to a constant isotherm, usually modeled to be in the range 1250–1350°C (McKenzie and Bickle, 1988). Depending on its age, this corresponds to a depth of 10–125 km below ocean floor or, in the case of continental lithosphere, to a thickness of 100–200 km. Old cratonic lithosphere may be up to 400 km in thickness (Jordan, 1975). The boundary between the lithosphere and the asthenosphere is a transition zone referred to as the lower thermal boundary layer. However, the lowermost part of the thermal lithosphere may deform in a ductile fashion over time and thus is not really part of the tectonic plate that moves as a mechanical unit in plate tectonics.

The movement of these plates on the surface of a sphere is best understood in terms of rotations by applying Euler's theorem. If one of two plates is taken to be fixed, the movement of the other plate corresponds to a rotation about some

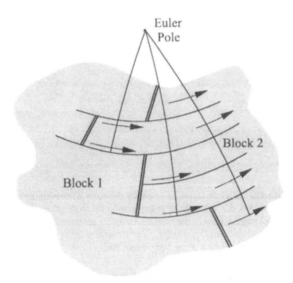

Fig. 5.3. On a sphere the motion of block 2 relative to block 1 must, according to Euler's theorem, be a rotation about some pole. Transform faults on the boundary between 1 and 2 must be small circles (lines of latitude) about the Euler pole. From Morgan (1968).

pole (Fig. 5.3) with angular velocity ω. Mathematically, ω is a vector pointing outwards along the rotation axis and is reckoned as positive when the rotation is clockwise looking outwards from the center of the sphere. However, it is commonly found more convenient to visualize these rotations as viewed from the surface of the Earth looking toward the center. The convention has therefore developed to regard rotations as being "clockwise" or "counterclockwise" when viewed from the surface of the Earth, corresponding mathematically to negative and positive angular rotations, respectively. If a is the radius of the Earth (Fig. 5.4) and the angular distance from a point S on one plate to the pole is φ, then the magnitude v_φ of the relative velocity at that point is given by

$$v_\varphi = a\omega \sin \varphi .$$ (5.1.1)

The relative velocity thus has a maximum at the "equator" and vanishes at the pole of rotation. The relative velocity vectors must lie along small circles or "latitudes" with respect to the pole, which has no significance other than being a construction point. When several plates are in relative motion, as at the present time (Fig. 5.1), then it is possible to use the property of angular velocities that they behave like vectors. Around any closed circuit crossing several plates, A, B, C, D, etc., the sum of the angular velocities must be zero, that is,

$$_A\omega_B +_B \omega_C +_C \omega_D +_D \omega_A = 0 ,$$ (5.1.2)

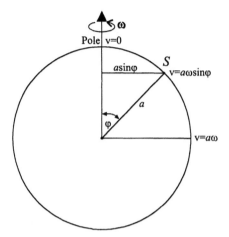

Fig. 5.4. Cross-section through the Earth showing the variation of spreading rate v with angular distance φ from the pole of rotation.

where $_A\omega_B$ is the angular velocity of the rotation that describes the magnitude and direction of the relative motion between plates A and B. The sense of the rotation is found by moving from plate A to plate B and so on. There are many points on the Earth's surface where three plates meet, called *triple junctions*. Triple junctions can arise from any combination of the three types of plate boundaries (ridges, trenches, transform faults) meeting at a point. Their stability conditions and evolution have been discussed by McKenzie and Morgan (1969).

The Euler poles of rotation and angular velocity vectors describing the average motion between the plates over the past few million years can be determined by combining ridge spreading rates, transform fault azimuths, and earthquake slip vectors determined from fault plane solutions. Since the original global calculations of Le Pichon (1968) and Morgan (1968), successive new global plate motion models have been constructed by Chase (1978), Minster and Jordan (1978), and DeMets *et al.* (1990, 1994) as new high-quality data have become available. In addition, current plate motions can be measured at the millimetre per year level using various geodetic techniques such as very long baseline interferometry (VLBI), satellite laser ranging (SLR), and global positioning system (GPS) interferometry. Global analyses using VLBI and SLR at 20 sites on five plates (Australia, Eurasia, Nazca, North America, and Pacific) show excellent agreement between the geodetically determined plate motions over the past two decades and those measured geophysically on the 0–2 Myr time scale (DeMets, 1995). The current instantaneous poles of rotation describing the motion between those pairs of plates that are mainly separated by mid-ocean ridges (Fig. 5.1) are summarized in Table 5.1 from the NUVEL-1A model of DeMets *et al.* (1994). The "equatorial spreading rate" is that which would occur

TABLE 5.1

Instantaneous Poles of Rotation and Magnitudes (ω) of Angular Velocities Describing the Present
Motion Between Pairs of Plates (Fig. 5.1)[a]

Plate Pairs	Pole Lat. (°N)	Pole Long. (°E)	ω (° Myr^{-1})	Equatorial spreading rate (km Myr^{-1})
North America – Pacific	48.7	-78.2	-0.75	-83.5
Cocos – Pacific	36.8	-108.6	-2.00	-222.6
Nazca – Pacific	55.6	-90.1	-1.36	-151.4
Antarctica – Pacific	64.3	-84.0	-0.87	-96.8
North America – Eurasia	62.4	135.8	0.21	23.4
North America – Africa	78.8	38.3	0.24	26.7
South America – Africa	62.5	-39.4	0.31	34.5
Antarctica – Australia	13.2	38.2	0.65	72.3
Antarctica – Africa	5.6	-39.2	0.13	14.5
Australia – India	-5.6	77.1	-0.30	-33.4
Africa – India	23.6	28.5	0.41	45.6
Africa – Arabia	24.1	24.0	0.40	44.5

Note. The second plate moves clockwise relative to the first plate.
[a]From DeMets *et al.* (1994).

when $\varphi = 90°$. For most ridges (e.g., the Cocos–Pacific ridge) φ never
approaches $90°$ and the fastest actual spreading rate today is about 150 km Myr^{-1}
along the East Pacific Rise separating the Nazca and Pacific plates.

5.1.2 Vine–Matthews Crustal Model

Newly erupted ocean-floor spreads smoothly away from the mid-ocean ridges.
Because these submarine basalts cool quickly from high temperatures in the sea
water environment, they contain fine-grained titanomagnetite of average
composition $Fe_{2.4}Ti_{0.6}O_4$ (TM60). This quenching process prevents high-
temperature deuteric oxidation, as generally occurs in subaerial basalts, so that
typical Curie temperatures lie in the range 150–200°C. Rapid cooling is also
responsible for the fine grain size, enabling oceanic basalts to acquire an intense
TRM and become excellent recorders of the paleomagnetic field.

Vine and Matthews (1963) and Morley and Larochelle (1964) proposed that
the lineated magnetic anomalies (magnetic stripes), which are observed parallel
to and on either side of the mid-ocean ridge, are records of past reversals of the
geomagnetic field. They therefore modeled the magnetic anomalies as being
produced by strips of crust of alternating polarity at increasing distances (and
therefore age) from the ridge crest (Fig. 5.5). For a uniform spreading rate the
strips, and their linear magnetic anomalies at the ocean surface, will have widths
related to the durations of the geomagnetic polarity epochs. In analyzing these
anomalies it is usual to use the effective susceptibility of the material according

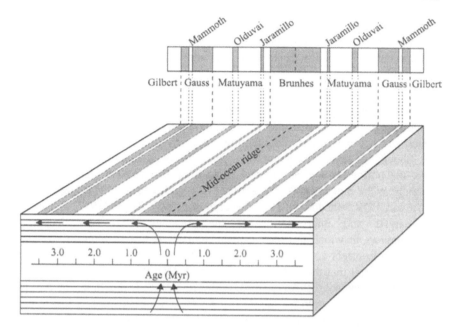

Fig. 5.5. Schematic representation of sea-floor spreading and the formation of linear magnetic anomalies due to reversals of the Earth's magnetic field as proposed by Vine and Matthews (1963). Normal polarity zones are shaded. Only major subchrons are shown. Updated after Allan (1969), with the permission of Elsevier Science.

to (2.1.3) given by the total magnetization (remanent plus induced) as in (2.1.2) divided by the present magnetic field strength. However, the magnetic properties of rock samples dredged from the ocean floor demonstrated the predominance of remanent magnetization since the actual susceptibility is comparatively low. For basalts the effective susceptibility is of the order of 0.1 SI units, corresponding to a magnetization of about 5 Am^{-1}.

5.1.3 Measurement of Marine Magnetic Anomalies

Marine magnetic anomalies are generally measured by towing a total-field magnetometer on the ocean surface behind a ship at sufficient distance so that the magnetic effects of the ship are below the noise level of the magnetometer. Anomalies may also be measured by flying an aircraft equipped with a suitable magnetometer at a known height above the ocean surface. Satellite measurements may also be used but the filtering effect of the height above the ocean floor makes the results less valuable because most of the detail is lost. Most ship-borne magnetometers are proton precession or fluxgate types that only measure the total intensity F of the geomagnetic field. At any point on the

Earth's surface the expected value of the total intensity (F_0) may be calculated using the International Geomagnetic Reference Field (IGRF) for the appropriate epoch, as defined in §1.1.3. The magnetic anomaly (ΔF) is then given by

$$\Delta F = F - F_0 . \tag{5.1.3}$$

It is worth noting that this does not necessarily give exactly the same value for the anomaly as would be obtained using vector observations and calculations.

Ocean-surface magnetometers measure the geomagnetic field up to 5 km or more above the ocean floor, but sometimes deep-tow magnetometers, which follow the bottom topography of the ocean floor at a fixed distance above it (generally in the range 50-200 m), are used (Luyendyk *et al.*, 1968; Greenewalt and Taylor, 1974; Macdonald, 1977; Sager *et al.*, 1998). Parker (1997) discussed the analysis and interpretation of measurements from two deep-tow magnetometers in which the second magnetometer is towed 300 m above the first, giving in effect a vertical gradiometer. Deep-tow magnetometers provide much more detail that those that measure at the ocean surface because they are nearer to the signal source. An example of this difference is shown in Fig. 5.6, in which the detail measured by the deep-tow near the ocean floor is lost when measured at the surface. In the early days of interpreting marine magnetic

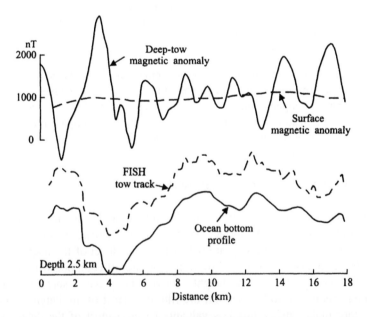

Fig. 5.6. Magnetic anomaly profile across the mid-Atlantic ridge at a height of about 100 m above the ocean floor using the deep-tow magnetometer package FISH (Fully Instrumented Submersible Housing) compared with that obtained at the ocean surface. After Greenewalt and Taylor (1974).

anomalies according to the Vine–Matthews crustal model the filtering effect of the height above the ocean floor actually provided some advantage in that it presented a simpler and cleaner data set. However, in their analysis of the Cenozoic reversal time scale (§4.5.2) McFadden and Merrill (1984) suggested that short subchrons (typically <30 kyr, see also Marzocchi, 1997) were missing from the reversal record. In a theoretical analysis of data from the Juan de Fuca Rise, Parker (1997) concluded that there is a maximum resolution of about 36 kyr. Note that this is not the detection limit; under favorable circumstances it is possible to detect events much shorter than 36 kyr. With the improved systems and analysis capability available today, it is important to obtain records with the maximum possible resolution. Autonomous underwater vehicles, AUVs, which operate without a tether or human intervention, represent an emerging technology that can be used in the deep oceans for this purpose.

5.1.4 Nature of the Magnetic Anomaly Source

Basaltic magma that rises from shallow depths in the upper mantle is extruded as pillow lavas in the region known as seismic layer 2A (approximately 0.5–1 km thick). The magma may be intruded as sheeted dikes and sills into the underlying layer 2B (about 1 or 2 km thick) or as slow cooling gabbros into layer 3 (about 2–5 km thick). This crustal structure is illustrated in Fig. 5.7 and was first inferred from studies of ophiolite suites, such as the Macquarie Island ophiolite

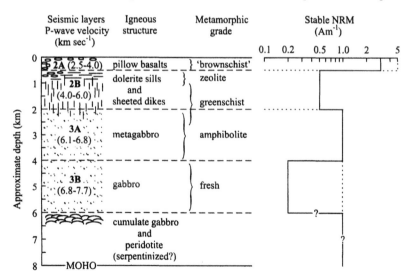

Fig. 5.7. The petrologic, seismic, and magnetic structure of typical oceanic crust. The solid and dashed lines indicate alternate possible variations in magnetization. From Dunlop and Özdemir (1997), with the permission of Cambridge University Press.

(Banerjee, 1980) or the Troodos ophiolite (Hall *et al.*, 1991). This was later confirmed by direct sampling of the ocean floor through the Deep Sea Drilling Project, DSDP, or the Ocean Drilling Project, ODP, (Johnson and Atwater, 1977; Prévot *et al.*, 1979; Smith and Banerjee, 1986).

The geothermal gradient beneath mid-ocean ridges can be as high as 100°C km^{-1}, and the Curie points of the primary titanomagnetites lie in the range 150–200°C (§5.1.2). Therefore, the depth of the Curie point isotherm is only a few kilometres and this is the main constraint on the thickness of the source layer for the magnetic anomalies (Fig. 5.8). The slower cooled sheeted dikes and gabbros of layers 2B and 3 generally have Curie points in excess of 500°C, but their magnetization is at least an order of magnitude less than the overlying pillow lavas. As the newly formed lithosphere spreads away from the ridge the deeper rocks cool more slowly than those near the surface. The cooling fronts, and thus the boundaries of the blocks of opposite polarity, are not vertical but curve away from the ridge with depth (Fig. 5.8).

Well before new ocean floor has spread 20 km from the ridge axis (~1 Myr at the relatively slow spreading rate of 20 km Myr^{-1}), sea water has penetrated along deep fissures and oxidized the primary titanomagnetites (Bleil and Petersen, 1983). This process of low-temperature oxidation produces titanomaghemite (see §2.2.2, Fig. 2.9, and §3.5.1, Fig. 3.18). The inverse spinel

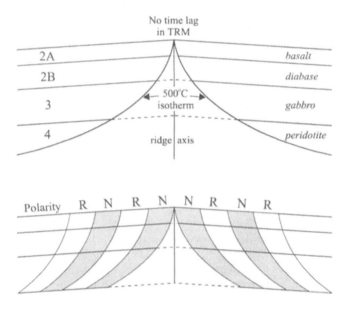

Fig. 5.8. Vertical structure and 500°C isotherm in oceanic crust near a spreading ridge. The normal (N) and reverse (R) blocks of magnetized ocean floor are not vertical but curve away from the ridge axis. After Dunlop and Özdemir (1997), with the permission of Cambridge University Press.

lattice of the titanomagnetite is preserved during oxidation except for high values of the oxidation parameter z. The change in the lattice parameters is <1% and as a result the exchange coupling will be unbroken across the titanomagnetite-titanomaghemite phase boundary. Therefore, it can be expected that the CRM of the titanomaghemite will inherit the original TRM direction of magnetization of the titanomagnetite. Even after a high degree of oxidation the CRM should be indistinguishable from the initial remanence except for a decrease in intensity due to the change in M_s. Both laboratory experiments (Özdemir and Dunlop, 1985) and observations of magnetizations in variably oxidized ocean floor pillow lavas (Soroka and Beske-Diehl, 1984) confirm this to be the case.

The variation of ocean crustal magnetization with distance from the mid-ocean ridge has been quantified by rock magnetic studies (Irving, 1970; Johnson and Atwater, 1977; Bleil and Petersen, 1983), magnetic anomaly surveys (Wittpenn *et al.*, 1989) including near-bottom surveys (Macdonald, 1977), and results from deep-sea drilling (Johnson and Pariso, 1993). Of particular interest is the fact that the central anomaly is much more intense than the subsequent anomalies and generally requires a magnetization source about twice as strong. Analyses of near-bottom profiles and rock magnetic studies of dredged samples have previously documented a decrease in magnetization by a factor of five within about 3 km of the ridge crest. The time constant of the magnetization decay (the time for the magnetization to reduce to 1/e of its initial value) has been inferred to be ~0.5 Myr. Seismic studies indicate that there is a rapid thickening of layer 2A within a few kilometres of the ridge crest where the extrusive layer increases from ~0.1–0.2 to 0.3–0.5 km through the addition of flows that may extend 1 or 2 km off the ridge crest. Paleointensity measurements indicate that prior to about 10 kyr ago the geomagnetic field strength was much less than it is now (Merrill *et al.*, 1996). A magnetic source layer modeled with such a variation would produce an overall axial anomaly low so that the only likely explanation of the magnetization contrast required for the central anomaly high is a more rapid alteration of the source layer and its magnetization than has previously been inferred.

The problem of explaining the central anomaly high has been resolved by remanence data from dredged basalts and from near-bottom magnetic profiles of the fast-spreading East Pacific Rise (Gee and Kent, 1994). Forward modeling incorporating the thickening of the source layer clearly demonstrates that the ~0.5 Myr time constant previously inferred for the magnetization reduction is too long by at least an order of magnitude (Fig. 5.9). The time constant of the magnetization reduction is estimated to be 20 kyr. The best fitting model for both the near-bottom and ocean surface magnetic profiles suggests layer 2A thickens from 0.2 to 0.45 km by 3 km from the ridge crest with a 0.3-km-wide neovolcanic zone at the center in which the magnetization is constant, followed by an exponential decay of the magnetization with distance. Further near-bottom magnetic profiles from the Juan de Fuca ridge (Tivey and Johnson, 1995) across a newly erupted lava flow show strong magnetic anomalies and a high crustal magnetization with >50 Am^{-1} contrast relative to the surrounding older lavas. There is also a clearly defined short wavelength magnetic low over the center of the flow as seen by the data from the East Pacific Rise (Fig. 5.9).

Although the spatial association of high-amplitude magnetic anomalies with iron-rich lavas is well documented, direct evidence of the link between high iron

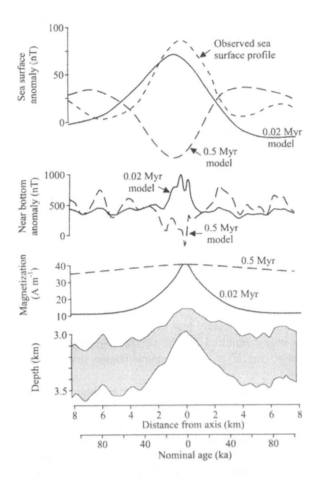

Fig. 5.9. Forward modeling of ocean-surface and near-bottom magnetic anomalies across the central anomaly on the East Pacific Rise. Seismic data indicate thickening of layer 2A from 0.2 to 0.45 km within 3 km from the ridge crest. A time constant of 0.5 Myr for the magnetization reduction from the ridge crest produces a negative anomaly in both cases. A 20-kyr time constant with a 0.3-km central neovolcanic zone of constant magnetization matches both observed profiles, including the short wavelength low within the central zone. From Gee and Kent (1994).

content and high remanent magnetization has been difficult to establish. Gee and Kent (1997) report the magnetization properties of axial lavas (age 0–6 ka) from the southern East Pacific Rise that provide the first strong support for the geochemical dependence of remanent intensity. The samples contain 8–16% FeO* (the asterisk indicates total iron content as FeO) and the analysis shows that saturation magnetization and saturation remanence are highly correlated with FeO* indicating that the more iron-rich lavas have higher abundances of otherwise similar titanomagnetite. These data indicate that the equatorial magnetization of submarine lavas has a range ~10–50 Am^{-1}, which corresponds with iron content in the range 8.5–16% FeO*.

5.2 Modeling Marine Magnetic Anomalies

5.2.1 Factors Affecting the Shape of Anomalies

The anomalies expected from the Vine–Matthews crustal model will vary depending on location and age. Consider the case of symmetrical anomalies about a spreading ridge at the north pole and about N–S or E–W oriented ridges at the equator as shown in Fig. 5.10. The crustal blocks will be magnetized in opposite directions as the polarity changes, but only the first magnetic reversal is

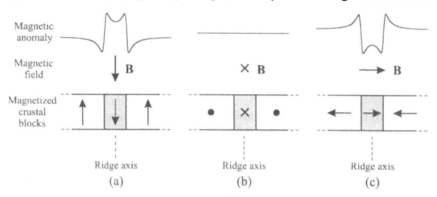

Fig. 5.1. Latitude variation of the shape of the magnetic anomaly due to magnetized crustal blocks adjacent to the axis of a spreading ridge. Normal polarity block is shaded. The external magnetic field **B** due to an axial geocentric dipole is shown for three cases. (a) At the north pole the intensity of the field is increased over the ridge axis. (b) At the equator a north–south oriented ridge spreading east–west produces no magnetic anomaly (crosses and dots indicate north–south horizontal fields perpendicular to the page, in and out respectively, along the ridge). (c) At the equator, for an east–west oriented ridge spreading north–south, the intensity of the field is decreased over the ridge axis.

shown. The central block will be magnetized in the same direction (normal polarity) as the present geomagnetic field and the adjoining blocks on either side will be magnetized in the opposite direction (reverse polarity). During normal polarity time the magnetic anomaly is positive at the pole (thus reinforcing the field intensity as in Fig. 5.10a). At the equator the magnetic anomaly depends on the ridge orientation and for a north–south ridge spreading east–west there is no magnetic anomaly (Fig. 5.10b). However, for an east–west ridge spreading north–south the magnetic anomaly is negative (thus reducing the field intensity as in Fig. 5.10c).

It should be noted that certain distributions of magnetization (e.g., that illustrated in Fig. 5.10b) will not produce an external field. Such distributions are known as *magnetic annihilators* (Parker, 1977, 1994) and clearly demonstrate the non-uniqueness of magnetic inversion. Of particular interest here is that an infinite slab with uniform magnetization is such an annihilator. Thus any anomaly observed over a large slab with constant thickness must arise from changes or discontinuities in the magnetization. In the ocean floor there are discontinuities at each change in polarity and it is these discontinuities that give rise to the anomalies as shown in Fig. 5.10. When the magnetized blocks are sufficiently narrow, as is often the case in the ocean floor, then these anomalies can become superimposed.

The shape and intensity of the observed magnetic anomalies will depend on several factors, including latitude as described above, the direction of the profile in relation to the orientation of the magnetized blocks, and the spreading rate. The effect of each of these factors is illustrated in Fig. 5.11. Figure 5.11a shows a more detailed pattern of the latitude effect described in Fig. 5.10 for north–south profiles at various latitudes. Note that at intermediate latitudes the symmetry of the anomaly about the ridge crest becomes obscured as individual blocks are no longer associated with a single positive or negative anomaly. Figure 5.11b shows the variation of the magnetic anomaly pattern with the direction of the profile for a location where the magnetic inclination is 45°. The variation arises because only the component of the magnetization vector lying in the vertical plane through the magnetic profile affects the anomaly. This component is a maximum for east–west ridges with north–south profiles and a minimum for north–south ridges. The spreading rate affects the detail shown in the observed magnetic anomalies and the ability to detect short subchrons as illustrated in Fig. 5.11c for slow- and fast-spreading ridges.

Because of the variation in the observed magnetic anomalies at different places, they obviously cannot simply be stacked to improve the signal to noise ratio. Blakely and Cox (1972) overcame this problem by applying the geocentric axial dipole assumption (§1.2.3) and the anomalies were transformed to what they would be if the observation site were at the pole. This process is referred to as *reduction to the pole*. After reduction to the pole and adjustment for different

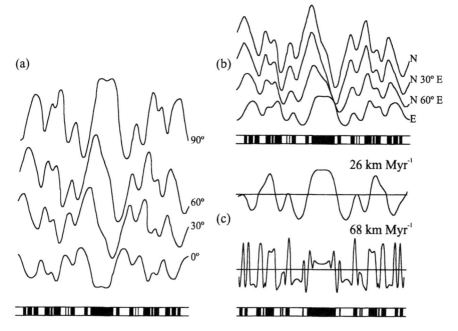

Fig. 5.11. The effects of latitude, profile orientation and spreading rate on the magnetic anomaly patterns. After Kearey and Vine (1996) and DeMets *et al.* (1994).

(a) Variation with geomagnetic latitude for north–south profiles. Angles refer to magnetic inclination.

(b) Variation with profile orientation at a fixed latitude, where the magnetic inclination is 45°.

(c) Variation with spreading rate. More detail can be obtained from the fast-spreading ridge.

spreading rates, magnetic anomalies may then be stacked so as to reduce the noise and enhance the signal.

The first applications of the Vine–Matthews crustal model to the interpretation of marine magnetic anomalies depended critically on the geomagnetic time scale available at the time. Improvements in the geomagnetic polarity time scale, especially the discovery of the Jaramillo event, led to improved analyses of the ridge anomalies. Four of these are shown in Fig. 5.12 from the analysis of Vine (1966) for four widely separated areas on the mid-ocean ridge system. In the first two profiles from the Juan de Fuca ridge and the East Pacific Rise, the profiles have been reversed for comparison. The remarkable symmetry displayed is noteworthy. Note that the three profiles from mid-latitudes have central positive anomalies, whereas the equatorial profile from the northwest Indian Ocean on the Carlsberg ridge has a central negative anomaly (c.f. Fig. 5.10). Spreading rates deduced varied from 15 to 44 km Myr^{-1} over the time interval 0–4 Ma, as far back as the land-based geomagnetic polarity time scale extended at that time.

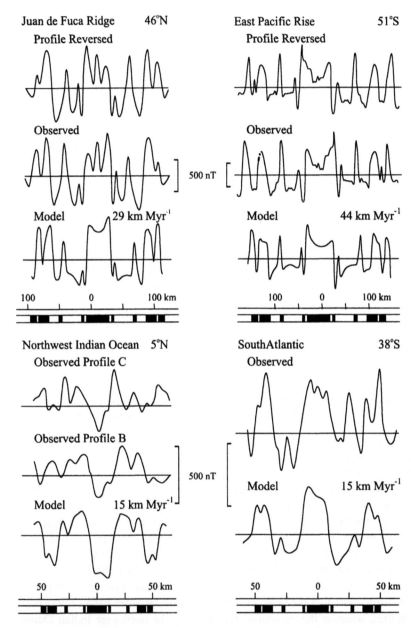

Fig. 5.12. Observed magnetic anomaly profiles at various points on the mid-ocean ridge system are compared with simulated profiles based on the polarity time scale and the Vine–Matthews crustal model. Reproduced with permission from Vine (1966). © American Association for the Advancement of Science.

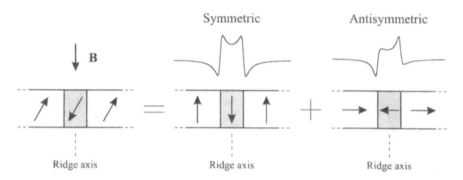

Fig. 5.13. Symmetry of magnetic anomalies about a ridge axis. Normal polarity block is shaded. A symmetrical magnetization of the crustal blocks (left) is not parallel to the present external field **B** (taken to be vertical) because the magnetization was acquired at another place and the magnetized blocks have subsequently been displaced to their present position. The magnetization can be divided into a component parallel (center) and perpendicular (right) to the external field. These components produce symmetric and antisymmetric anomalies, respectively.

Vine (1966) showed that the model could be used beyond the limits of the known polarity time scale. By constructing a configuration of blocks to match the observed profile and assuming a constant rate of spreading, the geomagnetic polarity time scale could be extrapolated to the limits of the area being surveyed (see §5.3.1).

The assumption that the anomalies are symmetric about a spreading axis if the magnetization is symmetric is not necessarily true and is not always observed. The symmetry may be partly an accident of the strike of the ridges under investigation. The observed symmetry in many marine magnetic anomalies over ocean ridges is a result of the predominant north–south orientation of the ridges at the present time. Consider the case of the symmetric magnetization of adjacent crustal blocks that have been displaced some distance from the spreading axis so that their magnetization is now not parallel to the external magnetic field. This situation is shown in Fig. 5.13, in which the external field is taken to be vertical. The magnetization can be divided into a component parallel to and perpendicular to the external field. These components produce symmetrical and antisymmetrical anomalies, respectively. If, after reduction to the pole, symmetric magnetic anomalies are observed, then this implies symmetric magnetization and vice versa.

5.2.2 Calculating Magnetic Anomalies

The theoretical foundations for modeling marine magnetic anomalies were established by Bott (1967), Schouten (1971), and Schouten and McCamy (1972). The magnetization of layers 2B and deeper is at least an order of magnitude less than that of the basalts making up layer 2A (see §5.1.4, Fig. 5.7). Therefore, it is assumed that the magnetization of the ocean floor can be modeled as two-dimensional horizontal planar bodies lying immediately below the ocean floor. Their thickness is defined by layer 2A, usually taken to be 0.5 km, which is much less than the typical ocean depth of about 3½ km. The direction of magnetization is assumed to be parallel to the normal or reverse geocentric axial dipole field.

When the magnetized blocks of the ocean floor have been displaced some distance from the spreading axis, their magnetization is not necessarily parallel to the present geomagnetic field. This changes the relative amplitude and shape asymmetry of the magnetic anomalies as discussed in §5.1.3 (Fig. 5.13). The shape asymmetry of an anomaly shows up in its Fourier transform as a phase shift θ (Bott, 1967; Schouten, 1971; Schouten and McCamy, 1972). Thus, phase-shifting the Fourier transform of the anomaly (a simple linear filter operation) can remove the asymmetry. The phase shift θ is referred to as the *skewness* of the magnetic anomaly and can be defined in two ways. Schouten and Cande (1976) defined skewness of an observed anomaly as the phase shift that transforms a symmetrical anomaly as seen in a downward vertical magnetic field into the observed anomaly. The inverse of this phase shift can then be applied to an observed anomaly to transform it so that the resulting shape is most similar to that seen at the pole and is the equivalent of the method of reduction to the pole

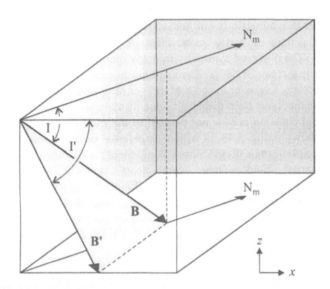

Fig. 5.14. The *x* axis is normal to the strike of the body (azimuth *A*). The effective inclination I' (and I'_R) is the inclination of the vector component in the *x–z* plane. N_m is the direction of magnetic north.

(see §5.2.1) of Blakely and Cox (1972). However, Petronotis *et al.* (1992) define skewness in the opposite sense, that is, by the phase shift θ that transforms an observed anomaly into the shape most similar to that expected for sea floor formed and observed in a downward vertical magnetic field. The definition of Schouten and Cande (1976) gives negative values of θ between 0 and 360°, whereas that of Petronotis *et al.* (1992) gives positive values of θ between 0 and 360°. In the discussion that follows, the definition given by Petronotis *et al.* (1992) is followed so that θ has positive values.

For total intensity magnetic anomalies the skewness θ is given by

$$\theta = 180 - I' - I'_R , \qquad (5.2.1)$$

where I' and I'_R are the *effective* inclinations of the ambient magnetic field and the remanent magnetization vectors, respectively. The skewness θ is therefore independent of depth, layer thickness, or magnetization. The effective inclination is the vector component in a plane normal to the strike of the body (Fig. 5.14). It is a useful concept because it reduces the five parameters D, D_R, A (the azimuth of the magnetic lineation at the site), I, and I_R, to the two parameters I' and I'_R. A is defined as the strike direction, measured from true north, that is, 90° clockwise from the direction in which the sea floor becomes younger. The effective inclinations are then defined by

$$\tan I' = \frac{\tan I}{\sin(A - D)} \qquad (5.2.2)$$

and

$$\tan I'_R = \frac{\tan I_R}{\sin(A - D_R)} . \qquad (5.2.3)$$

Since the source is two-dimensional, the component of the magnetic vector in the plane normal to the strike of the body is the only component that affects the shape of the anomaly. The amplitude of the anomaly depends on depth, layer thickness and magnetization, and the *relative amplitude factor C* given by

$$C = \frac{\sin I \sin I_R}{\sin I' \sin I'_R} \qquad (-1 \le C \le +1) . \qquad (5.2.4)$$

Figure 5.15 shows how an arbitrary magnetic anomaly profile alters appearance as θ is varied from 0 to 360° in 30° increments. When θ = 0 or 180°, the anomaly is symmetric and when θ = 90 or 270° it is antisymmetric. In all other cases it is asymmetric. Compare the anomalies labeled 1, 2, or 3 for skewness 0 to 90° with the central anomaly of Fig. 5.11a, in which the magnetic inclination varies from 0 to 90°.

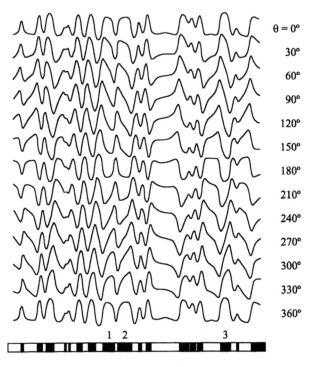

Fig. 5.15. Magnetic anomaly profiles produced by the same arbitrary model sea-floor magnetization but for various values of the skewness θ. Compare anomalies 1, 2, or 3 for skewness 0 to 180° with the central anomaly of Fig. 5.11a. Modified after Blakely (1995), with the permission of Cambridge University Press.

Weissel and Hayes (1972) first noticed that on many magnetic anomaly profiles the observed skewness differs systematically from that predicted from simple models. Cande (1976) introduced the term *anomalous skewness* to describe these discrepancies. He found that the effective remanent inclination determined from anomalies on either side of the Pacific–Antarctic Ridge were not equal but had an observed discrepancy of 20±8°. By attributing equal amounts of the discrepancy to the anomalies on each side of the ridge, the anomalous skewness was estimated to be 10±4°. An example of the effect of anomalous skewness is illustrated in Fig. 5.16 for two synthetic profiles covering the time interval 0–3 Ma. The amplitude of the sub-horizontal portion of an individual anomalously skewed anomaly tends to decrease more rapidly (or increase less rapidly) toward its younger end than do the anomalies with no anomalous skewness.

An age dependence of anomalous skewness has been established by Petronotis and Gordon (1989), and a dependence on spreading rate has been shown by Roest *et al.* (1992). Anomalous skewness increases as spreading rate decreases.

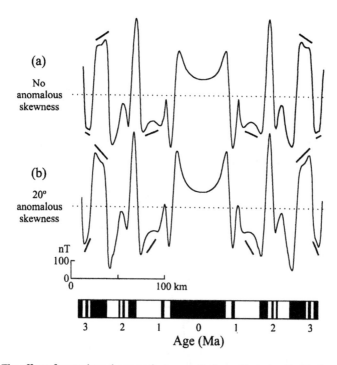

Fig. 5.16. The effect of anomalous skewness for two synthetic profiles using the block model shown covering the time interval 0–3 Ma across a spreading ridge. (a) An ideal anomaly sequence with no anomalous skewness. (b) The same anomaly sequence assuming 20° of anomalous skewness, represented here by phase-shifting the anomalies left of the axis of symmetry by −20° and those to the right of the axis by +20°. The amplitude of a subhorizontal portion of an individual anomalously skewed anomaly tends to decrease more rapidly (or increase less rapidly) than do the anomalies with no anomalous skewness. The bold lines above and below some of the wider anomalies show the differences in the slopes of the anomalies. From Petronotis *et al.* (1992).

Dyment *et al.* (1994) showed that anomalous skewness becomes negligible at spreading rates above 50 km Myr^{-1}. This dependence on spreading rate has been explained as being due either to systematic variations of the geomagnetic field intensity within each polarity interval (Cande, 1978) or to sea-floor spreading processes causing the magnetic properties of the oceanic lithosphere to depart significantly from the standard uniformly magnetized rectangular prism model (Cande and Kent, 1976; Cande, 1978; Verosub and Moores, 1981; Raymond and LaBrecque, 1987; Arkani–Hamed, 1989). Dyment *et al.* (1994) point out that obviously no geomagnetic field behavior can predict the observation, for the same time interval, of a significant anomalous skewness at slow-spreading centers and none at fast spreading centers so that this possible cause can be eliminated. Therefore, the spreading rate dependence of anomalous skewness most likely arises from sea-floor spreading processes.

5.3 Analyzing Older Magnetic Anomalies

5.3.1 The Global Magnetic Anomaly Pattern

In an analysis of profiles across the Pacific–Antarctic ridge, Pitman and Heirtzler (1966) extended the model as applied to the then known geomagnetic polarity time scale, GPTS, for the time 0–4 Ma. By constructing a configuration of blocks of alternating polarity and assuming the constant spreading rate of 45 km Myr^{-1}, as observed for the interval 0–4 Ma, they traced the reversals of the geomagnetic field back to 10 Ma. The model is shown in Fig. 5.17. A comparison of the profile and its reverse shows the remarkable symmetry observed over a distance of 1000 km. The agreement between the resulting computed anomaly profile with that observed convincingly demonstrated the validity of the extrapolation beyond the known GPTS at that time. When this

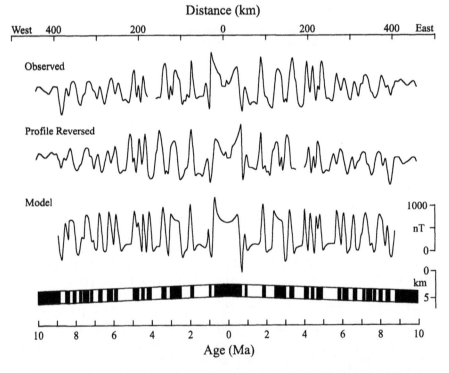

Fig. 5.17. The observed and reversed magnetic profiles across the Pacific–Antarctic ridge over a distance of 1000 km. The reversed sequence beyond 4 Ma is inferred from the magnetic anomalies by assuming a constant spreading rate in the interval 0–4 Ma. Reproduced with permission from Pitman and Heirtzler (1966). © American Association for the Advancement of Science.

model was applied to a profile across the Reykjanes Ridge on the opposite side of the globe (Heirtzler *et al.*, 1966), a good agreement was obtained if a spreading rate of 10 km Myr^{-1} was assumed.

By 1968 it was apparent that correlatable sequences of magnetic anomalies parallel to and bilaterally symmetric about the mid-ocean ridge system are present over extensive regions of the Pacific (Pitman *et al.*, 1968), Atlantic (Dickson *et al.*, 1968), and Indian (Le Pichon and Heirtzler, 1968) oceans. The system of magnetic anomaly numbers was initiated by assigning numbers from 1 (at the mid-ocean ridges) to 32 (for the oldest anomaly) to the most prominent positive magnetic anomaly peaks in the sequences. In a major review of these results, Heirtzler *et al.* (1968) assumed uniform spreading rates in the different oceans (the South Atlantic, the North and South Pacific, and the South Indian Oceans.) and extrapolated from the GPTS for the past few million years. Perhaps not surprisingly, this extrapolation led to different estimates of the polarity time scale for each of the four regions involved. Based on the then available information about the age of the sea floor, they took the bold step of resolving the problem by deciding that the South Atlantic Ocean had spread at a constant rate since the Late Cretaceous. Assuming a uniform axial spreading rate of 19 km Myr^{-1} at mid-latitudes in the South Atlantic, a magnetic reversal

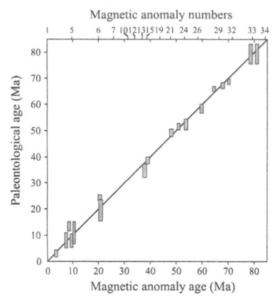

Fig. 5.18. Comparison of the paleontological ages of basal sediments in deep-sea drill holes with the basement ages predicted from the magnetic anomalies. The 45° line is that expected for perfect agreement. After Lowrie and Alvarez (1981), with the permission of the Geological Society of America.

chronology extending back to 80 Ma was then obtained. This chronology was subsequently found to be consistent with ages obtained by paleontological dating of basement sediments obtained by the DSDP (Maxwell *et al.*, 1970; LaBrecque *et al.*, 1977). One comparison of paleontological ages of basal sediments with magnetic anomaly ages is shown in Fig. 5.18.

The first substantial reanalysis of worldwide marine magnetic anomaly data since that of Heirtzler *et al.* (1968) was undertaken by Cande and Kent (1992a) to produce an updated time scale for the Late Cretaceous and Cenozoic. This was later refined by Cande and Kent (1995), who still used the South Atlantic as a reference ocean but assumed that its spreading rate had smooth variations that could be approximated by fitting a spline through a set of calibration points. Huestis and Acton (1997) argue that this assumption of smooth spreading at a reference ridge forces more erratic spreading rates at other ridges. To eliminate this problem, they proposed a formalism that penalizes nonsmooth spreading behavior equally for all ridges. The method then defines the time scale that has the best agreement with known chron ages and with anomaly-distance data from multiple ridges and allows the spreading rate for each ridge to be as nearly constant as possible. When applied to the data used by Cande and Kent (1992a) to establish their time scale (which singled out one ridge for the preferential assumption of smoothness) only modest changes, of <5%, were needed. A conceptual disadvantage of the Huestis and Acton (1997) approach is that it makes the unrealistic assumption that if any ridge changes spreading rate then all other ridges change their spreading rates at the same time.

Since the first major summary and interpretation by Heirtzler *et al.* (1968) in terms of the Vine–Matthews model and sea-floor spreading, detailed magnetic anomaly surveys and analyses have been made over most of the world's oceans. In the North Atlantic, analyses in terms of sea-floor spreading include those of Matthews and Williams (1968), Vogt *et al.* (1970, 1971), Rona *et al.* (1970), Vogt and Johnson (1971), Williams and McKenzie (1971), Pitman *et al.* (1971), Pitman and Talwani (1972), and Srivastava (1978). Sea-floor spreading in the Indian Ocean since the Late Cretaceous has been analyzed by McKenzie and Sclater (1971), Weissel and Hayes (1972), Sclater and Fisher (1974), and Norton and Sclater (1979) and reviewed by Schlich (1982). Recent analyses of sea-floor spreading between Australia and Antarctica include those by Cande and Mutter (1982), Veevers *et al.* (1990), and Tikku and Cande (1999). Sea-floor spreading between Madagascar and Africa has been analyzed by Rabinowitz *et al.* (1983). The Pacific Ocean has been investigated in most detail and significant analyses include those of Hayes and Heirtzler (1968), Pitman and Hayes (1968), Atwater

Fig. 5.19. (a) Isochron map of the ocean floor on a color-shaded relief map illuminated from the northwest. After Müller *et al.* (1997).

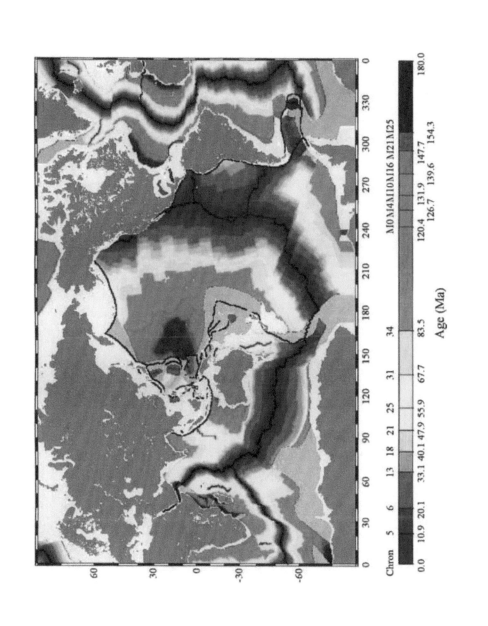

Chron 5 6   13 18 21 25 31 34 83.5

10.9 20.1 33.1 40.1 47.9 55.9 67.7 M0 M4 M10 M16 M21 M25

120.4 131.9 147.7
126.7 139.6 154.3

Age (Ma)

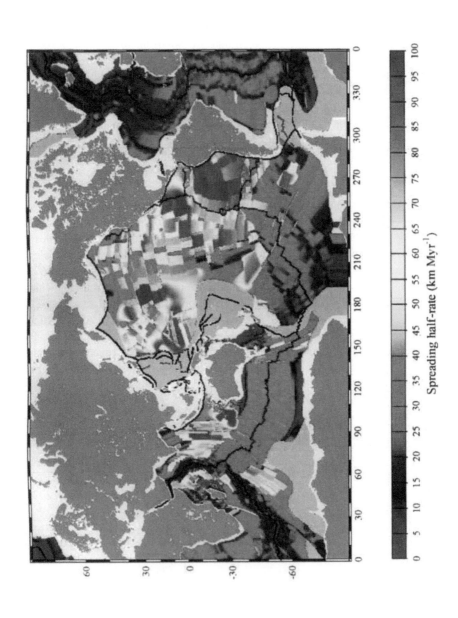

Spreading half-rate (km Myr^{-1})

(1970), Atwater and Menard (1970), Hayes and Pitman (1970), Larson and Chase (1972), Larson *et al.* (1972), Christoffel and Falconer (1972), Molnar *et al.* (1975), Larson and Hilde (1975), Hilde *et al.* (1976), and Weissel *et al.* (1977). A summary of the coverage of the Pacific Ocean from shipboard magnetic surveys is given by Isezaki (1988). Cande *et al.* (1989) provide a valuable summary in their map of magnetic lineations of the world's ocean basins, together with a comprehensive list of references.

The oldest magnetic anomalies occur in the western Pacific, the western North Atlantic, and off northwest Africa and are referred to as the M sequence anomalies (Larson and Chase, 1972; see §5.3.2). Larson and Pitman (1972) were the first to make a detailed worldwide correlation of these anomalies and establish their age. They realized that negative anomalies in the Pacific sequence correlated with positive anomalies in the North Atlantic because the Pacific anomalies were formed at or below the equator. Their reversal time scale for the Late Jurassic and Early Cretaceous involved anomalies M1–M22. In later studies of the western Pacific anomalies, Larson and Hilde (1975) added M0 and M23–M25 to the time scale and Hilde *et al.* (1976) recognized M26. However, Cande *et al.* (1978) resolved this into M26–M28 and added M29 to the sequence. The oldest anomalies observed in the oceans are M30–M41; small-amplitude anomalies in the western Pacific Jurassic quiet zone delineated by Handschumacher *et al.* (1988) initially to M38 and later by Sager *et al.* (1998) to M41. The Jurassic quiet zone anomalies are discussed further in §5.3.3.

Using the reversal time scale established from analyzing marine magnetic anomalies, it is possible to construct detailed maps of the age of the ocean floor worldwide. One of the most significant developments in our understanding of the structure of the ocean floor has come from the use of satellite altimetry data. The surface of the ocean bulges outwards and inwards according to variations in the topography of the ocean floor. These variations can be measured using satellite radar altimetry with closely spaced profiles and can then be converted to grids of vertical gravity gradient and gravity anomalies (Haxby, 1987; Smith and Sandwell, 1994; Sandwell and Smith, 1997). Maps produced from these measurements show many tectonic features in the oceans with more detail than was previously known, especially in areas not well covered by ship surveys. By combining these data with the latest magnetic anomaly identifications Müller *et al.* (1997) compiled a digital age grid of the ocean floor using a self-consistent set of global isochrons and associated plate reconstruction poles.

Figure 5.19a shows the isochron map of Müller *et al.* (1997) of the ocean floor and Fig. 5.19b shows the corresponding spreading rates at each age. The

---

Fig. 5.19. (b) Spreading rate map (kindly provided by D. Müller) corresponding to the isochron map of Fig. 19a.

TABLE 5.2

Estimates of the Timing of Continental Separation from Ages of the Oldest Magnetic Anomalies at
Various Continental Boundaries

Separating continents	Oldest magnetic anomaly	Age of separation (Ma)
Greenland – Europe	24+	60
North America – Greenland	34+	90
North America – Iberia	M0+	130
North America – Africa	M29+	175
South America – Africa		
North section: North of Benue Trough	M0	120
South section: Just north of Agulhas FZ	M10	130
Africa – Antarctica (+India)	M22–M24+	160+
Antarctica – India	M0–M10	120-130
Africa – Madagascar	34	85
Australia – Antarctica	34+	95
Australia – Lord Howe Rise	34+	90
New Zealand – Antarctica	33+	85
South America – Antarctica	M12	135
India – Australia		
North section: Argo Basin	M25+	155-160
South section: Perth Basin/Exmouth Plateau	M10	130

[a] A "+" indicates that sea floor older than the anomaly number exists.

isochron map has implications for continental drift and ocean evolution. The age
of the last anomaly observed between previously adjacent continents gives the
age at which the continents started to separate. Table 5.2 gives estimates of the
age of separation of various continents determined from the age of the oldest
anomalies observed in each case. In many cases sea floor still exists beyond the
oldest anomaly that can be identified. In Table 5.2 this is indicated by a "+" after
the anomaly number. In that case an arbitrary 5 Myr has been added to the age of
the oldest identifiable anomaly to give the age of separation. In some cases the
onset of sea-floor spreading anomalies is time transgressive, becoming
progressively younger along the margin.

5.3.2 Magnetic Anomaly Nomenclature

Polarity chrons older than 5 Ma are designated by numbers correlated with the
marine magnetic anomalies. For example polarity chron C26r represents the time
of reverse polarity between the normal polarity chrons correlated with magnetic
anomalies 26 and 27. Using the prefix M for Mesozoic, polarity chrons for the
pre-Aptian sequences are generally described in order of increasing age by the
designation M0 through M29. The situation is, however, somewhat confusing
because these M sequence anomalies were mainly (but not exclusively) assigned

to reverse polarity anomalies. For consistency, Opdyke and Channell (1996) used the prefix "C" (e.g., CM29) to distinguish polarity chrons as observed in magnetostratigraphy from the magnetic anomaly numbers.

As additional polarity chrons have been identified the nomenclature has had to evolve (LaBrecque *et al.*, 1977; Harland *et al.*, 1982, 1990; Cande and Kent, 1992a). The current nomenclature, following Cande and Kent (1992a), enables every chron and subchron to be identified uniquely. The longest intervals of predominantly one polarity are referred to by the corresponding anomaly number followed by the suffix n for normal polarity, or r for the preceding reverse polarity interval. When these chrons are subdivided into shorter polarity intervals, they are referred to as subchrons and are identified by appending, from youngest to oldest, .1, .2, etc., to the polarity chron identifier and adding an n or r as appropriate. For example, the three normal polarity intervals that make up anomaly 17 (chron C17n) are called subchrons C17n.1n, C17n.2n, and C17n.3n. Similarly, the reverse interval preceding (older than) C17n.1n is referred to as subchron C17n.1r. For more precise correlations, the fractional position within a chron or subchron can be identified by the equivalent decimal number appended within parentheses. For example, the younger end of chron C29n is C29n(0.0) and a level three-tenths from this younger end is designated C29n(0.3).

Cande and Kent (1992a) introduced the term *cryptochron* (see Table 4.3) to describe tiny wiggles in magnetic anomaly records that are clearly related to paleomagnetic field behavior (Cande and Kent, 1992b) but may not be short polarity subchrons or microchrons since they have not been confirmed in magnetostratigraphic section. They can be modeled either as short subchrons (or microchrons) or more likely as being due to longer period (50–200 kyr) global changes in the intensity of the Earth's magnetic field. This latter interpretation has been strongly supported by the similarity of ocean surface profiles of the central anomaly with synthetic profiles based on Brunhes age paleointensity records derived from deep-sea sediments (Gee *et al.*, 1996).

5.3.3 The Cretaceous and Jurassic Quiet Zones

Studies of marine magnetic anomalies have identified two quiet or smooth zones of Cretaceous and Jurassic age in which there appear to be either no anomalies or anomalies of extremely low amplitude with unclear interpretation. The Cretaceous Quiet Zone is the best defined and is bounded by magnetic anomalies M0 and M34 corresponding to the Cretaceous Normal (KN) Superchron originally identified by Helsley and Steiner (1969), as discussed in §4.2.3 and §4.3.6 and shown in Fig. 4.8. It seems clear that this quiet zone originated because the geomagnetic field remained in the normal polarity state from 118 to 84 Ma (Kent and Gradstein, 1986) and no magnetic anomalies were produced.

Larson and Pitman (1972) proposed that the KN Superchron represented a time during which there was a pulse of rapid spreading at all spreading centers in both the Atlantic and Pacific oceans. They related this pulse of rapid spreading to episodes of circum-Pacific intrusive and extrusive activity and orogenesis during this period. Plutonism on a large scale occurred in eastern Asia, western Antarctica, New Zealand, the southern Andes, and western North America during the mid-Cretaceous. This is best documented in western North America, where more than 50% of the exposed batholiths are dated between 115 and 85 Ma. If the granodiorites and granites that make up these batholiths are derived from underthrust oceanic lithosphere, then large-scale lithospheric subduction (a consequence of the rapid spreading in all the oceans during that time) is required. Although this pulse of rapid spreading was disputed by Berggren *et al.* (1975) on the grounds that the age limits of the quiet zone had been incorrectly assigned, the current state of the polarity time scale shown in Fig. 4.8 apparently still supports the original proposal (Larson, 1991). However, Heller *et al.* (1996) argue that there are sufficient uncertainties in reconstructing plate motions to question the reality of the rapid pulse in spreading rate.

Sclater *et al.* (1971) and Parsons and Sclater (1977) have shown that the depth of the ocean floor is primarily a function of its age, with the most recent age–depth curve being given by Stein and Stein (1992). This age–depth relation is a consequence of the fact that the lithosphere formed at a spreading ridge is hot and therefore elevated. As it moves away from the ridge axis it cools and subsides (Menard, 1969; McKenzie and Sclater, 1971). As a result there is an increase in the volume of spreading ridges with increase in spreading rate. Therefore the volume of any ridge is a function of its spreading rate history. Changes in the spreading rate cause changes in ridge volume together with associated rises and falls in sea level. Larson and Pitman (1972) proposed that a pulse of rapid spreading occurred during the time of the KN Superchron. Hays and Pitman (1973) demonstrated quantitatively that the worldwide great marine transgression and subsequent regression that occurred in the mid- to Late Cretaceous may have been caused by this contemporaneous pulse of rapid spreading deduced by Larson and Pitman (1972) to have occurred during the KN Superchron. However, Gurnis (1991) shows that there are problems with this linkage. Heller *et al.* (1996) point out that there are many other explanations for the timing and magnitude of long-term sea level changes, including plate reorganization with little or no change in spreading rates, such as the formation of a new ridge, a new ocean, a new continental rift, ridge jumps and reconfiguration of subduction zones. During the KN Superchron, plate reorganization in the Pacific shows evidence of ridge jumps, indicating that more of the ocean floor of this age may be preserved than has been presumed. Therefore, both the basis of and apparent need for the mid-Cretaceous pulse of sea-floor spreading is now in question (Heller *et al.*, 1996).

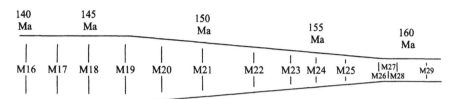

Fig. 5.20. Decreasing amplitude of magnetic anomalies in the Pacific observed from anomalies M19 to M26 approaching the Jurassic Quiet Zone. Ages are indicated according to the time scale of Kent and Gradstein (1986). The reduced amplitude from anomalies older than about 159 Ma corresponds with the minimum in the Mesozoic dipole low determined from paleointensities on land in the time interval 160–170 Ma. After Cande *et al.* (1978), with the permission of Elsevier Science.

The Jurassic Quiet Zone (JQZ) is observed off the east coast of North America, the northwest coast of Africa, and in the western Pacific. It corresponds to ocean floor of age older than M29 and is characterized by low-amplitude anomalies whose interpretation has been controversial. The known geomagnetic polarity time scale observed from sequences on land (Steiner and Ogg, 1988; Gradstein *et al.*, 1994; Opdyke and Channell, 1996) shows that, contrary to previous perception, there is no long period of a single polarity that could be correlated with this quiet zone. Several explanations have been offered, by Mascle and Phillips (1972), Poehls *et al.* (1973), Hayes and Rabinowitz (1975), Larson and Hilde (1975), Roots (1976), and Cande *et al.* (1978). The problem is that many of the quiet zones occur adjacent to continental margins and one suggestion is that they formed during the initial rifting of continents. Spreading rates are at first slow, producing narrow magnetic bodies. When these narrower bodies subside to continental margin depth, their anomalous magnetic field is then attenuated at the ocean surface.

However, one of the most important results not known until recently is that the intensity of the geomagnetic field was much reduced between 130 and 170 Ma, with a minimum of about one-third of its Cenozoic value between 160 and 170 Ma (Prévot *et al.*, 1990; Perrin *et al.*, 1991; Prévot and Perrin, 1992; Kono and Tanaka, 1995; Perrin and Shcherbakov, 1997; Kosterov *et al.*, 1997), as was illustrated in Fig. 1.15. An envelope of decreasing anomaly amplitudes with increasing age is observed from around anomaly M20 (~150 Ma) in the Pacific (Larson and Hilde, 1975). Typical amplitudes of 250–500 nT are observed for anomalies formed after M20, whereas those older than M25 have typical amplitudes of about 50–70 nT (Fig. 5.20). Hayes and Rabinowitz (1975), Larson and Hilde (1975) and Cande *et al.* (1978) suggested that this could be due to a systematic decrease in field intensity going backwards in time. There is also a remarkable linearity in the low amplitude magnetic anomalies that are seen in the JQZ, especially in the faster spreading Pacific regions (Handschumacher *et al.*, 1988; Sager *et al.*, 1998).

The frequency of reversals during the time of the JQZ is not observed to be remarkably high on land (Gradstein *et al.*, 1994; Opdyke and Channell, 1996), with rates of ~3 per Myr. However, Jurassic ocean floor is deep and anomalies measured at the surface are attenuated. Sager *et al.* (1998) investigated these anomalies using deep-tow profiles over the western Pacific JQZ. The anomalies show both short- and long-wavelength components. The short-wavelength components are preferentially attenuated when upward continued to the ocean surface. Sager *et al.* inferred that many of the short-wavelength anomalies represented geomagnetic intensity fluctuations and were therefore cryptochrons according to Cande and Kent (1992b). The polarity reversal model deduced from the long-wavelength anomalies then becomes reasonably consistent with that observed on land. This interpretation, coupled with the known decrease in global field intensity at that time, appears to be a reasonable explanation of the JQZ anomalies.

5.4 Paleomagnetic Poles for Oceanic Plates

5.4.1 Skewness of Magnetic Anomalies

The measurement of the skewness of magnetic anomalies as described in §5.2.2 provides a method for calculating paleomagnetic poles. The effective inclination I' is easily determined from the geomagnetic field direction at the site (generally calculated using the IGRF, see §1.1.3) and the azimuth A of the source using (5.2.2). Therefore, if the skewness θ is known, (5.2.1) can be solved for I'_R. Substituting into (5.2.3) the possible combinations of D_R and I_R, together with (1.2.9) and (1.2.10), defines a semi-great circle of paleomagnetic pole positions consistent with the observation θ. To further constrain the solution to a single pole position instead of a possible pole path, it is necessary to determine θ for two sets of contemporaneous anomalies on a plate. The two semi-great circles should intersect to give the paleomagnetic pole for the plate for the age of the anomalies (Fig. 5.21a). In practice the estimated value of θ in each case can only be determined at a level of confidence $\Delta\theta$ of ±5–10°. Therefore, the values $\theta\pm\Delta\theta$ define a lune of confidence for lineations of the same age. The intersection of two lunes of confidence for lineations of the same age at two places on a rigid plate then determines the zone of confidence for the paleomagnetic poles (Fig. 5.21b).

Schouten and Cande (1976) referred to the above method for determining paleomagnetic poles as the "theta method". It was first used by Larson and

Chase (1972) on three sets of lineations in the Western Pacific. They found θ by computing a series of models for their lineations phase-shifted in $10°$ increments. They then estimated errors by comparing the observed profiles with the models. Although this gives a good estimate of θ when the profiles are nearly symmetric (θ near $0°$), it is not as effective when the profiles are nearly antisymmetric (θ near $90°$). Schouten and Cande (1976) proposed that the skewing effect of θ be removed (deskewed) by inverse phase filtering (Schouten and McCamy, 1972), similar to the method of reduction to the pole of Blakely and Cox (1972). The profiles are then inverse phase-filtered in $5°$ increments of θ. To calculate θ and its confidence level $\Delta\theta$, suitable individual anomalies are compared with the model profile. This will then give a range of values of θ corresponding to the best fit for each chosen anomaly. These values of θ are then averaged to give a mean value for the observed profile and the confidence interval $\Delta\theta$ can be calculated.

The presence of anomalous skewness causes a systematic error in the effective remanent inclination, and this limits the accuracy of paleomagnetic poles determined from skewness data. Petronotis *et al.* (1992) proposed the following procedure to calculate paleomagnetic poles and to allow for the effect of anomalous skewness. The anomalous skewness θ_A is given by the difference between the *true* effective remanent inclination I'_R and the *apparent* effective remanent inclination I'_{AR}, according to

$$\theta_A = I'_R - I'_{AR} \quad (\theta_A \geq 0). \tag{5.4.1}$$

If λ_s and ϕ_s are the latitude and longitude of the site, and λ_p and ϕ_p are the corresponding coordinates of the pole position, then I_R and D_R can be calculated according to (1.2.9) and (1.2.10). Given a trial value for anomalous skewness θ_A and a trial pole position (λ_p, ϕ_p), a model apparent effective remanent inclination can be predicted at each site from (5.4.1) and then compared with the observations.

In the Schouten and Cande (1976) approach two distinct observations uniquely define a best fitting paleomagnetic pole (Fig. 5.21) because only two parameters λ_p and ϕ_p were adjusted. To allow for anomalous skewness θ_A must also be estimated; therefore, three distinct observations must be used in this case. For three or more observed values of skewness, a pole position will rarely fit all the data perfectly. A pole position is therefore calculated that minimizes the weighted squared differences between the predicted and observed apparent effective remanent inclination at each site.

The relative amplitude factor C also contains information about the remanent magnetization vector from (5.2.4). Since the overall amplitude of the anomalies also depends on the thickness of the source layer and the remanent magnetization, C cannot be resolved from the amplitudes of the observed

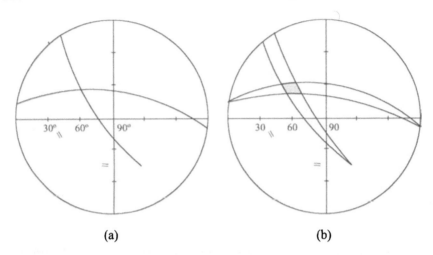

(a)   (b)

Fig. 5.21. Estimating a paleomagnetic pole position from the skewness of magnetic anomalies. (a) Semi-great circles of possible pole positions, determined from unique values of θ from two sets of magnetic lineations located at =, intersect at the pole position. (b) The zone of confidence for the pole position is determined by the intersection of two lunes of confidence, shown here with $\Delta\theta = 10°$ in each case. From Schouten and Cande (1976).

anomalies. It is possible, however, to determine a locus of paleomagnetic poles from a ratio of C for two regions. Assuming that the thickness of the source layer is the same for the two regions and its magnetization only varies as a geocentric axial dipole field, then the ratio of the amplitudes equals the ratio of C times the ratio of the dipole paleointensity. Although the resolution of the ratio of amplitudes is far less than the resolution of θ, the locus of paleomagnetic poles can also be used as an independent source of information (Schouten and Cande, 1976).

5.4.2 Magnetization of Seamounts

Vacquier (1963) showed that if a combined magnetic and bathymetric survey of a seamount is made, it is possible to calculate the direction of magnetization of the seamount under the assumption of uniform magnetization. These directions of magnetization are an equivalent of those determined paleomagnetically by conventional methods and it is possible to calculate a paleomagnetic pole for the seamount. The growth of a small seamount takes place over 10^4-10^5 years. Secular variation will tend to be averaged out in the computation of its average direction of magnetization. Larger seamounts may remain volcanically active over a period of 10^4-10^7 years, and the problem arises that several polarity changes may have taken place during this time. The seamount will not then be uniformly magnetized as will be found when the computed anomaly fails to match the observed anomaly.

The original method proposed by Vacquier (1963) used a linear least squares inversion of the magnetic anomaly constrained by the shape of the seamount. The seamount is assumed to be uniformly magnetized, to be bounded by an upper surface equal to its bathymetric surface, and to have a known lower surface that is usually flat and assumed to be the same depth as the surrounding ocean floor. This method has been improved or extended using more efficient algorithms (Talwani, 1965; Plouff, 1976). Vacquier (1963) modeled the shape as an aggregate of rectangular blocks, whereas Talwani (1965) used a stack of laminas and Plouff (1976) a stack of layers. Attempts have been made to allow for non-uniform magnetization in seamounts by dividing the topography into discrete blocks. For example, McNutt (1986) divided the topographic shape into two or three blocks, assuming each was uniformly magnetized and used least squares methods to solve for the direction of magnetization within each block.

In order to allow for nonuniform magnetization the boundaries between the uniformly magnetized blocks were determined subjectively, usually by inspection of the residual magnetic anomaly. Parker *et al.* (1987) proposed an entirely different approach and developed a method using linear inverse theory in a Hilbert space. The method is particularly useful where seamounts have magnetization inhomogeneities. The derived magnetic model maximizes the uniformly magnetized part and minimizes the inhomogeneities. Hildebrand and Parker (1987) applied this method to some seamounts in the Pacific Ocean whose magnetization had previously been determined by the linear least squares method. In most cases the differences between the results obtained by the two methods were small. When modeling seamount magnetization it is assumed that there are no induced or viscous components. This assumption is difficult to verify because few seamount interiors have been sampled. However, it is surprising that there is a predominance of seamounts modeled to have normal polarity magnetization. Gee *et al.* (1989) sampled an uplifted seamount and found that one-sixth of the magnetization may arise from induced magnetization. Despite this, it seems that as long as the induced and viscous components are less than about 15% ($Q \geq 7$), the calculated paleomagnetic pole will be in error by $\leq 5°$ (Sager and Pringle, 1988). A useful review of these methods together with appropriate theory is given by Blakely (1995).

There has been considerable discussion as to the validity of the methods of modeling seamount magnetization using the methods described above. Parker (1988) has shown that the standard least squares method is unsatisfactory because, in the case of nonuniform magnetization, the unfitted field is not due to random contamination as required. He therefore proposed a statistical theory that overcame this problem, showing that the paleomagnetic pole derived in this way can be completely incompatible with that deduced from standard least squares inversion. Also, the magnetization of dredged and drilled samples from the ocean floor shows that both basalts (Lowrie, 1974) and gabbros (Kent *et al.*, 1984) have log-normal distributions of magnetization. Limited sampling of seamounts (Kono, 1980a; Gee *et al.*, 1988, 1989) also shows similar log-normal distributions that vary according to lithology and grain size. This means that the model of uniform magnetization usually assumed is unrealistic. Parker (1991) therefore developed a model for seamount magnetization in which the direction of magnetization is fixed but the intensity of magnetization can vary with no upper limit on magnitude. Application of this model to several seamounts suggests that many published results using the assumption of uniform magnetization may not be as well determined as has previously been thought.

It is useful to calculate some parameter that indicates how well the calculated magnetic anomaly approximates to the observed anomaly. The most widely used parameter is the *goodness-of-fit ratio* (GFR) defined as

$$\text{GFR} = \frac{\text{Mean observed magnetic anomaly}}{\text{Mean residual anomaly}}, \qquad (5.4.2)$$

where the mean residual anomaly is the mean difference between the observed and calculated anomaly. A GFR <2.0 is usually regarded as an unreliable result (Sager and Pringle, 1988).

Most seamount magnetizations have been determined from the Pacific plate. However, to be useful for paleomagnetic studies the ages of the seamounts must be known. The first attempt at plotting an apparent polar wander path (APWP, see chapter 6) for the Pacific plate was carried out by Francheteau *et al.* (1970b). They showed that the pole position determined from 17 seamounts in the vicinity of the Hawaiian islands with ages 85–90 Ma lay at 61°N, 16°E with $A_{95} = 8°$. The Pacific plate has thus drifted approximately 30° northwards since the Cretaceous. Sager (1987) indicates that there are now more than 90 results with GFR >2.0 from this region. Sager and Pringle (1988) conclude that only 22 of these seamounts can be used for determining the apparent polar wander path for the Pacific plate. There are 17 that have radiometric ages, with all but 3 of these being $^{40}\text{Ar}/^{39}\text{Ar}$ total fusion or incremental heating ages, which are considered the most reliable for dating basalts that have undergone submarine alteration. The other 3 ages are high-quality K–Ar ages. Other means, including fossils dredged from the reef cap or magnetostratigraphic arguments (Gordon and Cox, 1980a), have been used to date the remaining 5 seamounts.

5.4.3 Calculating Mean Pole Positions From Oceanic Data

Oceanic plates present the following types of data that provide information about paleomagnetic pole positions.

(i) Paleomagnetic measurements of inclinations (but not declinations) obtained from deep-sea drilling cores (see §3.2.3 for analysis of inclination-only data). This constrains the pole to lie on a small circle, centered on the sampling site, at a distance given by the paleocolatitude (1.2.5).

(ii) Effective inclinations, calculated from the skewness of marine magnetic anomalies, which constrain the pole to lie on a specific half great circle (§5.4.1).

(iii) Ratio of relative amplitude factors from two sets of magnetic lineations of the same age but created at different locations on the same plate. This constrains the pole to lie on a locus defined by the intensity structure of the geocentric axial dipole field (§5.4.1). Naturally these provide only a low-resolution result.

(iv) Paleomagnetic poles determined from magnetic anomalies over seamounts (§5.4.2).

(v) Paleoequators determined from geological analysis of marine sediment cores (van Andel, 1974). In this case the equatorial zone of upwelling is also a region of high biological productivity so that this zone is characterized by high sedimentation rate and higher biogenic content than at other latitudes. It is possible in some deep-sea cores to identify the age at which the drill site crossed the equator. A datum from a single site can thus be considered to be a paleocolatitude of 90°, defining a great circle of possible pole positions.

Gordon and Cox (1980b) presented a method to yield a best fit pole position and confidence limits for a combination of data of the types listed above. The two fundamental assumptions in their method are that (a) each of the observations arises from a common paleomagnetic pole, and (b) the errors from each datum are random and mutually independent. It is then a question of identifying the pole position that is most compatible with all the observations. Gordon and Cox (1980b) chose maximum likelihood as the appropriate measure of compatibility. That is, given the probability distribution for each observation, the most appropriate pole is the one that gives the maximum joint probability (likelihood) for the actual observations. They further assumed that for small errors the error distribution for each datum is Gaussian. With this assumption, maximum likelihood reduces to a weighted least squares estimation. Hence, as their test parameter they used

$$\chi^2 = \sum \left(\frac{\delta_i}{\sigma_i} \right)^2, \tag{5.4.3}$$

where the summation is over all observations, σ_i is the error in the ith observation, and δ_i is the deviation of the observation from the calculated value for the current test position of the pole.

For example, for inclination-only data, the datum is considered to be the equivalent paleocolatitude. Thus, for these data

$$\delta_i = p_i^{obs} - p_i^{model}, \tag{5.4.4}$$

where p_i^{obs} is the observed paleocolatitude and p_i^{model} is the colatitude of the site for the current test pole position. Seamount poles constrain two degrees of freedom (the latitude and longitude of the pole) in the analysis. The same constraint can be achieved by having two great circles intersecting at right angles at the seamount pole position. Therefore, seamount data with elliptical confidence limits from one site are computationally equivalent to paleoequator data from two sites with different errors. However, paleoequators are in effect just a paleocolatitude with a nominal value of 90°. Therefore, the analyses of

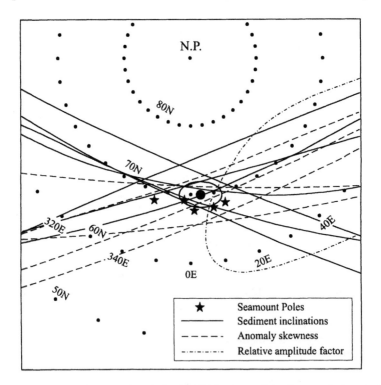

Fig. 5.22. The 72-Ma paleomagnetic pole for the Pacific plate and the data used in its calculation. The solid dot is the location of the pole and the ellipse is its 95% confidence region. The stars indicate paleomagnetic poles from seamounts, solid lines give the locus of possible pole positions determined from paleocolatitudes, dotted lines show the half great circle locus of paleomagnetic poles determined from the skewness of magnetic anomalies, and the dashed line represents the locus of possible pole positions determined from the relative amplitude factor. From Sager and Pringle (1988).

seamount poles, inclination-only data, and paleoequator data are all reduced to the analysis of paleocolatitudes from a computational viewpoint. For anomaly skewness the datum is the effective inclination and so the difference between observed and model effective inclination is used for δ_i in (5.4.4). The best estimate of the pole position is that which gives the minimum value of χ^2.

Confidence limits in the estimated pole position are obtained by standard linear propagation of errors. With these errors it is possible to calculate the *importance* of each datum, bearing in mind that a datum such as a seamount pole provides two constraints and therefore two importances, which should be summed. Data importances are a measure of the contribution that any particular datum makes in constraining the final estimate. If the importance of one datum is substantially greater than that of the others, then there are three possibilities that should be considered: (a) The uncertainty in that datum may have been grossly underestimated, (b) the remainder of the data genuinely do not provide a useful constraint on the result, and (c) the datum in question may be incompatible with the rest of the data. Naturally, (c) is the most serious because it suggests that the datum with the highest "importance" is in error either because of incorrect measurement or because the age of the datum has been incorrectly assigned. Bearing in mind the problems discussed in §5.4.2 regarding the modeling of seamount magnetization, it is critical that seamount poles be accurately (as distinct from precisely) determined. It is inevitable that seamount poles will have the greatest importance since they constrain two dimensions. This is especially the case if the great circles from the other data are subparallel because then the result will be controlled entirely by the seamount data.

An example of the application of this method to a 72-Ma paleomagnetic pole determined for the Pacific plate by Sager and Pringle (1988) is illustrated in Fig. 5.22. As listed in Table 5.3 the analysis combines data from paleocolatitudes

TABLE 5.3

Paleomagnetic Poles for the Pacific Plate Derived from the Analysis of Various Oceanic Data

Age (Ma)	Methods[a]	Pole Position Lat.(°N)	Long.(°E)	Major semi-axis (azimuth)(°)	Minor semi-axis (°)	Ref.[b]
26	I	81.1	2.4	7.1 (80)	1.2	1
39	I,M	77.6	7.6	3.5 (91)	2.4	2
57	S	77.6	3.6	6.5 (91)	4.0	3
65	I,S,C,M,E	71.6	7.9	2.9 (75)	1.8	4
72	S,I,C,M	70.0	3.6	3.2 (91)	1.9	5
82	I,M	58.4	359.0	2.9 (91)	2.7	5
88	I,M	56.6	330.7	3.8 (41)	3.1	5
125	I,S,M	50.9	322.6	10.0 (60)	3.6	6

[a] I, inclinations from deep-sea cores; S, skewness of magnetic anomalies; C, relative amplitude factor from magnetic anomalies; M, seamount magnetization; E, equatorial sediment facies.
[b] 1, Acton and Gordon (1994); 2, Sager (1987); 3, Petronotis *et al.* (1994); 4, Acton and Gordon (1991); 5, Sager and Pringle (1988); 6, Petronotis *et al.* (1992).

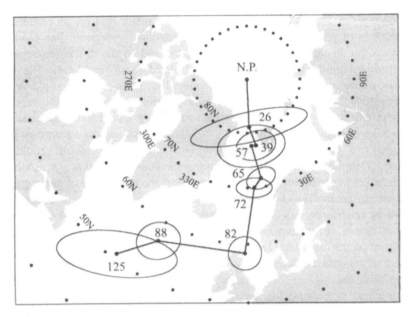

Fig. 5.23. Apparent polar wander path for the Pacific plate for the past 125 Myr, based on the data of Table 5.3. From Petronotis *et al.* (1994).

determined from 7 deep-sea sediment cores, 5 analyses of skewness, 1 analysis of the relative amplitude factor and 5 seamount paleomagnetic poles. These 18 measurements combine to give a paleomagnetic pole at 70.0°N, 3.6°E with confidence limits having semimajor axis of 3.2° with an orientation of 91° and semiminor axis of 1.9°.

The most recently determined apparent polar wander path for the Pacific plate is that given by Petronotis *et al.* (1994), based on the data listed in Table 5.3 and illustrated in Fig. 5.23 (see §6.4.1 for details regarding plotting apparent polar wander paths). The apparent polar wander path shows that the Pacific plate drifted northwards from 82 Ma to the present, although not at a steady rate. Prior to 82 Ma there is little northward motion and the east–west trend of the poles could have been caused by a rotation of the Pacific plate between 125 and 82 Ma. Most of this period coincides with the Cretaceous Normal Superchron, following which the plate drifted northwards. Several authors have suggested from studies of the evolution of the North Pacific that a reorganisation of plate boundaries occurred near the end of the KN Superchron (Rea and Dixon, 1983; Sager and Pringle, 1987; Mammerickx and Sharman, 1988).

Unfortunately, analyses as described above have not been applied to plates other than the Pacific. However, paleomagnetic inclinations have been measured using the sediments and basalts from the many DSDP cores that have been

TABLE 5.4
Paleomagnetic Pole Positions for the Indian Plate Derived from DSDP Core Inclinations[a]

Age range (Ma)	Mean age (Ma)	No. of sites	Pole position Lat.(°N)	Long.(°E)	A_{95} (°)
5-22.5	20	2	78.7	294.7	–
25-35	30	4	72.5	315.8	3.5
35-45	40	3	67.5	312.2	12.0
45-55	50	4	57.7	301.3	9.1
55-65	60	5	40.7	299.9	10.7
65-75	70	2	29.0	291.4	–

[a]Declinations are simulated following Klootwijk (1979).
Note. A_{95} is the radius of the circle of 95% confidence about the mean pole position.

drilled at various sites in the Indian Ocean. Data from the eastern Indian Ocean have been analyzed by Peirce (1976, 1978). Averaging inclination data alone consistently underestimates the true mean value and correction must be made for this effect (Briden and Ward, 1966; Kono, 1980b; McFadden and Reid, 1982; see §3.2.3 for details). In addition, secular variation is often not averaged out sufficiently when averages are taken from basalt horizons or from rapidly deposited sediments. Peirce (1976) proposed a method that takes all these factors into account and derived mean paleolatitudes as a function of time that described the northward drift of the Indian plate for the past 70 Myr (Peirce, 1978). His results were in good agreement with movements derived from sea-floor spreading data.

Klootwijk (1979) extended the analysis of Peirce (1976, 1978) to include other data from DSDP cores in the northwest Indian Ocean (see results listed in Table 5.4 and Fig. 5.24 for the apparent polar wander path). By setting $D = 0°$ at the present time and using the rotation parameters for the Indian plate as given by Powell *et al.* (1980), the paleodeclinations at any site for a given age could be simulated. The method enabled the calculation of paleomagnetic pole positions for the past 70 Myr for the Indian plate derived solely from oceanic data. There is excellent agreement with the corresponding apparent polar wander path determined from the Indian subcontinent itself (Klootwijk and Peirce, 1979).

5.5 Evolution of Oceanic Plates

5.5.1 The Hotspot Reference Frame

Most of the basaltic volcanism on the Earth's surface occurs at plate boundaries, either as mid-ocean ridge basalts or as volcanic arc basalts at subduction zones.

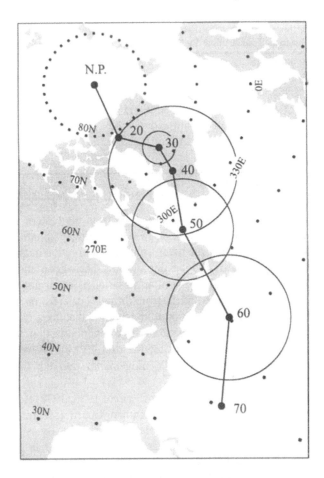

Fig. 5.24. Apparent polar wander path for the Indian plate for the time interval 70–20 Ma derived from DSDP cores as listed in Table 5.4.

Basaltic volcanism also occurs within plates on a much smaller scale and is characterized by linear volcanic chains (hotspot lineaments) that grow older in the direction of plate motion. The best known of these are the Hawaiian islands and other island chains in the Pacific Ocean. Wilson (1963) proposed that these resulted from limited upper mantle zones of melting called hotspots that remain stationary as the lithospheric plates move over them.

Morgan (1971, 1972) proposed that hotspots were maintained by material rising from the lower mantle through narrow mantle plumes [see Stacey and Loper (1983), Loper and Stacey (1983), and Loper (1984) for theoretical work on plumes] that formed part of the upward flow of a long-lived stable pattern of whole-mantle convection. It was supposed that these plumes do not change their relative positions once they have become established.

The possibility that such hotspots might be fixed with respect to the mantle is of interest in paleomagnetic studies because it suggests that they might form a fixed reference frame for plate motions. In plate tectonics all plate boundaries are in continuous motion with respect to one another and only relative motions can be measured between plates. Paleomagnetic measurements are only able to describe motions with respect to the Earth's spin axis and thus determine changes in latitude and orientation but not east–west motions (see further discussion in chapter 7). Any possible motion between hotspots can be measured by comparing the geometry and age distribution of volcanism along hotspot tracks with past plate movements reconstructed from relative motion data. Morgan (1972) first noted that the geometry of the Hawaiian–Emperor, Tuamoto–Line and Austral–Gilbert–Marshall island and seamount lineaments can be well matched by rigid plate motions over fixed hotspots presently located at Hawaii, Easter Island, and Macdonald seamount, respectively. Subsequent work has confirmed the fixed position of Pacific hotspots over the past 65 Myr. Morgan (1981) and Duncan (1981) confirmed the fixed position of hotspots on the African plate in the central and South Atlantic and the western Indian Ocean over the past 120 Myr. Therefore, there appears to be good evidence that the motion between hotspots *within* a given plate can remain fixed over long periods of time (Duncan and Richards, 1991).

Measuring the possible motion between hotspots on different plates requires that the relative motions between plates be well determined. Initial analyses by Minster *et al.* (1974) and Minster and Jordan (1978) found no detectable motion between hotspots globally for the past 5–10 Myr. Duncan and Richards (1991) suggested that the maximum relative motion between hotspots has been only 2–5 km Myr^{-1} over the past 120 Myr. However, the fixed hotspot reference frame has been seriously questioned by Tarduno and Gee (1995) who determined paleolatitudes from the paleomagnetism of basalts from seamounts drilled in the central, western, and northern Pacific Ocean during the Ocean Drilling Program. Comparison of these direct measurements of paleolatitude with those predicted from plate reconstructions provides a first-order test of the validity of fixed hotspots. The data suggest only minor latitude shifts of Pacific hotspots during the Cretaceous but require relative velocity between Atlantic and Pacific hotspots of 30 km Myr^{-1}. Similarly, Tarduno and Cottrell (1997) measured the paleolatitude of the Detroit seamount (81 Ma) in the northern part of the Hawaiian–Emperor chain. If the hotspot has remained fixed then the paleolatitudes of any part of the volcanic chain should be equal to that of present-day Hawaii. The measured paleolatitude of 36.2° is clearly different from the present-day latitude of Hawaii of 19°. Previous paleomagnetic data from the Suiko seamount (65 Ma and also part of the Emperor chain) gave a paleolatitude of 27° (Kono, 1980a), a value that is significantly different from that of present-day Hawaii and of the Detroit seamount.

Originally the bend in the Hawaiian–Emperor chain was regarded as the best example of a change in plate motion recorded in a fixed hotspot reference frame. An alternative is that the bend merely records differences in motion of the Hawaiian hotspot relative to the Pacific lithosphere. The paleomagnetic data from the Detroit and Suiko seamounts now clearly support the latter view. Tarduno and Cottrell (1997) therefore suggest that the Pacific hotspots may have moved at rates comparable to those of lithospheric plates (>30 km Myr^{-1}) in the time interval 81–43 Ma. McNutt *et al.* (1997) also questioned the validity of the hotspot plume theory to explain the origin of the Cook–Austral chain of volcanoes. New radiometric dates demonstrate that the southern Austral volcanoes are actually composed of three distinct volcanic chains with a range of ages spanning 34 Myr and with inconsistent age progressions; the technique of cumulative volcano amplitudes (Wessel and Kroenke, 1998) may help resolve this problem. It is possible that the volcanism in this region is controlled by stress in the lithosphere rather than the locus of narrow plumes rising from the deep Earth. A similar situation is apparent in relation to the Joban seamount chain in the northwestern Pacific (Desiderius *et al.*, 1997).

Results from seismic tomography suggest that much of the return convective flow that balances plate motions occurs in the lower mantle. However the hypothesis of a fixed hotspot reference frame assumes that the lower mantle is almost rigid and thus has minimal internal relative motion. Steinberger and O'Connell (1998) modeled the advection of plumes in a large-scale mantle flow. Their results explain how it is possible for plumes under a single plate to show little relative motion while at the same time moving relative to plumes under a different plate. Therefore, it seems unlikely that plumes can be firmly anchored in a convecting mantle and so the idea of a fixed hotspot reference frame can be dispelled.

5.5.2 Evolution of the Pacific Plate

The Pacific is currently the largest oceanic plate. It contains no continents and the magnetic anomalies extend back to 180 Ma. It should therefore be possible by careful analysis to be able to deduce the evolution of the Pacific plate since

---

Fig. 5.25. Reconstructions of the Pacific basin at 20, 37, and 65 Ma in a fixed hotspot reference frame after Engebretson *et al.* (1985). Double heavy lines are ridge boundaries between oceanic plates (dashed when inferred). Single heavy lines are transform faults (dashed when inferred). Arrows show the motion of the plates, with the length of each arrow indicating 10 Myr of motion. Pattern (x-x-x) indicates the extinct Pacific–Kula ridge. Diagonal shading indicates lithosphere that could be either Kula (KU) or Farallon (FA) plate.

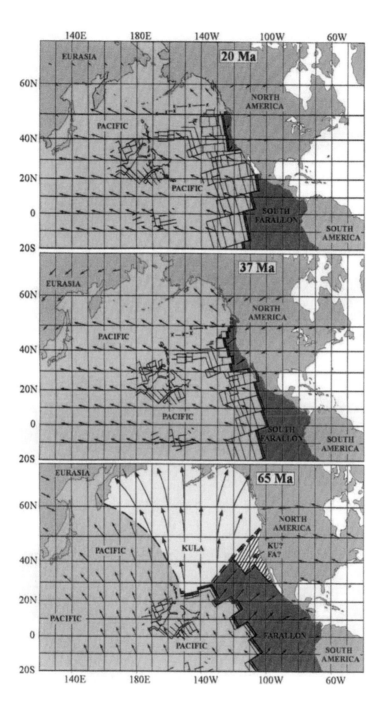

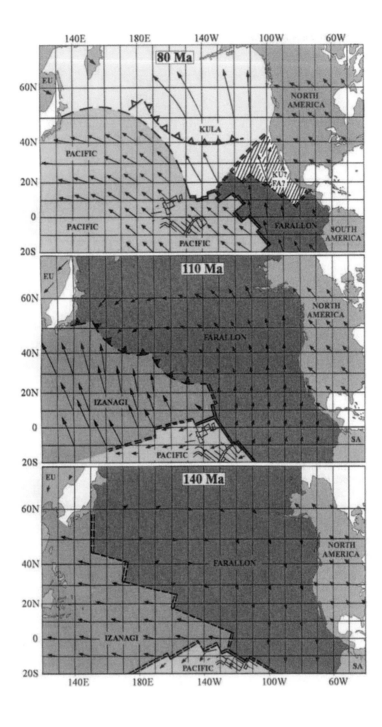

that time by using all the information that is still preserved both in the magnetic anomaly sequence and in the ocean floor features. Grow and Atwater (1970) deduced, backwards in time, that there were at least two plates present in addition to the Pacific plate. The present Juan de Fuca and Cocos plates are relics of the Farallon plate, named after the Farallon islands located off the coast of central California (McKenzie and Morgan, 1969). There was another plate north of the Farallon plate, named the Kula plate by Grow and Atwater (1970). The remnants of the Kula plate (Kula means "all gone" in an Athabascan Indian dialect) are currently being subducted beneath the Aleutian arc. There was a Pacific–Kula spreading center that became extinct at about 43 Ma. Other important early studies include those of Francheteau *et al.* (1970a), Coney (1972), and Atwater and Molnar (1973).

Many models have been presented for the former locations of ridges between oceanic plates in the Pacific basin (Larson and Chase, 1972; Hilde *et al.*, 1976; Alvarez *et al.*, 1980; Carlson, 1982; Woods and Davies, 1982). Engebretson *et al.* (1984) derived relative motion Euler poles describing the displacement histories between the Pacific plate and other plates that were once adjacent during the late Mesozoic and Cenozoic (Farallon, Kula, Izanagi, and Phoenix plates). Mainly using these reconstruction models, Engebretson *et al.* (1985) produced a model for the displacement history between western North America, eastern Eurasia and the adjacent oceanic plates for the past 180 Myr (Fig. 5.25). The model was based on the assumption that the hotspots in the Atlantic region have remained fixed relative to the hotspots in the Pacific basin (but not necessarily relative to the spin axis). Despite the problems that now appear to be present in sustaining the fixed hotspot assumption (§5.5.1), the model presented by Engebretson *et al.* (1985) is based on the only complete analysis that has yet been made of Pacific basin history over the entire time interval covered by the Pacific magnetic anomalies.

The color diagrams of Fig. 5.25 show how the Pacific basin has evolved over time with reconstructions at 20, 37, 65, 80, 110, and 140 Ma. These reconstructions are shown in a reference frame in which the hotspots are held fixed. Known plate boundaries based on isochrons and fracture zones now preserved in the Pacific plate are depicted with solid double lines (ridges) or solid single lines (transforms). Speculative ridges and transforms in regions where neither of a plate pair is preserved are depicted with dashed lines drawn perpendicular and parallel to tangents of small circles about the appropriate relative motion Euler pole for the time of the reconstruction. The locations of

---

Fig. 5.25 continued. Reconstructions of the Pacific basin at 80, 110, and 140 Ma in a fixed hotspot reference frame after Engebretson *et al.* (1985). Barbed arcs are inferred island arcs, solid barbs when active and open barbs when inactive. EU, Eurasia; FA, Farallon; KU, Kula; SA, South America. Reproduced with the permission of the Geological Society of America.

these boundaries are regarded as highly uncertain. Engebretson *et al.* (1985) have estimated that the combined uncertainty in location for the reconstruction at 100 Ma is about 900 km.

At 20 Ma the reconstruction shows the Farallon plate in two parts, the northern part being the Juan de Fuca plate of today and a southern part separated from the Phoenix plate by a ridge system to the south. The location of the extinct Pacific–Kula spreading center, assumed to have died at 43 Ma, is shown by the (x-x-x-x) pattern. This extinct ridge arrived at the Aleutian trench at ~10 Ma. At 37 Ma the Pacific plate first came into contact with North America. At 65 Ma the location of the Kula–Farallon boundary is based on the interpretation of Woods and Davies (1982) that the Kula plate broke from the Farallon plate at about 85 Ma. The shaded areas in the 80- and 65-Ma reconstructions could be either Kula or Farallon plate. The location of the Kula–Pacific boundary is not constrained by ocean floor data and here the model proposed by Hilde *et al.* (1976) has been followed.

For times older than 85 Ma, Woods and Davies (1982) proposed there was an older plate to the west that they called the Izanagi plate, and this interpretation is followed here for the 110-Ma reconstruction. The reconstructions for 110 and 80 Ma show that a major change took place in the plate geometry of the northwest Pacific during this interval. At 140 Ma the Pacific–Izanagi and Pacific–Farallon boundaries are known to have been spreading centers because of the presence on the Pacific plate of magnetic anomalies produced along these ridges. In the Cretaceous the Pacific Ocean was much more extensive than it is today because India, Australia, and New Zealand were then joined to Antarctica. Spreading was occurring from at least five spreading centers joined at two triple junctions (Larson and Chase, 1972). At 110 and 140 Ma there was another triple junction to the south and spreading centers to the south of the Pacific and Farallon plates bordering the Phoenix plate.

As one moves forward in time from 140 Ma to the present, the Pacific plate as shown in Fig. 5.23 has moved progressively northwards with respect to North America. This is in total agreement with the northward motion shown by the apparent polar wander path illustrated in Fig. 5.23.

Chapter Six

Continental Paleomagnetism

6.1 Analyzing Continental Data

For ages younger than about 180 Ma the sea floor provides a wealth of evidence that can be used to understand the evolution of the Earth's crust as discussed in chapter 5. However, in the absence of an oceanic record for earlier times, recourse must be made to evidence from the continents. This then demands the application of continental paleomagnetism to plate tectonics, for which this chapter lays the foundations.

There are four steps that have to be undertaken to establish the necessary foundation. Techniques for the acquisition and analysis of continental paleomagnetic data have evolved with time. This has led to the existence of a large amount of data of highly variable quality, with the quality and quantity generally improving with time. Despite this general trend, there is no one-to-one relationship between when the data were acquired and their quality; some of the earlier data may still be regarded as being of high quality, whereas some more recently acquired data are of poorer quality. Therefore, the first step is to establish a set of reliability criteria and then to apply these to the selection of appropriate data.

The second step is to test the validity, throughout Earth history, of the basic assumption used in the application of paleomagnetism to tectonics. This assumption is that the geocentric axial dipole (GAD) model, described in §1.2.3, is valid and therefore the data can be described in terms of the equivalent paleomagnetic poles. Even though suitable reliability criteria have been used to select an appropriate set of data, the paleomagnetic poles for the same epoch for a given region will often have been derived from widely varying numbers of sites, yet they must be combined in a sensible manner. The third step, then, is to establish a methodology for combining these poles, with the correct weighting, into composite poles over suitable time intervals and then to apply this methodology to the selected data. The fourth step is to make use of these composite paleomagnetic poles to determine the motion of continental blocks with respect to the geographic pole by determining apparent polar wander paths as described in §6.4.2.

The results of these analyses provide the foundation at the continental level for the plate tectonic syntheses that are discussed in chapter 7.

6.2 Data Selection and Reliability Criteria

6.2.1 Selecting Data for Paleomagnetic Analysis

The laboratory and analytical techniques used by paleomagnetists have evolved and improved substantially with time. Cryogenic magnetometers with their greater sensitivity and increased speed of measurement have led to larger and more elaborate field collections. The use of computers has enabled more sophisticated data analyses (such as principal component analysis) to be carried out on large data sets. These developments have led to a general improvement in the quality and quantity of data, which in turn has led to the currently available data being of highly variable reliability. One cannot therefore merely take all paleomagnetic measurements and use them indiscriminately as if they were equally valid representations of the paleomagnetic field at the time a given rock unit was formed. Thus, it is necessary to set up generic selection criteria that can be used to eliminate those data that may be suspect. Naturally, these reliability criteria are not foolproof and additional careful selection may still be required.

6.2.2 Reliability Criteria

With the improving quality of the data several different schemes for judging the reliability of data have been suggested (see e.g., Irving, 1964; McElhinny, 1973a; Khramov, 1987; Van der Voo, 1993). Some hierarchical schemes have been proposed (Briden and Duff, 1981; Li *et al.*, 1990) but these have now been abandoned because results that failed the first hierarchical level were automatically rejected even though they had other excellent attributes that would have survived subsequent levels. The problem that arises in setting up any selection scheme is that even if some criteria are not satisfied, the pole position may still be a valid record of the ancient magnetic field, and conversely those results that satisfy more than the minimum criteria may sometimes turn out to be seriously in error. In practice it is often easier to know when a pole position has been well determined than it is to know with any certainty that it is flawed. Van der Voo (1990a) proposed a useful set of seven reliability criteria that can be used for the selection of paleomagnetic data for tectonic studies. The selection criteria are simple to use and overcome the objections to the hierarchical schemes.

There are three basic criteria for a good paleopole determination: structural control, age of the pole, and the appropriate laboratory treatment of sufficient samples. In the scheme proposed by Van der Voo (1990a) one point is scored for each of the criteria that is judged to have been satisfied in any study. A quality factor Q in the range 0–7 is then assigned to each study and the data can be accepted or rejected according to the overall point score. A summary of the seven criteria is given in Table 6.1.

Van der Voo (1990a) suggests the following standards for the different criteria. For criterion 1 the age for Phanerozoic rock units should be constrained to be within a half-period, such as Late Jurassic or Early Silurian, or to be within a numerical age range of ±4%, whichever is larger. For Precambrian rocks the age limits should be ±4% or ±40 Myr, whichever is smaller. It is important that both the minimum number of samples *and* the minimum precision parameter *and* the minimum error be satisfied (No. 2). The demagnetization procedure must appropriately isolate the various magnetic components (No. 3). Blanket demagnetization treatment is not sufficient for this criterion to be satisfied. Only when vector subtraction is performed as illustrated by orthogonal vector diagrams, by the use of stereonets giving change in direction combined with M/M_0 intensity decay plots, or by principal component analysis can one be

TABLE 6.1
Reliability Criteria for Paleomagnetic Data[a]

No	Brief description
1	Well-determined rock age and a presumption that the magnetization is the same age
2	Sufficient number of samples ($N \geq 24$), k (or K) ≥ 10, and α_{95} (or A_{95}) $\leq 16°$
3	Adequate demagnetization that demonstrably includes vector subtraction information
4	Field tests that constrain the age of magnetization
5	Structural control and tectonic coherence with craton or block involved
6	Documented evidence of the presence of reversals
7	No resemblance to paleomagnetic poles of younger age (by more than a period)

[a]After Van der Voo (1990a).

assured that the magnetic components are isolated as much as possible. Field tests may not always be possible as required by (No. 4), but if such tests are positive they satisfy this criterion. The area studied should clearly belong to the craton or tectonic block involved (No. 5). For orogenic belts, results from intrusives with ages older than the last tectonic phase, or results from thrust sheets that may have been rotated, will generally not satisfy this criterion. The presence of reversals (No. 6) is a powerful test that enough time has elapsed for secular variation to be averaged. Antipodal reversals also preclude any systematic bias caused by an overprint. There should be no suspicion of remagnetization (No. 7), a criterion satisfied when a pole position does not resemble results from rocks of much younger age. A pole position based on a remagnetization is only viable if the age of remagnetization is constrained by independent means.

Few results satisfy all seven criteria; however, Van der Voo (1990a) suggests that a data set satisfying on average most of the criteria ($Q \geq 4$) can be described as more robust than one with $Q = 2$.

6.2.3 The Global Paleomagnetic Database (GPMDB)

The Paleomagnetism Working Group of the International Association of Geomagnetism and Aeronomy, IAGA, has established various databases in paleomagnetism. The most important of these from the perspective of this book is the *Global Paleomagnetic Database*, GPMDB, which contains details of all

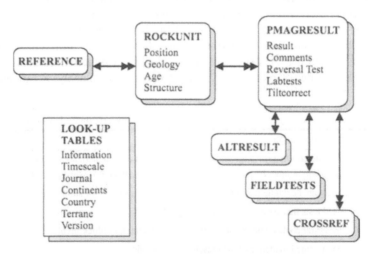

Fig. 6.1. The structure of the Global Paleomagnetic Database (GPMDB). Tables are linked as either one to one (single-headed arrows) or one to many (single- to double-headed arrows) relationships. From McElhinny and Lock (1996), with permission from Kluwer Academic Publishers.

paleomagnetic pole positions determined worldwide since the work of Chevallier (1925). The database was first established in 1991 using the Oracle™ relational database management system for personal computers (McElhinny and Lock, 1990a, b; Lock and McElhinny, 1991; McElhinny and Lock, 1993) and is updated at least every two years and sometimes more frequently. In 1994 it was moved to the new Microsoft Access™ relational database system, which was much better suited to running on a personal computer (McElhinny and Lock, 1996). The database files are available on the internet from World Data Center A in Boulder, Colorado, U.S.A. (*http://www.ngdc.noaa.gov*), or can be queried directly from the internet site at the Norwegian Geological Survey in Trondheim, Norway (*http://dragon.ngu.no/Palmag/paleomag.htm*).

The GPMDB is a fully relational database and it enables researchers to search for data according to various selection criteria. Its structure is shown in Fig. 6.1. However, no judgment or opinion is given regarding any individual entry in the database, except where data are superseded by later studies of the same formation and at the same localities as previous studies. Therefore, it is important that, to assist with data selection, careful use be made of the parameters included in the database.

Most of the important parameters required by the reliability criteria of Van der Voo (1990a), discussed in §6.2.1, are included in the tables ROCKUNIT and PMAGRESULT (McElhinny and Lock, 1990b). The rock age and its estimated error (required for *criterion 1*) are given in ROCKUNIT. The parameters k (*KD* in the database), N, and α_{95} (*ED95* in the database), whose minimum values are required by *criterion 2*, are given in PMAGRESULT. In the same table is a parameter called the Demagcode (*DC*), which summarizes the demagnetization procedures carried out and has integer values 0–5 as explained in Table 6.2. *Criterion 3* can only be satisfied when $DC \geq 3$. A summary of the various tests carried out is given in the column *TESTS* using symbols F (fold test), F* (synfold test), Fs (fold test with strain removal), G (conglomerate test), G* (intra-formational conglomerate test), C (baked contact test), C* (inverse contact test), R (reversals test) and M (rock magnetic tests such as those described in §3.5). When the symbol for the field test is followed by a + sign this indicates that the test is positive as required for *criterion 4*, with a – sign indicating a negative test.

TABLE 6.2

Integer Values of the Demagcode (*DC*) in the Global Paleomagnetic Database

DC	Description
0	No demagnetization carried out. Only NRM values reported.
1	Pilot demagnetizations on some samples suggest stability. Only NRM values reported.
2	Bulk demagnetization carried out on all samples, but no vector diagrams shown.
3	Vector diagrams or stereoplots with M/M_o justify demagnetization procedures used.
4	Principal component analysis (PCA) carried out from analysis of Zijderveld diagrams.
5	Magnetic vectors isolated using two or more demagnetization methods with PCA.

Ra, Rb, and Rc indicate a positive reversal test at levels A, B, and C, respectively, and Ro indicates the test is indeterminate, as discussed in §3.3.6. The presence of reversals as required to satisfy *criterion 6* is therefore quite simply indicated by the symbol R. To show that *criteria 5 and 7* are satisfied requires further study by the investigator.

The table ALTRESULT gives details of the calculation for the mean of the VGPs when available. The FIELDTESTS table provides information and parameters relating to any field tests indicated under the column *TESTS* in PMAGRESULT and CROSSREF gives the cross reference to the result if it has also been detailed in any of the catalogs of paleomagnetic data (such as the Pole Lists that used to be summarized in the *Geophysical Journal of the Royal Astronomical Society*).

6.3 Testing the Geocentric Axial Dipole Model

6.3.1 The Past 5 Million Years

For many years it was customary to illustrate the validity of the GAD model by plotting all paleomagnetic poles determined for the past few million years on the present latitude–longitude grid and showing that they center about the geographic pole (Cox and Doell, 1960; Irving, 1964; McElhinny, 1973a). However, the validity of the GAD model is not correctly deduced from the above observation. Worldwide data from a paleomagnetic field consisting of a series of zonal harmonics (g_1^0, g_2^0, g_3^0, etc.), when plotted in this way, will always produce a set of paleomagnetic poles that center about the geographic pole. The paleomagnetic field is a *time-averaged field* and when this time averaging is performed over time intervals much longer than the longest periods of secular variation it might be supposed that the nonzonal components of the magnetic potential V of (1.1.4) will be eliminated, leaving only the zonal terms ($m = 0$) in the potential. At the Earth's surface $r = a$, so (1.1.4) reduces to

$$V = \frac{a}{\mu_0} \sum_{l=1}^{\infty} g_l^0 P_l^0 (\cos \theta) . \tag{6.3.1}$$

Opdyke and Henry (1969) showed that inclinations obtained from deep-sea sediment cores worldwide were consistent with the GAD model for the past few million years. However, this was only a first-order solution and showed that the GAD was the dominant term. Wilson and Ade-Hall (1970) first noted that the

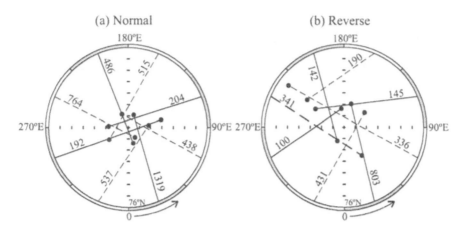

Fig. 6.2. Analysis of global paleomagnetic results for the past 5 Myr by averaging over 45° longitude sectors: (a) normal and (b) reverse. The paleomagnetic poles tend to plot too far away, which typically puts them over the geographic pole. The number of spot readings of the field is indicated for each average. Polar stereographic projection at 76°N. After McElhinny *et al.* (1996).

paleomagnetic poles from Europe and Asia for the past few million years tended to plot too far away from the observation site along the great circle joining the site to the geographic pole. This has been referred to as the *far-sided effect*. Successive analyses by Wilson (1970, 1971), McElhinny (1973a), Wilson and McElhinny (1974), Merrill and McElhinny (1977, 1983), Quidelleur *et al.* (1994), Quidelleur and Courtillot (1996) and McElhinny *et al.* (1996) confirmed that this occurs on a global scale. Whereas in tectonic studies the selection criteria given in Table 6.1 would require $DC \geq 3$ (Table 6.2), for data in the time range 0–5 Ma it has been considered that data with $DC \geq 2$ are acceptable. The far-sided effect is illustrated in Fig. 6.2 from the analysis of McElhinny *et al.* (1996) in which global data for the past 5 Myr have been averaged in 45° longitude sectors. The position of the mean pole is related to the sector in each case, but the average of all eight sector poles still gives the geographic pole within a degree. Note that the reverse data tend to plot further over the pole from the sampling region than do the normal data.

Wilson (1971) introduced the concept of the common-site longitude pole position as a convenient way to analyze the overall far-sided effect. The method is to place all observers at zero longitude by replacing the pole longitude with the common-site longitude given by the difference between the pole and site longitudes. The data shown in Fig. 6.2 for the past 5 Myr have been analyzed in this way and the global mean pole positions are summarized in Table 6.3 either as an overall mean or as a common-site longitude mean. The mean common-site longitude poles for the normal and reverse global data are shown in Fig. 6.3.

TABLE 6.3

Mean Paleomagnetic Pole Positions for the Time Interval 0–5 Ma Both as the Overall Mean and as
the Common-Site Longitude Mean[a]

Data Set	Polarity	N	Overall Pole Position			Common-Site Longitude Pole Position		
			Lat N	Long E	A_{95}	Lat N	Long E	A_{95}
All Data	N+R	6943	88.9	172.8	0.8	86.4	143.7	0.7
	N	4455	89.4	105.4	0.8	87.2	143.9	0.8
	R	2488	87.3	195.7	1.5	85.0	143.6	1.4
Igneous Rocks	N+R	4408	89.0	143.4	0.8	87.4	148.3	1.4
	N	2986	89.4	107.1	0.9	87.6	152.4	0.9
	R	1422	87.7	162.3	1.5	86.8	142.0	1.5

[a]After McElhinny *et al.* (1996). N is the number of sites. A_{95} is the radius of the circle of 95%
confidence about the mean.

When all the data are used (sedimentary and igneous rocks), the means are
discernibly different at the 95% confidence level (Fig. 6.3a), whereas they are
not when only igneous rocks are considered (Fig. 6.3b). Note that the mean poles
always fall to the right of the line joining the zero common-site longitude to the
geographic pole (particularly for the reverse polarity data). Wilson (1971, 1972)
referred to this as the *right-handed effect*. Egbert (1992) has shown that, for
simple statistically homogeneous models of secular variation, the distribution of
VGP longitudes peaks 90° away from the sampling locality. The bias is not
large, but it may contribute to the right-handed effect.

Wilson (1970, 1971) modeled the far-sided effect as originating from an axial
dipole source that is displaced northward along the axis of rotation rather than
being geocentric but was unable to explain the right-handed effect satisfactorily.

(a) All Data

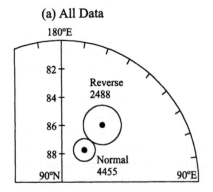

(b) Igneous Rocks Only

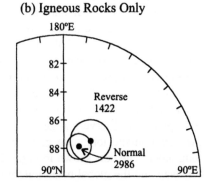

Fig. 6.3. Common-site longitude representation of the normal and reverse data shown in Fig. 6.2
and listed in Table 6.3 with circles of 95% confidence about each mean. Polar stereographic
projection at 80°N. After McElhinny *et al.* (1996).

However, such modeling of the sources of the geomagnetic field is nonunique and there are many possible sources that could equally satisfy the data. It is thus always more appropriate to use spherical harmonics as suggested by (6.3.1). Following James and Winch (1967), the magnetic potential V at the Earth's surface of a dipole of strength m displaced a distance x along the axis of rotation is given by

$$V = \frac{m}{4\pi a^2}\cos\theta + \frac{mx}{4\pi a^3}(3\cos^2\theta - 1) + \dots\dots . \tag{6.3.2}$$

If this is related to the expansion of (6.3.1), namely

$$V = \frac{a}{\mu_0}g_1^0\cos\theta + \frac{a}{\mu_0}g_2^0\left(\frac{3\cos^2\theta - 1}{2}\right) + \dots\dots, \tag{6.3.3}$$

then equating (6.3.2) with (6.3.3) gives

$$g_1^0 = \frac{\mu_0 m}{4\pi a^3} \quad \text{and} \quad g_2^0 = \frac{2\mu_0 mx}{4\pi a^4} . \tag{6.3.4}$$

For a small offset, therefore, Wilson's offset-dipole model is (nonuniquely) equivalent to a geocentric axial dipole (g_1^0) plus a geocentric axial quadrupole (g_2^0), and the displacement x of the offset dipole can be expressed in terms of the zonal coefficients (Wilson, 1970) as

$$x = \frac{g_2^0}{2g_1^0}a . \tag{6.3.5}$$

McElhinny *et al.* (1996) examined in some detail what second-order terms, beyond the geocentric axial dipole term, can realistically be deduced from paleomagnetic data for the past 5 Myr (see also Merrill *et al.*, 1996). Their analysis suggests that there are no discernible nonzonal terms present and that only a geocentric axial quadrupole term can be distinguished with the present data set. This conclusion has been confirmed independently by Quidelleur and Courtillot (1996). However, some authors (Johnson and Constable, 1995; 1997; Gubbins and Kelly, 1993; Kelly and Gubbins, 1997) have attempted a full spherical harmonic analysis of the data and claim that significant nonzonal terms are present. The second-order terms are small in relation to the geocentric axial dipole term and their determination is sensitive to a variety of artifacts of the data that can be shown to be present. A significant problem seems to be the incomplete magnetic cleaning of Brunhes-age overprints in reversely magnetized rocks (Merrill and McElhinny, 1983; McElhinny *et al.*, 1996). Until a significant data set becomes available with the demagnetization procedures and directional analyses represented by $DC = 4$ (Table 6.2), it seems unlikely that further progress can be made. In the absence of such a data set, it seems likely that the apparent differences between the normal and reverse data given in Table 6.3 and shown in Figs. 6.2 and 6.3 arise as a consequence of the current data being inappropriate for the resolution of such second-order problems.

For the time interval 0–5 Ma the best estimate of the geocentric axial quadrupole term is given by the ratio $g_2^0/g_1^0 = 0.038\pm0.012$ (McElhinny *et al.*, 1996). That is, the geocentric axial quadrupole present in the time-averaged paleomagnetic field is less than 4% of the geocentric axial dipole and has the same value for the normal and reverse fields. The presence of such a geocentric axial quadrupole would mean that paleomagnetic poles would be in error by no more than 3–4° if only a geocentric axial dipole field is assumed. This is much less than the typical 95% confidence limits determined for paleomagnetic pole positions. Therefore, the GAD is an acceptable model for paleomagnetic data in the range 0–5 Ma.

6.3.2 The Past 3000 Million Years

Attempts have been made to measure the ratio g_2^0/g_1^0 for epochs older than 5 Ma, but these analyses suffer even more from data artifact problems than the analysis of data for the time interval 0–5 Ma discussed in §6.3.1. It is reasonable to assume that continental drift has been small over the past few million years and that the relation of present continents to the axis of rotation has remained unchanged. For older times it is necessary to reconstruct the continents using sea-floor spreading data and then analyze the paleomagnetic data for each configuration. For Cretaceous and younger epochs these analyses suggest that it is unlikely that the geocentric axial quadrupole term has ever been more than a few percent of the geocentric axial dipole term (Coupland and Van der Voo, 1980; Livermore *et al.*, 1984).

If paleomagnetic poles for a given geological epoch (e.g., the Permian of Europe or North America) are consistent for rocks sampled over a large geologically stable region such as a craton of continental extent, this provides compelling evidence that the *dipole* assumption used in calculating the poles is essentially correct. However, this does not in itself provide any information about whether the dipole is *axial*. Testing the *axial* nature of the field requires some independent measure of paleolatitude as will be described in §6.3.4.

Evans (1976) suggested a test of the dipolar nature of the geomagnetic field throughout the past. For any given magnetic field a definite probability distribution of magnetic inclination, *I*, exists for measurements made at geographical sites uniformly distributed around the globe. This is easily obtained by simply estimating the surface area of the globe corresponding to any set of $|I|$ classes. For example, the dipole field is horizontal at the equator and has $|I| = 10°$

at latitude 5.0°. At the poles the field is vertical and has $|I| = 80°$ at latitude 70.6°. The surface areas of these two zones imply that if sampling is sufficient and geographically random the $0° \leq |I| \leq 10°$ band would make up 8.8% of results and the $80° \leq |I| \leq 90°$ band would make up only 5.7% of results.

The present uneven distribution of land on the Earth's surface and the existence of areas of relatively intense study means that in present-day terms a random geographical sampling of paleomagnetic data has not been undertaken from this uniform distribution. However, over the past 600 Myr or more, considerable movement of continents relative to the pole has taken place and it might be assumed that this has been sufficient to render the paleomagnetic sampling random in a paleogeographical sense. Evans (1976) therefore compared the observed frequency distribution of $|I|$ from paleomagnetic data over the past 600 Myr with that expected for the first four axial multipole fields using 1271 observations. This analysis was later updated by Piper and Grant (1989) using 4787 observations covering the past 3000 Myr. Kent and Smethurst (1998) analyzed 6419 mean inclinations for data covering the past 3500 Ma. The dependence of $|I|$ on colatitude θ for axial multipoles can be obtained from the relationship

$$\tan I_l = \frac{-(1+l)\,P_l}{(\partial P_l / \partial \theta)} \ , \tag{6.3.6}$$

where P_l is the Legendre polynomial of degree l and I_l is the inclination arising from an axial multipole of degree l. The expected frequency distributions of $|I|$ for the first three multipoles are illustrated in Fig. 6.4a.

In their analysis of global paleomagnetic data Kent and Smethurst (1998) used the data listed in the Global Paleomagnetic Database (§6.2.3). They excluded only those results described by the authors in each case as representing secondary magnetization. This rather liberal acceptance criterion allowed a large data inventory, but Kent and Smethurst (1998) argued that any systematic errors introduced through the selection of data that have been remagnetized can be modeled and evaluated statistically. To minimize any bias from the overconcentration of data from any region, the observed values of $|I|$ were averaged within $10° \times 10°$ latitude/longitude areas for eleven geological periods in the Phanerozoic (Neogene, Paleogene, Cretaceous,..., Ordovician, and Cambrian) and in 50-Myr intervals in the Precambrian. This technique results in fewer but more evenly distributed data points and is similar to that used by Evans (1976). These data were then grouped to construct frequency distributions of $|I|$ for the Cenozoic (253 points), Mesozoic (342 points), Paleozoic (352 points), and Precambrian (531 points), a total of 1478 binned data points, or about one-quarter of the 6419 discrete inclinations. For the Phanerozoic the 5142

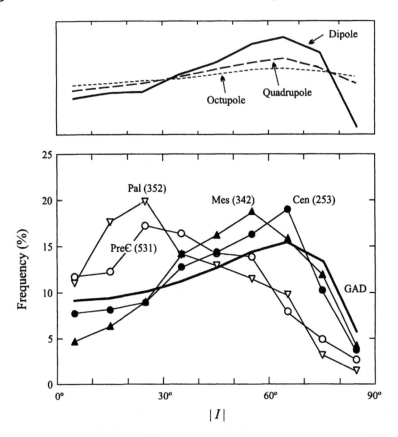

Fig. 6.4. (a) Comparison of the probabilities of observing magnetic inclination |*I*| within 10° latitude bands for the first three axial multipole fields. (b) The observed frequency distribution of |*I*| for the Cenozoic, Mesozoic, Paleozoic and Precambrian compared with that expected for the GAD. The numbers in brackets are the numbers of 10° × 10° latitude/longitude areas averaged in each case. After Kent and Smethurst (1998), with permission from Elsevier Science.

inclination values analyzed by Kent and Smethurst (1998) are represented by 947 bins, compared with the 1271 values analyzed by Evans (1976) in 430 bins.

The observed frequency distributions of |*I*| for the Cenozoic, Mesozoic, Paleozoic, and Precambrian calculated by Kent and Smethurst (1998) are shown in Fig. 6.4b. Data for the Cenozoic and Mesozoic closely resemble the GAD model and a χ^2 test confirms that there is no statistical reason to reject the hypothesis that these data conform with the predicted GAD distributions. However, the data for the Paleozoic and Precambrian are decidedly skewed toward lower values and a χ^2 test confirms that these distributions differ significantly from the expectations of the GAD model. Kent and Smethurst (1998) argue that if the bias to low inclination values for the Paleozoic and

Precambrian is due to contamination by younger magnetizations then one would expect such overprinting to produce frequency distributions more like the GAD-like patterns of the Mesozoic and Cenozoic. They therefore conclude that the paleomagnetic field in the Paleozoic and Precambrian can best be explained by a geomagnetic source model which includes a relatively modest (~25%) contribution to the geocentric axial dipole from a zonal octupole field and an arbitrary zonal quadrupole contribution. An alternative, and perhaps more likely, explanation is that the underlying assumption of random sampling of the globe through continental drift during the Paleozoic and Precambrian is invalid. In this case, if the GAD model is assumed, the results may reflect a tendency for continental lithosphere to have been continuously cycled into the equatorial belt.

6.3.3 Global Paleointensity Variations

The intensity of the Earth's magnetic field varies from the equator to the pole and for a geocentric axial dipole field has a latitude variation given by (1.2.7). This variation may be rewritten as

$$F = \frac{\mu_0 m}{4\pi a^3}(1 + 3\sin^2 \lambda)^{\frac{1}{2}} = F_0(1 + 3\sin^2 \lambda)^{\frac{1}{2}} \quad , \tag{6.3.7}$$

where F_0 is the equatorial value of the intensity of the geocentric axial dipole field, m is the Earth's dipole moment and a is the radius of the Earth (see §1.2.4). If the GAD model is valid in the past, then worldwide paleointensity measurements should show an equator to pole variation given by (6.3.7), in which the intensity at the pole is twice the equatorial value. For more details on the determination of paleointensities in the geological past, readers are referred to Merrill *et al.* (1996).

The determination of paleointensities is a much more difficult problem than the determination of paleomagnetic field directions. This is because the method that is generally considered most reliable requires that the sample be heated and cooled in successively higher temperatures up to the Curie temperature and this can result in chemical changes taking place. Thus often many measurements are made that produce no reliable result (e.g. Kosterov *et al.*, 1998), and the number of determinations that have been made globally is relatively few. Also, it has been found that paleointensities for any epoch can vary quite widely with standard deviations of ~50% of the mean value to be expected (McElhinny and Senanayake, 1982; McFadden and McElhinny, 1982; Kono and Tanaka, 1995; Perrin and Shcherbakov, 1997). Therefore, in order to determine reasonably well-defined mean values for any latitude, it is necessary to make a large number of measurements.

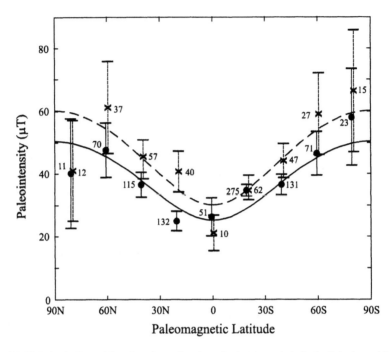

Fig. 6.5. Global paleointensities plotted as a function of paleomagnetic latitude. Paleointensities are averaged over 20° latitude bands. Vertical bars show the 95% confidence limits. The numbers of units used in averaging are indicated. Crosses with dashed vertical error bars are for 0–10 Ma, and solid circles with solid vertical error bars are for 0–400 Ma. The curves represent the best fits for a geocentric axial dipole field; dashed curve for 0–10 Ma, solid curve for 0–400 Ma. From Perrin and Shcherbakov (1997).

The most recent summaries of the latitude variation of paleointensities in the past are given by Tanaka *et al.* (1995) for the time interval 0–10 Ma, and by Perrin and Shcherbakov (1997) for 0–400 Ma. Mean values averaged over 20° latitude bands given by these authors are summarized in Fig. 6.5 and the best fit curves to each data set for a geocentric axial dipole field are drawn through the data. Although all the mean values for each band are shown in Fig. 6.5, it should be noted that Perrin and Shcherbakov (1997) take the view that when the number of units being averaged is less than 20 then the calculated mean value cannot be considered reliable. A chi-square test indicates that the data are consistent with the latitude variation to be expected from a geocentric axial dipole field, so the current data set gives no statistical reason to reject the GAD model. This provides useful confirmation of the acceptability of the GAD model for the past 400 Myr. A subset of the global data set for the time of the Mesozoic dipole low (120–260 Ma) is also consistent with the GAD model (Perrin and Shcherbakov, 1997). Therefore, it appears that the GAD remains a reasonable first-order model irrespective of the variation in the Earth's mean dipole moment. The best fitting curves shown in Fig. 6.5 give the mean geocentric axial dipole moment as 7.8×10^{22} Am^2 for 0–10 Ma and 6.5×10^{22} Am^2 for 0–400 Ma. These may be compared with the present value of 7.9×10^{22} Am^2 for the best fitting dipole and 7.7×10^{22} Am^2 for the axial dipole (g_1^0 term).

6.3.4 Paleoclimates and Paleolatitudes

At the present time the mean annual equatorial temperature is about $+25°C$, and the polar value is about $-25°C$. Although the range in temperature from the equator to the pole may have varied in the past, the simplest model used by paleoclimatologists relates to the fact that the net solar flux reaching the surface of the Earth has a maximum at the equator and a minimum at the poles. Temperature therefore can be expected to follow the same pattern. The density distribution of many climatically sensitive sediments (*climatic indicators*) at the present time shows a maximum at the equator and either a polar minimum or a high-latitude zone from which the indicator is absent (such as reefs, evaporites and carbonates). A less common distribution, seen today in the distribution of glacial phenomena and some deciduous tress for example, has a maximum in polar and intermediate latitudes. Paleoclimatologists have demonstrated over the past few decades that the distribution of *paleoclimatic indicators* can be related to the present-day climatic zones that are roughly parallel with latitude (e.g. Parrish *et al.*, 1982; Witzke, 1990). Therefore, if the expected latitude distribution of various paleoclimatic indicators can be generalized, it becomes possible to use statistical methods to estimate pole positions over geological time (Scotese and Barrett, 1990).

Irving (1956) first suggested that comparisons between paleomagnetic results and geological evidence of past climates could provide a test for the GAD hypothesis over geological time. The essential point is that both paleomagnetic and paleoclimatic data provide evidence for past latitudes and the factors controlling climate are quite independent of the Earth's magnetic field. The determination of the paleomagnetic pole for any region on the GAD assumption enables paleolatitude lines to be drawn across the region (§1.2.3). These are then compared with the occurrences of various paleoclimatic indicators. These comparisons can be made in four distinct ways.

(i)   The time variation of paleolatitude for any single place may be compared with paleoclimatic evidence from that place (Blackett, 1961; Irving, 1956).

(ii)   The space variations of paleolatitude over a given region for a given time may be represented as maps with the paleolatitudes and the occurrences of paleoclimatic indicators compared (Runcorn, 1961; Opdyke, 1962; Irving and Briden, 1962; Drewry *et al.*, 1974).

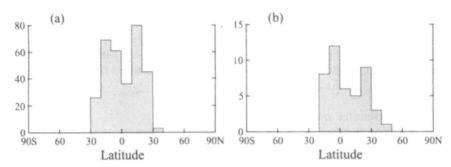

Fig. 6.6. Equal angle latitude histogram for organic reefs. (a) Present latitudes of modern reefs. (b) Paleolatitudes of fossil reefs. After Briden and Irving (1964), with permission from John Wiley & Sons.

(iii) The latitude (or paleolatitude) range of the main occurrences of paleoclimatic indicators may be plotted against geological time (Briden and Irving, 1964).

(iv) The paleolatitude values for a particular occurrence may be compiled in the form of equal angle or equal area histograms to give the paleolatitude spectrum of the particular indicator (Irving and Gaskell, 1962; Irving and Briden, 1962; Briden and Irving, 1964).

Method iv is the most useful approach to adopt. Three of the four examples that follow are from the compilations of Briden and Irving (1964). Although these examples are based on paleomagnetic data available at that time, for the purposes of the comparison they are still reasonably valid because the analyses are based on averaging over $10°$ latitude bands.

The distribution of modern coral reefs is symmetrical about the equator (Fig. 6.6a), the maximum frequency occurring between 10 and $20°$ latitude with most occurrences lying within $30°$ of the equator. The present latitude of fossil reefs does not show this distribution, but when referred to their paleolatitudes (Fig. 6.6b) the spectrum is similar to that of modern reefs, with over 95% of the occurrences falling within $30°$ of the equator. The deposition of evaporites is thought to require high temperatures, the most notable deposits being in the tropical and temperate deserts or semideserts, associated with the dry trade wind belts and with the arid centers of large continents. There are no occurrences recorded north of $53°N$ or south of $43°S$ (Fig. 6.7a) and there are only two regions, one in east Africa and the other in Peru, in which evaporite deposition occurs within $10°$ of the equator. The present latitudes of fossil evaporites through the geological column show a spread from over $30°S$ to over $80°N$, but when referred to their paleolatitudes the distribution condenses into one in which 75% lie within $30°$ of the paleoequator (Fig. 6.7b).

An example that relates to plant distribution is that of coalfields. Coal deposits indicate the existence of moist conditions that could arise from either heavy

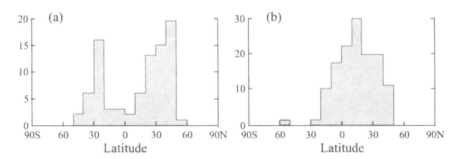

Fig. 6.7. Equal-angle latitude histogram for evaporites. (a) Present latitudes of modern evaporites. (b) Paleolatitudes of fossil evaporites where paleomagnetic data are available. After Briden and Irving (1964), with permission from John Wiley & Sons.

precipitation in warm environments or modest rainfall in cold climates. Coal forms where the accumulation of vegetation exceeds its removal or decay. This occurs either in hot rain forests where, although decay is rapid, growth rates are high, or in cold environments where, although growth may be less rapid, decay is inhibited by cold winters. The paleolatitude spectrum of coalfields shows this effect in Fig. 6.8, where the hemispheres are plotted together using a histogram in which the class intervals vary as sin λ_p (where λ_p is the paleolatitude) and therefore contain equal areas of the Earth's surface. The low paleolatitude group is mainly the Carboniferous coals of western Europe and North America, whereas the high paleolatitude group is mainly the Permian and younger coals from Canada, Siberia, and the southern continents. The two groups also contain distinct fossil floras.

An example of the paleolatitude distribution of a faunal group has been given by Irving and Brown (1964) for labyrinthodont reptiles. For large groups of widely distributed organisms, taxonomic diversity should be a maximum at or near the equator with a decrease into higher latitudes (Irving and Brown, 1964). The study of paleowind directions determined from cross stratification in eolian sandstones is another paleoclimatic indicator that can be compared with paleolatitudes derived from paleomagnetism. Opdyke and Runcorn (1960) and Opdyke (1961) have shown that in the upper Paleozoic eolian sandstones of Europe and North America, the directions observed correspond to a trade wind belt relative to the paleomagnetic equator.

The trade wind belts of the present day lie within 30° of the equator. The present day latitudes of Phanerozoic eolian deposits show a much wider scatter than this. However their paleolatitudes place most of them within 30° of the paleoequator (Drewry *et al.*, 1974). Other paleoclimatic indicators that have been studied include oil fields (Irving and Gaskell, 1962), organic siliceous ooze or its diagenetic equivalent radiolarian chert (Drewry *et al.*, 1974), and Phanerozoic glacial deposits (Blackett, 1961; Drewry *et al.*, 1974). These glacial deposits

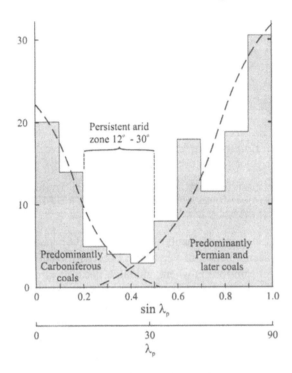

Fig. 6.8. Equal-area histogram of coal deposits. Paleolatitude values are plotted irrespective of sign. The dashed lines show the possible division into two groups centered near the equator and poles, respectively. From Irving (1964), after Briden and Irving (1964), with permission from John Wiley & Sons.

show the expected high paleolatitude concentration, although it has been suggested that some Precambrian glacial deposits were formed near the equator (Schmidt *et al.*, 1991; Schmidt and Williams, 1995; Park, 1997; Evans *et al.*, 1997).

The basic assumption made by paleoclimatologists is that ancient climatic belts run essentially parallel to ancient lines of latitude. Today, however, there are wide departures from such a pattern caused by perturbations in planetary atmospheric circulation produced by the distribution of land and sea, the so-called monsoon effect. Because the distribution of land and sea was markedly different in the past, there could be quite considerable departures from the present-day climatic belts and from simple parallelism. For example, if the most commonly accepted Triassic and Permian reconstructions are assumed (see chapter 7), then the low-pressure equatorial belt responsible for the monsoon would have been substantially displaced (Robinson, 1973).

The occurrence of phosphate deposits is an example of the way in which the distribution of land and sea plays a critical part in their formation. For a

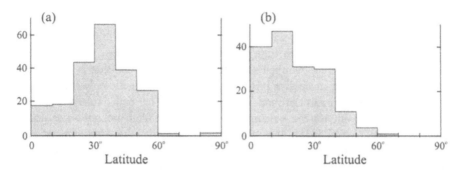

Fig. 6.9. Equal-angle latitude histogram for Phanerozoic phosphate deposits. (a) Present latitudes of all deposits. (b) Paleolatitudes of those deposits where paleomagnetic data are available. Updated from Cook and McElhinny (1979).

phosphorite to form, the coastal portion of a continent must drift into a low-latitude location and therefore the formation of phosphorites is not uniform as a function of time (Cook and McElhinny, 1979). Sheldon (1964) and Cook and McElhinny (1979) have examined the paleolatitudes of Phanerozoic phosphate deposits and have shown that they occur mainly within $40°$ of the equator (Fig. 6.9) as is observed for young phosphorites.

Thus, at least through Phanerozoic time, the paleolatitudes determined paleomagnetically under the assumption of the GAD hypothesis have been shown to be consistent with the paleolatitudes derived from a wide variety of independent indicators. This provides compelling evidence of both the axial *and* the dipole nature of the paleomagnetic field.

6.4 Apparent Polar Wander

6.4.1 The Concept of Apparent Polar Wander

From a tectonic perspective one of the purposes of paleomagnetic investigations is to provide information relating to the way in which parts of the Earth's crust have moved apart or come together during Earth history. In order to achieve this, it is necessary to have paleomagnetic data from localities well distributed both in space and in time and to have a method by which these data can be presented and analyzed.

There are two principal ways of presenting paleomagnetic information for a given region over several epochs. For each epoch a paleogeographic map of the region can be drawn showing the directions of the paleomeridians and lines of latitude across the region. These maps are generally found to be most useful by climatologists for comparison with relevant information, but a simple view of the overall variation from time to time is not easy because several maps, one for each epoch, are required. A much simpler approach, and one that is inherently more useful, is to plot successive positions of the paleomagnetic pole from epoch to epoch on the present latitude–longitude grid. The path, relative to the continent, traced out by the paleomagnetic pole is called the *apparent polar wander path* (APWP) for that region. To provide the optimum visual discrimination, three projections have been used here when plotting APWPs: equal angle, equal area, and Mollweide. It should be noted that the procedure adopted is to *assume* the region fixed and plot the movement of the pole. This relative polar movement could represent polar wandering or continental drift and the data from a single region cannot distinguish between the two. In general, it is necessary to have information for the entire globe to isolate polar wander from continental drift (§7.5.1).

6.4.2 Determining Apparent Polar Wander Paths

The first estimates of apparent polar wandering were made by Creer *et al.* (1954) relative to the British Isles. To draw apparent polar wander paths it is simplest to consider a continent presently lying in the northern hemisphere. For the late Cenozoic a paleomagnetic pole will be determined lying closest to the north or south geographic pole. Obviously the pole lying closest to the present north pole is regarded as the north paleomagnetic pole, even though this is strictly indeterminate due to reversals of the geomagnetic field. During the early Cenozoic the pole might lie $20°$ away from the north geographic pole, and its antipole will lie $160°$ away. The principle of minimum movement is applied. It is always assumed that the path is the one found by joining successive poles through the shortest distance, going back in time, from the present geographic pole. There are thus two paths, one starting at the north geographic pole and the other starting at the south geographic pole representing the path of the south pole. If the north (south) pole wanders through an angle greater than $90°$ it could cross the present equator and lie in the southern (northern) hemisphere. In these circumstances, if the path were not known it might then quite wrongly be assumed that the pole that lies in the northern (southern) hemisphere was, by definition, the north (south) pole. This would result in the polarity of the rock magnetization being wrongly defined.

To plot the apparent polar wander path for a continent or region requires some method to average the paleomagnetic poles that have been determined for that region for the same epoch. A problem that arises is that each of the paleomagnetic poles will have been determined from different numbers of sites. How does one decide whether the poles have each been determined from a sufficient number of sampling sites such that the secular variation has been averaged out so that the GAD model is truly valid in each case, or that they are determined accurately enough to be included in any analysis? This is a problem that has been the subject of much debate over the years. In the past most workers have taken the simple approach and taken the mean of all the paleomagnetic poles that pass some minimum sampling and statistical criteria such as those of criterion 2 in Table 6.1.

Each paleomagnetic pole can be considered to be a single group mean determined from the paleomagnetic directions or VGPs at several sites. McFadden and McElhinny (1995) suggested a method by which the results from m groups of mean paleomagnetic directions or poles can be combined where the groups have widely different numbers and where the resultant vectors for the individual groups are not known. The advantage of the method is that one does not have to consider whether each group by itself is sufficient to average secular variation, and even a single VGP or direction can be included as one of the groups.

Assuming a Fisher (1953) distribution for the underlying population, suppose that n_i is the number of sites at the ith of m groups of sites, such that the total number of sites is $N = \Sigma n_i$ and let R be the (unknown) length of the vector resultant of the N unit vectors. An unbiased estimate of the precision κ is the value of k given by (3.2.8). Alternatively, given k, one can calculate an equivalent resultant length R_k, where

$$R_k = N - \frac{N-1}{k} \; . \tag{6.4.1}$$

If the individual values r_i (the resultant lengths of the n_i unit vectors for each of the m groups) are known and their corresponding direction cosines are given by x_i, y_i, z_i, it follows that

$$R^2 = \left(\sum_{i=1}^{m} r_i x_i\right)^2 + \left(\sum_{i=1}^{m} r_i y_i\right)^2 + \left(\sum_{i=1}^{m} r_i z_i\right)^2 \; . \tag{6.4.2}$$

Although it is not true for the individual group observations that

$$r_i = n_i - \frac{n_i - 1}{k} \; , \tag{6.4.3}$$

it is true when averaged over the populations. Therefore, by setting

$$(R_{sum})^2 = \left(\sum\left(n_i - \frac{n_i-1}{k}\right)x_i\right)^2 + \left(\sum\left(n_i - \frac{n_i-1}{k}\right)y_i\right)^2 + \left(\sum\left(n_i - \frac{n_i-1}{k}\right)z_i\right)^2$$

a consistent estimate is obtained if k is chosen such that

$$R_{sum} = R_k .$$

(6.4.4)

This can be achieved by numerical iteration using some starting value of k until (6.4.4) holds. Having determined k (equivalent to determining R), the circle of 95% confidence may then be determined from (3.2.13).

   In the following sections of this chapter the above method is used to calculate mean paleomagnetic poles to enable APWPs to be determined for each of the major blocks for any given time interval. Other methods have been tried in the past, such as using a moving window of fixed width (Irving, 1977) or using cubic splines (Thompson and Clark, 1981) on the assumption that the path can be fitted by a smoothly varying function. When there are many data available the differences between methods will be minimal, but both moving windows and cubic splines are smoothing techniques that can oversimplify the resulting path when there are few data, when data are contaminated from remagnetization, or when the path is complex.

   Gordon *et al.* (1984) noted that APWPs generally consist of long, gently curved segments they termed *APWP tracks* linked by short segments with sharp curvature (cusps). The tracks correspond to time intervals when the direction of plate motion was essentially constant and the cusps correspond to time intervals when the direction of plate motion was changing. APWP tracks, like hot spot tracks, mark the motion of a plate with respect to some fixed point. In the case of hot spot tracks this point is the position of a rising mantle plume, and in the case of APWP tracks the point is the paleomagnetic pole, representing the geographic axis. In both cases the nature of the motion means that they tend to lie along small circles. APWP segments can therefore be described by rotation about some pivot point lying on the normal to the small-circle plane and a rotation rate measured in degrees per million years. This pivot point was termed a *paleomagnetic Euler pole* (PEP) by Gordon *et al.* (1984). They noted that parts of the then known APWP for North America could be represented in such a way over periods of 80–100 Myr. Van der Voo (1993) has subsequently noted that significant departures from such small circle trends are now evident. Therefore, increasingly smaller small-circle segments will ultimately be needed for a complete description of APWPs by this method. Modern-day plate motions, as revealed by sea-floor spreading, have not been constant for such long times. If intervals of constant plate motion are only 20 Myr, the usefulness of the PEP method will be much reduced because the resolution of paleomagnetic data is much less than that required to measure the trends of APWP tracks of short duration.

6.4.3 Magnetic Blocking Temperatures and Isotopic Ages

In paleomagnetism the concept of blocking temperature (§2.3.3) was introduced by Néel (1949, 1955) and has played a fundamental role since those early days of the development of the subject. It was introduced into geochronology somewhat later in order to assist in the interpretation of mineral ages (Armstrong, 1966; Harper, 1967; Jäger *et al.*, 1967). In geochronology the concept corresponds exactly to that in paleomagnetism (York, 1978). As the system cools through a critical blocking temperature it becomes blocked against thermal agitation at lower temperatures. Above this temperature all evidence of its history is destroyed. However, in geochronology confusion arises from the use of the term "blocking temperature" because the closure of a new system during initial cooling and the opening of an established system during a heating event need not necessarily occur at the same temperature. Therefore, it is usual to adopt the term *closure temperature* to discuss the cooling of isotopic systems (Dodson and McClelland Brown, 1985).

Of particular importance in paleomagnetic studies is the relation between magnetic blocking temperature and isotopic closure temperature because at least in principle, one should be able to relate isotopic ages determined in different ways with paleomagnetic directions determined over differing temperature intervals during thermal demagnetization. Unfortunately, the blocking temperature spectrum of a single magnetic mineral can vary widely up to the Curie temperature depending on the shape and size of the grains. In isotopic systems the closure process is not instantaneous. In a hot, fully open system, the daughter product of the radioactive decay process is lost as soon as it is formed. On cooling, the fully open system changes gradually to a fully closed system passing through a transitional temperature range in which the daughter product is partly lost and partly retained. The definition of closure temperature is therefore the temperature of the system at the time represented by its apparent age (Dodson, 1973).

Theoretical analysis of the closure process (e.g., Dodson and McClelland Brown, 1980) has shown that, in a wide range of kinetically controlled systems, the closure temperature T_{cl} is related to the cooling time constant τ_i by the equation

$$\tau_i = \frac{1}{Af_i}\exp\left(\frac{E}{\gamma T_{cl}}\right), \qquad (6.4.5)$$

where A is a geometry-dependent factor, f_i is a frequency factor, E is the activation energy, and γ is the universal gas constant. The cooling time constant τ_i is the time required for the value of $\exp(-E/\gamma T)$ to reduce to $1/e$ of its value in the transitional temperature range. Thus, τ_i given by (6.4.5) is similar to the

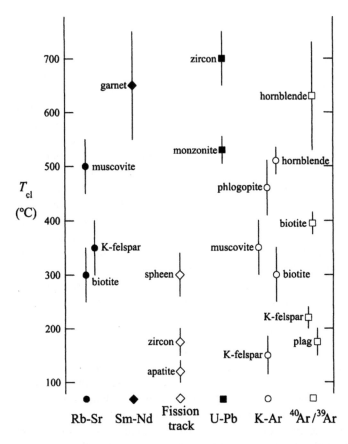

Fig. 6.10. Estimates of closure temperature T_{cl}, using different isotopic dating methods, following Berger and York (1981) for $^{40}Ar/^{39}Ar$ dating and Dodson and McClelland Brown (1985) for other methods.

paleomagnetic relaxation time τ given by (2.3.14). In isotopic systems the closure temperature T_{cl} depends on the particular mineral species, grain size, and dating method used. In magnetic systems the blocking temperature T_B depends on the mineral species and is extremely sensitive to grain size.

Dodson and McClelland Brown (1985) have summarized possible estimates of closure temperatures in various isotopic systems (adjusted for a cooling rate of 30°C Myr^{-1}) and Berger and York (1981) made estimates for the $^{40}Ar/^{39}Ar$ method (using a cooling rate of 5°C Myr^{-1}). These values are plotted in Fig. 6.10, in which it should be noted that the $^{40}Ar/^{39}Ar$ values need to be revised upwards slightly for comparison with other methods. Uncertainties in diffusion size and strong compositional dependence mean that much of the data should be regarded as tentative. However, the data shown in Fig. 6.10 provide some idea of the possible relation between magnetic blocking temperatures and isotopic ages determined using different minerals and dating methods. Ages determined from Sm–Nd, U–Pb on zircon or ^{40}Ar/^{39}Ar on hornblende have closure temperatures above the Curie temperature of magnetite but similar to that of hematite. In all other cases the closure temperatures fall below these Curie temperatures. These factors need to be considered when relating isotopic ages to the various magnetic components determined from thermal demagnetization experiments and using them to plot apparent polar wander paths. It should be noted that the application of isotopic closure temperatures as a means of placing ages on secondary components of magnetization is still in its infancy and considerable caution is necessary in attempting to use this method. This point is pursued in §7.4.2.

6.5 Phanerozoic APWPs for the Major Blocks

6.5.1 Selection and Grouping of Data

The method described in §6.4.2 has been applied to data published up to the end of 1997 and summarized in the GPMDB (§6.2.3). The data are selected using the criterion that $Q \geq 3$, where Q is the quality index of Van der Voo (1990a). Because of space limitations a list of the data used in these analyses is not provided in this book, but Van der Voo (1993) provides such a list up to about the years 1990/91. The mean poles listed for each of the major blocks are essentially an update of those determined by Van der Voo (1993). The resulting APWPs can be considered as reference APWPS for these major blocks. One important difference is that the large amount of data available from the former Soviet Union is now summarized in the GPMDB and these data are incorporated in the analyses that follow. The Phanerozoic time scale used is a modified version of the *Geologic Time Scale 1989* of Harland *et al.* (1990). It takes into account the change in the age of the Precambrian/Cambrian boundary from 570 to 545 Ma, and the resulting modifications to the boundaries of the periods from Cambrian through to the Devonian as given by Tucker and McKerrow (1995). A summary of the period boundaries using this modified time scale is given in Table 6.4.

Each paleomagnetic pole is assigned a numerical age range based on either the available isotopic ages or the time scale of Table 6.4 when fossils and stratigraphic considerations are used. A median age corresponding to the middle point of the age range is then assigned to each pole. The time groupings for the paleomagnetic poles are somewhat arbitrary and are slightly different from those

TABLE 6.4
Modified version of the *Geologic Time Scale 1989* used in the GPMDB[a]

Epoch	Top (Ma)	Base (Ma)
Quaternary	0.0	1.8
Tertiary	1.8	65
Cretaceous	65	146
Jurassic	146	208
Triassic	208	245
Permian	245	290
Carboniferous	290	363
Devonian	363	417
Silurian	417	443
Ordovician	443	495
Cambrian	495	545

[a]Harland *et al.* (1990) modified using early Paleozoic period boundaries of Tucker and McKerrow (1995).

of Van der Voo (1990b, 1993). Rates of apparent polar wander are generally in the range 0.1–0.2° Myr^{-1} so that over 30 Myr changes of between 3 and 6° in the mean pole position can be expected. Van der Voo (1990b) therefore grouped poles in intervals of about 25 Myr or about half a period. This reasoning has been followed here, with the choice of groups being based on the most obvious changes in polar position.

6.5.2 North America and Europe

North America and Greenland
North America and Greenland together once formed part of a combined North America–Greenland continent called *Laurentia* (Fig. 6.11). Greenland is an old Precambrian craton that separated from North America when the Labrador Sea opened at about 90 Ma (see Table 5.2). The main North American craton is surrounded by a disturbed cratonic margin and by displaced (or displacing) terranes in the west, east, and south. The disturbed cratonic margin includes the fold and thrust belts of the Appalachians, the Ouichita Mountains in Arkansas and Oklahoma, and the Rocky Mountains. Structural control and tectonic coherence with the main craton as required by criterion 5 in Table 6.1 is generally not satisfied. Data from the displaced terranes have only been included in calculating mean poles when relative movements associated with these terranes are considered to have ceased. Movements of the displaced terranes along the western margin and Alaska (e.g., *Alexander, Peninsular, Stikine,* and *Wrangell* as shown in Fig. 6.11) and in Mexico to the south continued well into the Tertiary. For the displaced terranes to the northeast (e.g., *West Avalon*), data have been included for Devonian and younger times, whereas for those to the southeast (e.g., Outer Piedmont) data have been included for Late Carboniferous

Fig. 6.11. Tectonic framework of North America and Greenland showing the Laurentian craton and its disturbed margin (diagonal lines) and some of the surrounding displaced terranes (A, Alexander; F, Florida; OP, Outer Piedmont; P, Peninsular; S, Stikine; W, Wrangell; WA, West Avalon). CP is the Colorado Plateau.

and younger times. Note that the Avalon terrane is usually referred to as West Avalon because East Avalon is found in southern Britain (see Fig. 6.13). Further discussion of some of these displaced terranes and their history and evolution with respect to the major blocks is given in §7.3. The data selection from North America essentially follows that of Van der Voo (1990b) and the results presented here may be considered as a straightforward update from that time.

The Colorado Plateau (see Fig. 6.11) is an isolated rigid block whose coherence with the main craton has been disputed. However, comparison of paleomagnetic data from the Colorado Plateau with corresponding data from the Laurentian craton indicates that the plateau has rotated a few degrees clockwise with respect to the main craton as originally proposed by Hamilton (1981). Steiner (1986, 1988) and Bryan and Gordon (1986, 1990) have made differing estimates of the parameters for this rotation. Kent and Witte (1993) apparently resolved this difference through a comparison of Norian age poles between the plateau and the main craton. However, Molina Garza *et al.* (1998) have shown that the ages of the poles used from the Colorado Plateau are not as accurate as was supposed by Kent and Witte (1993). A comparison of the entire Colorado Plateau data set from Late Carboniferous through Late Jurassic with corresponding poles from the Laurentian craton gives a best estimate of the Colorado Plateau rotation as 7.4 ± 3.8°. Since the paleomagnetic poles from the Plateau form a significant part of the North American database, all poles from the Plateau and its surrounding areas in Colorado, Arizona and New Mexico have been rotated using an Euler pole at 34N, 105W through an angle of +7.4° (counterclockwise; see §5.1.1 for Euler rotation conventions). Kent (1988) made an estimate of the rotation of thrust sheets in the northern limb of the Pennsylvania Salient in the Appalachians. Following Kent (1988) paleomagnetic poles from the northern limb have therefore been rotated about an Euler pole at 41N, 77W through an angle of +23° (counterclockwise).

Unfortunately, there are so few data from Greenland that it is not possible to construct a separate APWP. Therefore, these data have been included with the North American data to construct the APWP for Laurentia after allowing for the closure of the Labrador Sea following Bullard *et al.* (1965) (rotation about an Euler pole at 65.8N, 92.4W through an angle of -19.7° clockwise). The full

TABLE 6.5

Phanerozoic APWP for Laurentia with Paleomagnetic Poles in Each Time Interval Averaged Using the Method that Takes into Account the Varying Number of Sites on Each Pole[a]

Geological period	Age range (Ma)	Mean (Ma)	m	N	Mean pole position Lat N	Long E	A_{95} (°)
Eocene	35-56	45	14	401	80.0	158.0	3.7
Paleocene	56-65	60	19	395	75.2	168.4	3.9
Late Cretaceous	65-97	80	12	225	72.3	194.8	3.7
Early Cretaceous 2	97-130	115	7	192	72.7	191.1	2.1
Early Cretaceous 1	130-146	135	5	144	64.0	203.4	3.9
Late Jurassic	146-160	155	6	60	66.5	155.1	8.3
Middle Jurassic	160-180	170	7	221	64.5	111.9	6.6
Early Jurassic	180-208	195	25	438	62.9	90.3	2.8
Late Triassic	208-225	215	17	299	55.9	93.9	2.1
Middle Triassic	225-240	235	11	333	54.4	104.4	1.7
Late Permian/Early Triassic	240-256	250	13	100	48.9	117.5	2.9
Early Permian	256-280	270	18	211	43.6	124.9	2.0
Late Carboniferous/Early Permian	280-300	290	15	144	37.3	132.3	3.5
Late Carboniferous	300-330	315	23	250	32.8	128.9	4.1
Early Carboniferous	330-363	345	13	107	29.9	130.1	4.7
Late Devonian	363-377	370	7	35	29.1	121.4	3.6
Middle Devonian	377-391	385	7	51	21.6	117.3	7.6
Late Silurian/Early Devonian	391-423	405	6	38	2.8	97.5	6.3
Early/Middle Silurian	423-443	435	3	32	21.3	131.6	10.4
Late Ordovician	443-458	450	4	29	22.2	129.1	6.4
Early/Middle Ordovician	458-495	475	6	37	17.7	154.2	9.0
Late Cambrian	495-505	500	8	59	-1.9	165.0	8.6
Early Cambrian	525-545	535	7	92	4.3	154.7	10.1

[a]From McFadden and McElhinny (1995). *m*, number of paleomagnetic poles; *N*, total number of sites; A_{95}, radius of circle of 95% confidence about the mean pole position. The mean age in each case is given to the nearest 5 Myr.

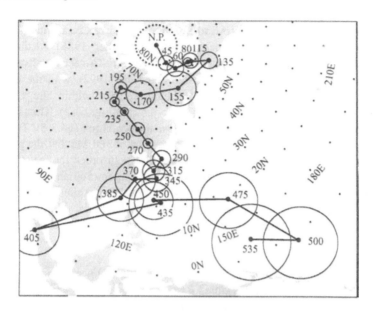

Fig. 6.12. Phanerozoic APWP for Laurentia using the mean pole positions given in Table 6.5. Each mean and its circle of 95% confidence is labeled with the mean age in millions of years.

rotation applies to data for times before 90 Ma, and after that time the rotations from Olivet *et al.* (1984) have been used. Triassic poles from the Fleming Fjord and Gipsdalen Formations of Greenland are discordant with those from North America after closing the Labrador Sea, differing by more than 15° from their well-grouped North American equivalents. Van der Voo (1990b) considered these poles as unreliable, but new results (Kent and Clemmensen, 1996) confirm the validity of the old data, so it appears that there must be some unresolved tectonic problem associated with the sampling region. Therefore these results have not been included in the analysis that follows.

The choice of time intervals for calculating mean pole positions is somewhat arbitrary but is made on the basis of the consistency of poles within any given time interval. For the most part these time intervals correspond to a half period or less, although this is increased to whole periods in the early Paleozoic. The resulting paleomagnetic poles have been averaged in the time groupings summarized in Table 6.5 and the corresponding APWP for Laurentia for the Phanerozoic is shown in Fig. 6.12. The analysis of 253 studies representing data from 3893 sites is summarized in Table 6.5 providing a substantial data set.

The general shape of the APWP is still similar to that originally determined by Irving (1956) and Runcorn (1956), and over the subsequent 25 years much more detail has become available (Van der Voo, 1981). There is a cusp in the APWP in the Early Jurassic (195 Ma) and this is followed by rapid apparent polar wander during the time interval Middle Jurassic to Early Cretaceous (170–115 Ma) as has previously been documented (May and Butler, 1986). At the end of this period of rapid polar wander there is a mid-Cretaceous "standstill" (Globerman and Irving, 1988; Van Fossen and Kent, 1992), with the circle of confidence for the 80-Ma mean pole being entirely within the circle of confidence of the 115-Ma mean pole (Table 6.5 and Fig. 6.12). This standstill appears to coincide with the Cretaceous Normal Superchron (§4.3.6). The mean pole position at 465 Ma returns close to its position at 360 Ma, but the poles are significantly different at the 95% confidence level. In order for the APWP to move from its 415-Ma position to its Cambrian position (500 and 535 Ma), the path must inevitably pass close to the previous 360-Ma position. Thus, it seems unlikely that remagnetization has occurred. In the Cambrian, the mean poles at 500 and 535 Ma are not discernibly different at the 95% confidence level.

Europe
Europe is a composite continent as illustrated in Fig. 6.13. The main European craton during Paleozoic times is called *Baltica*. It consists of Scandinavia and the Russian Platform, which lies between the Urals and the Paleozoic or younger mobile belts to the west. The rocks of these belts either occur at the surface or are buried in Denmark, Germany, Poland, western Podolia (Ukraine), Romania, and Bulgaria. In Scandinavia its disturbed margin includes most of the Caledonian belt of Norway, whereas in other parts of western Europe either there is no apparent disturbed margin or else it is mostly buried. The northern part of the British Isles (Scotland and the northern part of Ireland) is believed to have formed part of Laurentia during the Paleozoic, with the old Precambrian Lewisian basement being formerly contiguous with Greenland. During the early Paleozoic (Ordovician–Silurian times) Baltica and Laurentia collided so that pre-Late Silurian results from the northern British Isles are not necessarily representative of Baltica. *North Britain* may thus be considered as an early Paleozoic displaced terrane (Fig. 6.13), which can be reconstructed against Greenland by closure of the Atlantic Ocean. Therefore, when constructing the APWP for Europe for the purposes of testing the Mesozoic opening of the Atlantic Ocean, for pre-Late Silurian times the results from the North Britain terrane are the only ones that can be considered.

   The region to the south of the North Britain terrane, including almost all of England, Wales, and southern Ireland, represents another displaced terrane or group of terranes that was last affected by the Caledonian orogeny of Silurian to Early Devonian age. The late Precambrian basement of this region is sometimes referred to as the *Midland craton* and is now mostly buried beneath shallow younger rocks (Fig. 6.13). It had a drift history independent of Baltica and Laurentia until possibly the Late Silurian, when it collided with Laurentia. Therefore, only latest Silurian and younger results (<425 Ma) from the Midland

Fig. 6.13. Tectonic framework of Europe, showing Baltica (Baltic Shield and the Russian Platform), the North Britain terrane that was formerly part of Laurentia, together with various terranes in the Caledonian, Hercynian and Alpine belts. Note that the Alpine belt contains older Hercynian massifs, just as the Caledonian and Hercynian belts contain older Precambrian elements. AM, Armorican Massif; AR, Ardennes–Rhenish; BM, Bohemian Massif; EA, East Avalon (southern Britain); MC, Massif Central; V, Vosges.

craton and its margins are included in the results analyzed to produce the European APWP.

Another group of displaced terranes occurs to the south and east of the Midland craton that were last affected by the Hercynian orogeny of Carboniferous age. The southern part of Britain is referred to as the *East Avalon* terrane and is considered to be an easterly section of the West Avalon terrane of North America (Fig. 6.11). These Hercynian displaced terranes have been referred to collectively as *Armorica* following Van der Voo (1982, 1988). This designation does not imply that together they formed a coherent block, rather it implies that these terranes were formerly closely associated with one another. Results from the Armorican Massif will be discussed in more detail in §7.3.3. The Hercynian front runs from southern Ireland through southernmost England to Belgium, Germany, and southern Poland (Fig. 6.13). Therefore, only latest

Carboniferous and younger results (< 300 Ma) from these terranes have been
included in the European APWP. South of Hercynian Europe, across the Alpine
front, lie displaced terranes that form part of Alpine Europe. Alpine Europe
contains large Hercynian basement massifs (e.g., Iberian Meseta and Corsica–
Sardinia) that have not been affected by Alpine deformation but have themselves
been displaced with respect to the rest of Europe. These are discussed further in
§7.3.4. Because displacements in Alpine Europe often occurred well into the
Tertiary, results from this region are not considered in establishing the APWP
for Europe.

The first APWP for Europe was compiled by Creer *et al.* (1954) based on data
from Great Britain and, until the advent of plate tectonics in 1968, was
continually updated in successive reviews assuming that Europe was a single

TABLE 6.6

Phanerozoic APWP for Stable Europe (Separating North Britain and Baltica for >425 Ma)[a]

Geological period/epoch	Age range (Ma)	Mean (Ma)	m	N	Mean pole position Lat N	Long E	A_{95} (°)
Eocene	35-56	45	10	947	82.0	168.9	2.0
Paleocene	56-65	60	11	1085	77.3	164.4	3.1
Late Cretaceous	65-97	80	9	79	71.5	151.8	4.1
Early Cretaceous	97-146	120	6	55	70.5	206.5	7.3
Late/Middle Jurassic	146-176	160	11	97	73.8	144.0	3.8
Early/Middle Jurassic	176-208	195	10	125	66.1	128.0	8.6
Late Triassic	208-225	215	6	39	53.2	136.3	11.1
Early/Middle Triassic	225-245	235	13	106	50.7	150.0	5.7
Late Permian	245-270	255	48	535	46.9	163.2	1.6
Early Permian	270-290	280	37	308	44.9	165.5	1.9
Permian (Western Europe only)	(245-290)	(265)	53	434	46.6	163.0	1.8
Permian (Russian Platform only)	(245-290)	(265)	32	409	45.7	165.1	1.6
Late Carboniferous	290-323	305	19	268	36.3	168.1	4.8
Early Carboniferous	323-363	345	15	146	27.2	157.8	7.1
Late Devonian	363-377	370	9	77	23.0	161.4	4.9
Middle Devonian	377-391	385	12	64	20.1	157.1	7.5
Early Devonian	391-417	405	26	377	2.7	142.1	4.2
Late Silurian	417-425	420	11	155	2.6	164.0	4.9
North Britain							
Early/Middle Silurian	425-443	435	12	126	19.5	164.2	5.1
Late Ordovician	443-458	450	4	46	10.1	186.8	5.1
Early/Middle Ordovician	458-495	475	3	28	9.1	205.5	7.2
Baltica							
Early/Middle Silurian	425-443	435	7	78	16.7	177.2	9.3
Late Ordovician	443-458	450	2	15	2.9	174.4	13.6
Middle Ordovician	458-470	465	3	22	-8.5	231.6	9.1
Early Ordovician	470-495	480	4	52	-20.5	230.1	5.8
Middle/Late Cambrian	495-530	510	1	10	-31.0	266.0	-

[a]Symbols and method of averaging are as given in Table 6.5.

block. The realization that plate tectonics changed what was then the conventional view of the tectonic history of Europe to that outlined above led to the first experiments in the British Isles to try to delineate terrane boundaries during the Paleozoic. Paleomagnetic traverses across the British Isles produced a series of papers culminating in the review by Briden *et al.* (1973) that was the benchmark for paleomagnetic studies in Europe. Since that time further important reviews include those of Briden and Duff (1981), Torsvik *et al.* (1990), Torsvik and Trench (1991), and Trench and Torsvik (1991) and Torsvik *et al.* (1992).

Using the tectonic constraints described above, mean poles for "Stable Europe" for successive time intervals through the Phanerozoic are listed in Table 6.6 and the resulting APWP is illustrated in Fig. 6.14. For times >425 Ma the data are split into separate sets for North Britain and Baltica. As pointed out by Van der Voo (1990b, 1993), the distribution of poles for the Late/Middle Jurassic of Europe is strongly bimodal for reasons that are not understood, so they have been averaged to produce the mean at 160 Ma. Note that for times prior to 425 Ma, data from North Britain and from Baltica are listed separately. The data from North Britain are relevant when its reconstruction with Laurentia is being considered and those from Baltica are relevant to the past relationship between Baltica and Laurentia to be discussed in §7.4. The analysis of 289 studies representing 4840 sites is summarized in Table 6.6, a substantial data set that is somewhat larger than that from North America and Greenland.

The mean pole positions listed in Table 6.6 contain many data for the Russian Platform as summarized in the GPMDB. A point of concern has always been how these data may be compared with their western counterparts. Russian data have generally not been published in full and are only reported in summary reports and catalogs compiled by A.N. Khramov at the VNIGRI Institute in St. Petersburg. It is the data from these compilations that are included in the GPMDB and in this form it is not easy to assess their quality index Q according to the seven criteria set out in Table 6.1 following Van der Voo (1990a). Therefore, in attempting to assess these data, the important criterion first considered has been the presence of reversals (criterion 6 of Table 6.1). Many of the results are derived from the use of bulk demagnetization procedures, so the presence of reversals becomes a strong criterion for inferring a primary magnetization. The results from such data are then compared with others where reversals are not present, and when there is general agreement these are also deemed to be acceptable. Such data are then considered to be at least at the level of the minimum acceptance criterion of $Q = 3$. Many of the Russian results only indicate the number of samples collected rather than the number of sites, which is required for the method of combining data (§6.4.2). In these cases a conservative estimate of 10 samples per site has been used to determine the number of sites investigated in each study.

The Permian of Europe is the period which contains the most data and between the Early and Late Permian apparent polar wander is only 2.0°. Therefore, to

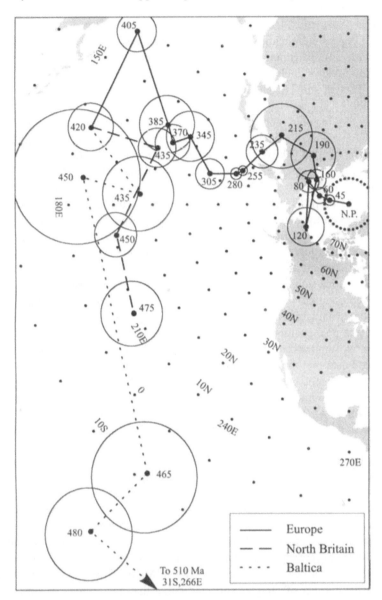

Fig. 6.14. Phanerozoic APWP for Stable Europe (separating North Britain and Baltica for >425 Ma) using the mean pole positions given in Table 6.6. Each mean and its circle of 95% confidence is labeled with the mean age in millions of years.

obtain a good data set enabling a comparison of data from the Russian platform and western Europe it is convenient to combine the data for the whole of the Permian. As is evident from Table 6.6, the mean pole positions differ by only 1.7°. These means are not discernibly different at the 95% confidence limit (angle of separation would need to exceed 2.4°). Therefore, the large amount of Russian data has been included in this analysis with some confidence. As a result, for pre-Permian times the number of pole positions averaged is now more than double that listed by Van der Voo (1993).

The separated paths for North Britain and Baltica prior to 420 Ma became widely separated in the Middle Ordovician (470- and 465-Ma poles, respectively). For the Ordovician and Cambrian of Baltica it appears that many of the data from Russia have been remagnetized. Therefore, those poles for Baltica identified by Torsvik *et al.* (1992) as being reliable have been mainly used in determining the extension of the APWP for Europe to Baltica for pre-425 Ma. Note that when the number of groups $m = 1$, A_{95} cannot be determined.

6.5.3 Asia

Asia is a composite continent made up of blocks that have accreted during most of the Phanerozoic (Hamilton, 1970; Zonenshain *et al.*, 1990). The major cratonic nucleus is *Siberia* (Fig. 6.15). To the west the early Paleozoic *Kazakhstan* sub-continent consists of a mosaic of displaced terranes that, together with Siberia, collided with Baltica during the late Paleozoic along the site of the Urals (Urals Orogenic Belt). Between the Kazakhstan sub-continent and Siberia lies the Central Asia Orogenic Belt formed during the Paleozoic collisions between Siberia, Kazakhstan in the north, and *Junggar*, *Tarim* and *North China* to the south. To the north lies the Precambrian *Kara* block separated from Siberia by the Taimyr Orogenic Belt. To the southeast of Siberia the Mongol–Okhotsk Orogenic Belt formed through the late Paleozoic and early Mesozoic during the collision of *Mongolia* with Siberia. Mongolia, which is referred to as *Amur* by Zonenshain *et al.* (1991), is probably made up of several terranes. The Mongol–Okhotsk suture extends to the east along the Okhotsk–Chukotka Orogenic Belt for a combined length of 5000 km into far-eastern Asia. Further to the south, *South China* amalgamated with North China during the Mesozoic.

To the east of Siberia the Verkhoyansk–Kolyma Orogenic Belt formed mostly in the Early and mid-Cretaceous when numerous displaced terranes were accreted to the Asian margin during rapid convergence of the Kula, Izanagi and Eurasian plates (see Fig. 5.25). The *Kolyma* block to the east and *Sikhote Alin* terrane to the southeast (Fig. 6.15) formed during this amalgamation. The Kolyma block is a loose grouping of several terranes, notably the *Omolon* and *Chukotka* terranes, which, together with Sikhote Alin, are discussed in §7.3.5.

Fig. 6.15. Simplified tectonic map of Asia showing the major blocks (shaded) whose paleomagnetic data are analyzed in the text (§6.5.3 or §7.2.2 and §7.3.5). The positions of other smaller terranes are indicated by names in italics. Outlines in northern Asia are after Zonenshain *et al.* (1991).

India, which collided with Asia during the Cenozoic, is considered in the context of the Gondwana continents in §6.5.4. The associated *Lhasa* and *Qiangtang* terranes to the north are discussed further in §7.3.5. *Iran* is here considered as a composite terrane comprising several stable blocks of Iran and central Afghanistan (e.g., Alborz–Great Kavir, Lut and Sistan–Helmand), which most workers argue have a common paleogeographic setting. *Iran* may have been a northern part of Gondwana as discussed in §7.3.5. The tectonic evolution of this region has been summarized by Boulin (1991).

Siberia

Russian scientists have carried out extensive paleomagnetic studies in Siberia since the early 1960s. Data for the former Soviet Union were regularly summarized in catalogs prepared by A.N. Khramov and made available through the World Data Centers. Reviews of these data in English in terms of APWPs for the various blocks in eastern Europe and Asia have been given by Khramov *et al.* (1981) and Khramov (1987). Besse and Courtillot (1991) derived a reference APWP for Eurasia (meaning Siberia in the present context) on the basis of worldwide data and rotation parameters derived from ocean magnetic anonmalies. Smethurst *et al.* (1998) reviewed paleomagnetic data for the Neoproterozoic and Paleozoic to construct an APWP.

Gurevich (1984) and Pavlov and Petrov (1996) have proposed that Siberia can be divided into two parts referred to as *Anabar* (northern Siberia) and *Aldan* (southern Siberia) in Fig. 6.15 and separated by the Viljuy Basin. The basin encloses the Ygyatta and Viljuy grabens which began to form in the Middle Devonian resulting in an increasing amount of extension toward the northeast where buried oceanic crust has been found (Zonenshain *et al.*, 1990). Pavlov and Petrov (1996) have proposed that the northern Anabar block rotated counterclockwise with respect to the southern Aldan block about a pole at 60N, 100E during Siluro–Devonian times. Therefore, it is necessary to rotate results from the Anabar block clockwise for pre-Devonian times to bring them into agreement with the Aldan block. Smethurst *et al.* (1998) conclude that the paleomagnetic data sets from the two blocks are insufficient to determine a rotation angle but suggest that a rotation of -20° (clockwise) is a good approximation.

The method for judging the reliability criteria for Russian data for Europe given in §6.5.2 has been used to compile the data set for Siberia giving the mean pole positions listed in Table 6.7. For pre-Devonian times the five mean poles in Table 6.7 were analyzed by rotating results from the northern Anabar block through angles. of -10, -15 and -20° (clockwise) about an Euler pole at 60N, 100E. Four of the five mean poles in Table 6.7 improve their grouping with a rotation of -10° compared with three for a rotation of -15° and only two for -20°. Therefore, in determining an APWP for Siberia, data from the Anabar block have been rotated about an Euler pole at 60N, 100E through an angle of -10° (clockwise) for pre-Devonian times. Data from the Taimyr Orogenic Belt have been included for post-Permian times. These data are those for the Siberian Traps that also occur on the Siberian craton and this is the reason why the Early

TABLE 6.7

Phanerozoic APWP for Siberia[a]

Geological period/epoch	Age range (Ma)	Mean (Ma)	*m*	*N*	Mean pole position		A_{95} (°)
					Lat N	Long E	
Early Cretaceous	97-146	130	4	25	71.0	161.0	8.4
Late Jurassic	150	150	1	4	54.0	123.0	-
Late Triassic	208-235	225	6	102	50.9	145.8	7.0
Early Triassic	241-245	245	22	627	48.7	139.7	4.4
Permian	245-290	265	6	146	51.5	150.5	6.1
Late Carboniferous	290-323	310	3	19	44.9	130.2	13.2
Early Carboniferous	323-363	345	3	23	34.8	122.5	10.6
Late Devonian	363-377	370	4	36	33.7	103.4	5.4
Middle Devonian	377-391	385	5	30	15.2	122.1	12.3
Silurian	417-443	440	3	18	-4.1	104.3	8.6
Middle/Late Ordovician	443-470	460	10	54	-24.7	130.9	5.9
Early Ordovician	470-495	485	4	30	-38.4	136.6	5.8
Middle/Late Cambrian	495-518	505	11	189	-36.0	131.7	3.7
Early Cambrian	518-545	530	8	29	-44.6	149.8	7.8

[a]Symbols and method of averaging are as given in Table 6.5.

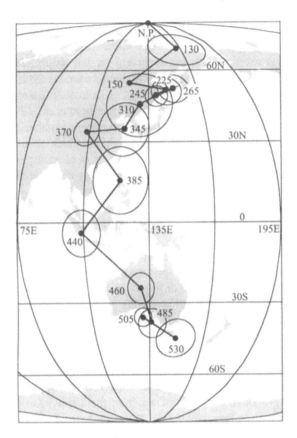

Fig. 6.16. Phanerozoic APWP for Siberia using the mean pole positions given in Table 6.7. Each mean and its circle of 95% confidence is labeled with the mean age in millions of years.

Triassic result includes so much more data than any of the other mean poles listed in Table 6.7. There are data from 90 studies representing 1332 separate sites, most of which are of Paleozoic age. This is only about one-third of the data available from either North America or Europe, but the remoteness and difficulty of access to the region has clearly restricted the acquisition of paleomagnetic data, and this is likely to be the case for some time.

The APWP for Siberia for the Phanerozoic represents the longest of these paths for any of the world's major cratons. It extends back from the present pole through an angle of nearly 135° to its Early Cambrian position. Note that the paleomagnetic pole crosses the equator during Siluro–Devonian times and by the Early Cambrian is located at 43°S (Fig. 6.16). Unlike the data for North America and Europe, the data for Siberia in the late Paleozoic suffer from poor age control. That is, most results are stated to be broadly Permian or Carboniferous and for this reason it has not been possible to subdivide the data for these two epochs in any finer detail.

North China, South China, and Tarim

The North and South China blocks consist of Precambrian basement but are not really stable cratons because the Phanerozoic cover rocks in most parts have been deformed by subsequent orogenies. North China is often shown as including Korea and in that case has been referred to as Sino–Korea. Faunal and other similarities suggest a close affinity between Korea and North China certainly in Mesozoic times and probably also during the Paleozoic. However, paleomagnetic data from Korea suggest that there has been relative rotation of Korea with respect to North China in pre-Cretaceous times (Van der Voo, 1993). Therefore, most analyses of paleomagnetic data usually regard Korea as a separate entity from North China. The eastern boundary of North China block is represented by the Tanlu fault running north–south and to the south the boundary with the South China block is the Qilian–Qinling–Dabie Shan (mountains). There appears to be no clear suture between the North China block and Tarim and consequently many geologists have considered them to have formed a single block since the late Precambrian. However, paleomagnetic data suggest that the two blocks could not have attained their present configuration until at least the Mesozoic (McFadden *et al.*, 1988b; Li, 1990). The collision between Tarim and Junggar occurred along the Tianshan Mountains and is inferred to have occurred sometime during the Permian. A full discussion of the paleomagnetic data from these three major blocks of China in relation to the timing of their amalgamation with Eurasia is given in §7.2.2.

Only limited paleomagnetic studies were carried out in China pre-1980 and then mostly on Mesozoic and Cenozoic rocks (McElhinny, 1973a). Many Western scientists visited China in the early 1980s and this resulted in an increased interest in paleomagnetic studies in China. The first publication resulting from this work was that of McElhinny *et al.* (1981), who reported new results for the Permo–Triassic of the North and South China blocks. In the 1980s cryogenic magnetometers and the methods of principal component analysis became widely used in western laboratories. As a result the database for China is overall of a much higher standard than for most other parts of the world and it is mainly the early studies from pre-1980 that have quality index $Q < 3$.

The first APWPs for North and South China were given by Lin *et al.* (1985). Zhao and Coe (1987), Enkin *et al.* (1992) and Zhao *et al.* (1990, 1996) discussed the significance of data acquired for the timing of the collision and suturing of these blocks to each other and to Asia.

TABLE 6.8

Phanerozoic APWP for North China[a]

Geological period/epoch	Age range (Ma)	Mean (Ma)	m	N	Mean pole position Lat N	Long E	A_{95} (°)
Paleocene/Eocene	35-65	50	1	6	86.3	200.1	-
Late Cretaceous	65-97	80	4	21	79.0	192.0	9.5
Early Cretaceous	97-146	120	7	48	79.3	213.7	5.5
Late Jurassic	146-157	150	6	40	72.9	220.0	2.5
Middle Jurassic	157-178	170	6	37	74.6	238.4	4.8
Early Jurassic	178-208	195	1	10	82.4	286.0	-
Middle/Late Triassic	208-235	220	8	87	59.8	11.0	5.9
Early Triassic	235-245	240	9	47	58.6	353.1	3.7
Late Permian	245-256	250	11	56	48.1	3.8	7.4
Early Permian	256-290	275	4	25	40.9	353.7	11.0
Late Carboniferous	290-323	305	3	14	35.3	339.5	21.4
Middle/Late Devonian	363-391	375	1	16	-34.2	48.7	-
Middle/Late Silurian	417-428	420	1	20	-26.2	48.4	-
Late Cambrian/Early Ordovician	470-505	485	1	8	-28.8	130.9	-
Early/Middle Cambrian	505-545	525	1	10	-21.2	155.2	-

[a]Symbols and method of averaging are as given in Table 6.5.

The analysis of the data for North China is summarized as the mean pole positions listed in Table 6.8 and the corresponding APWP is illustrated in Fig. 6.17. There are only 64 studies listed giving data from 445 sites, a data set

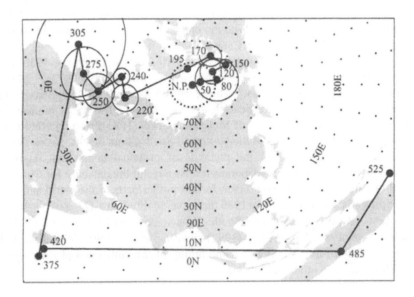

Fig. 6.17. Phanerozoic APWP for North China using the mean pole positions given in Table 6.8. Each mean and its circle of 95% confidence is labeled with the mean age in millions of years.

TABLE 6.9

Phanerozoic APWP for South China[a].

Geological period/epoch	Age range (Ma)	Mean (Ma)	*m*	*N*	Mean pole position Lat N	Long E	A_{95} (°)
Paleocene/Eocene	35-65	50	3	13	77.2	278.5	10.2
Late Cretaceous	65-97	80	11	86	74.0	199.7	7.3
Early Cretaceous	97-146	120	8	69	73.9	208.0	9.2
Late Jurassic	146-157	150	3	31	74.9	198.9	6.5
Middle Jurassic	157-178	165	2	8	64.4	186.4	5.2
Early Jurassic	178-208	195	3	15	64.8	190.0	3.4
Middle/Late Triassic	208-235	220	10	76	31.5	197.1	8.6
Early Triassic	235-245	240	16	79	46.8	216.5	4.9
Late Permian	245-256	250	20	132	51.5	238.8	5.0
Early Permian	256-290	275	1	11	31.4	224.1	-
Carboniferous	290-363	325	2	6	23.5	223.0	4.8
Devonian	363-417	390	1	7	-8.9	190.4	-
Silurian	417-443	430	1	15	4.9	194.7	-
Early Ordovician	470-495	480	1	5	-38.9	235.7	-
Early/Middle Cambrian	505-545	525	3	20	-4.3	204.6	14.5

[a]Symbols and method of averaging are as given in Table 6.5.

that is much smaller than that for Siberia. All the mean positions for Late Permian and younger times are based on multiple results, whereas for older times only single pole positions are available. The APWP lies in the southern hemisphere in pre-Permian times but is not as long as that for Siberia. Apparent polar wander during the Paleozoic appears to have been quite large.

The data for South China and Tarim are summarized in Tables 6.9 and 6.10 and the corresponding APWPs are illustrated in Fig. 6.18. In the case of South China data have been included from Mongolia for post-Early Jurassic times. The APWP lies in the southern hemisphere during the early Paleozoic. The data set for South China is slightly larger than that for North China with 85 studies

TABLE 6.10

Phanerozoic APWP for Tarim[a]

Geological period/epoch	Age range (Ma)	Mean (Ma)	*m*	*N*	Mean pole position Lat N	Long E	A_{95} (°)
Paleocene/Eocene	35-65	50	1	11	63.7	320.3	-
Late Cretaceous	65-97	80	4	31	66.9	219.4	6.2
Early Cretaceous	97-146	120	5	44	70.0	220.6	6.7
Middle/Late Jurassic	146-178	160	4	25	64.2	189.6	13.7
Late Triassic	208-235	220	1	8	52.1	166.8	-
Late Permian/Early Triassic	235-256	245	3	57	67.7	185.5	3.6
Early Permian	256-290	275	4	69	57.2	178.9	6.6
Late Carboniferous	290-323	305	3	15	47.4	170.9	7.0
Devonian	363-417	390	3	68	9.3	161.1	16.3

[a]Symbols and method of averaging are as given in Table 6.5.

(a) South China   (b) Tarim

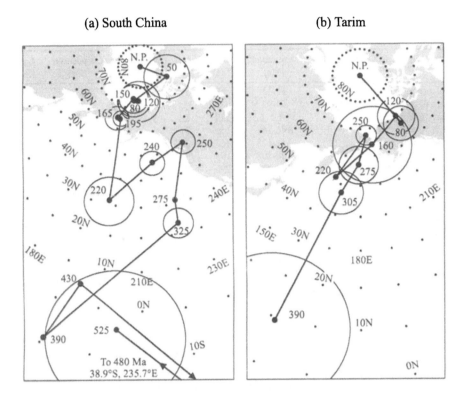

Fig. 6.18. Phanerozoic APWP for (a) South China using the mean pole positions given in Table 6.9 and (b) Tarim using the mean pole positions given in Table 6.10. Each mean and its circle of 95% confidence is labeled with the mean age in millions of years.

giving data from 573 sites. Paleomagnetic data for Tarim have only become available since the first APWP for the late Paleozoic was published by Bai *et al.* (1987). This was followed soon afterwards by the first studies involving Western scientists (Li *et al.*, 1988; McFadden *et al.*, 1988b).

Over the past decade sufficient data have become available to be able to determine the APWP back to the Devonian. From the Devonian to the present time the APWP is restricted to the northern hemisphere. The data set is much smaller than that for North and South China with 28 studies giving data from 328 sites. Although there are relatively few data, they are all of high quality, having been determined using all the modern methods and analytical techniques.

The timing of the amalgamation of Eurasia involves a combined analysis of the paleomagnetic data for all the major blocks and this is considered in detail in §7.2.2. Although the results for the major blocks as outlined above appear to be restricted compared with those for Laurentia and Europe, the fact that they have been acquired over the past 15 years results in a data set of higher reliability.

6.5.4 The Gondwana Continents

West Gondwana

Africa and South America are referred to as West Gondwana because it is thought that the constituent parts of these continents assembled during the Brasiliano (600–530 Ma) and Pan-African (≈500 Ma) orogenies. West Gondwana then merged with East Gondwana, which had remained as a unit for very much longer (see §7.2.3). A simplified outline of the major blocks of Africa and South America is shown in Fig. 6.19. The three major cratons of Africa are the *Congo*, *Kalahari*, and *West Africa* cratons. These cratons are surrounded by the Pan-African Orogenic Belts of age ≈550 Ma relating to their time of amalgamation. In the northern part of Madagascar there is a small remnant of the India craton (see Fig. 6.22).

For the purpose of analyzing paleomagnetic data, the subsequent rift history of the African continent has significance because these rifts have caused internal deformations that need to be taken into account. When Africa and South America separated in the Early Cretaceous, a failed third arm of a triple junction was formed in the region of the Benue Trough. This created an Early Cretaceous

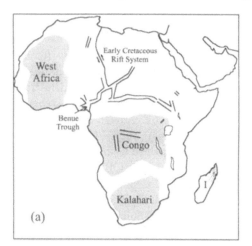

Fig. 6.19. (a) The major cratons of Africa (shaded) and their relationship to the Early Cretaceous rift system of northwest and northeast Africa following Pindell and Dewey (1982). The Benue Trough represents the failed third arm of a triple junction that formed when Africa and South America separated in the Early Cretaceous. The resulting rift system produced internal extension within the African continent. The northern part of Madagascar includes a small former part of the India craton labeled I. (b) Simplified tectonic framework of South America showing the major cratons (shaded), together with the Phanerozoic orogenic belts along the west and southern margin. The São Francisco craton was formerly a westerly extension of the Congo craton and WA is a former southern part of the West Africa craton. The Arequipa Massif is a former part of North Britain when it was joined to Laurentia.

TABLE 6.11

Phanerozoic APWP for Africa[a]

Geological period/epoch	Age range (Ma)	Mean (Ma)	m	N	Mean pole position Lat N	Long E	A_{95} (°)
Paleocene/Eocene	35-65	50	6	65	82.3	175.4	4.6
Late Cretaceous	65-97	80	16	162	69.2	239.7	4.5
Early Cretaceous	97-146	120	7	72	54.6	263.7	3.2
Middle/Late Jurassic	146-178	160	7	44	55.3	247.0	10.9
Early/Middle Jurassic	178-208	195	15	219	73.9	245.8	3.5
Late Triassic	208-235	220	4	42	59.5	239.8	12.7
Late Permian/Early Triassic	240-250	245	2	5	62.5	245.5	12.0
Early Permian	256-290	275	10	66	31.4	247.1	5.4
Late Carboniferous	290-323	305	5	37	27.9	236.4	3.5
Early Carboniferous	323-363	345	3	24	5.2	236.1	8.7
Middle/Late Devonian	363-391	375	3	9	17.1	199.3	12.9
Early Devonian	391-417	405	1	12	43.4	188.6	-
Late Ordovician	443-485	465	3	22	-33.3	179.3	24.0
Early Ordovician	485-495	490	2	5	-50.0	213.9	15.0
Late Cambrian	495-505	500	1	3	-17.7	157.2	-
Early Cambrian	520-530	525	3	10	-29.7	164.4	7.3

[a]Symbols and method of averaging are as given in Table 6.5.

extensional rift system from the Benue Trough into central and northern Africa as shown in Fig. 6.19a.

Lottes and Rowley (1990) compiled rotation parameters that allow for the differential rotation between northwest Africa, southern Africa and northeast Africa. Data from the main Arabia craton have been included with data from Africa after allowing for the opening of the Red Sea by rotation of Arabia to northeast Africa about an Euler pole at 31.3N, 25.5E through an angle -8.5° (clockwise; see §5.1.1 for Euler rotation conventions). Therefore, in analyzing paleomagnetic data from Africa, pole positions are rotated to the co-ordinates of West Africa for times before 130 Ma. For times between 130 and 100 Ma the paleomagnetic poles have been partially rotated, and for times after 100 Ma they have not been rotated. The rotation parameters are as follows.

(i)   Southern Africa to West Africa: rotation about an Euler pole at 9.3N, 5.7E through an angle -7.8° (clockwise).

(ii) Northeast Africa to West Africa: rotation about an Euler pole at 19.2N, 352.6E through an angle -6.3° (clockwise).

(iii) Arabia to West Africa: rotation about an Euler pole at 26.2N, 11.2E through an angle of -14.2° (clockwise).

Mean pole positions for Africa are given in Table 6.11 and the corresponding APWP is illustrated in Fig. 6.20a. The earlier paleomagnetic studies in Africa were mainly carried out in southern Africa with reviews of these data being given by Gough et al. (1964), McElhinny et al. (1968), and Brock (1981). Significant data were later acquired from northern Africa, mainly by workers

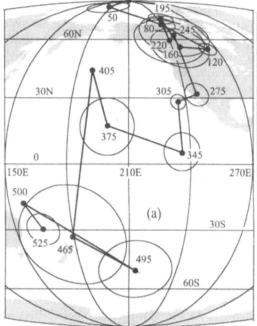

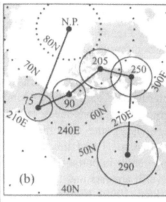

Fig. 6.20. Phanerozoic APWPs with each pole position and its circle of 95% confidence labeled with the mean age in millions of years.
(a) Africa, using the data listed in Table 6.11.
(b) Madagascar, using the data listed in Table 6.12.

from French and German laboratories. Currently, 90 pole positions representing 807 sites are summarized for Africa in Table 6.11.

Reviews of data from Africa over the past two decades have generally been made in the context of the reconstruction of Gondwana (Morel and Irving, 1978; Kent and Van der Voo, 1990; Bachtadse and Briden, 1990, 1991; Schmidt *et al.*, 1990; Li *et al.*, 1990; Van der Voo, 1993). This is because no single Gondwana continent by itself has unambiguous data through the Paleozoic to define each APWP with certainty. Of particular interest is the section of the path from Late Ordovician through Early Carboniferous times. Part of the problem has been due to the fact that most of the results in this time interval have been derived from the Tasman Fold Belt in eastern Australia, and it has been argued that these could be derived from displaced terranes and are not therefore related to the cratonic part of Australia. This problem will be discussed more fully in relation to the Australian APWP.

Bachtadse and Briden (1991) revised their previous view (Bachtadse and Briden, 1990) of the Gondwana Paleozoic APWP on the basis of new data from Devonian ring complexes in the Bayuda Desert, Sudan. Two possible primary components of magnetization were distinguished and it was argued (quite reasonably) that the high-temperature component was primary, and that the intermediate-temperature component was secondary. This scenario relegated all eastern Australian poles to the displaced terrane category and produced a simple APWP. However, new data from central Australia now clearly define the trend of the Gondwana APWP during the time interval Early Devonian to Early Carboniferous (Chen *et al.*, 1995). They are compatible with those from eastern Australia and the pole path is in the opposite sense to that proposed by Bachtadse and Briden (1991). It turns out that, if the interpretation of the high- and intermediate-temperature components is reversed (i.e., the intermediate-temperature component is primary and the high-temperature component is secondary), the data then become compatible with the Australian data. An alternative is that the age of the ring complexes is incorrect. Therefore this apparently important result has not been included in the analysis of the APWP here. It appears that the basic structure of the paths proposed by Bachtadse and Briden (1990), Kent and Van der Voo (1990), Schmidt *et al.* (1990), and Li *et al.* (1990) is probably correct. The Gondwana path for the Phanerozic is discussed more fully in §7.2.3.

The discussion of the Late Ordovician to Early Carboniferous poles for Africa and Gondwana is further complicated by the Late Carboniferous and Early Permian poles from Africa because the pole from the Late Carboniferous Dwyka varves of central Africa (McElhinny and Opdyke, 1968) is often taken to be a key pole in discussion of the Gondwana APWP. However, this pole differs by 30.7° from five well-grouped Late Carboniferous poles from northern Africa. Two of the sites in the Dwyka varves (those with the largest applied structural correction) are from the K1 beds of the Galula coalfield in Tanzania. These are located adjacent to the Tertiary Rukwa rift where the tectonic activity associated with the rifting relates to the structural corrections applied. Similarly, the pole from the Early Permian K3 beds located in the same section of the Galula coalfield differs by 41° from the pole determined from the K3 beds in the Songwe–Kiwira and Ketewaka–Mchuchuma coalfields to the east (Opdyke, 1964). Recent data from these same beds to the east (Nyblade *et al.*, 1993) confirm the previous result of Opdyke (1964). There are 8 Early Permian poles from northern Africa and these agree well with those from the eastern localities of the K3 beds in Tanzania. However, the K3 Galula pole differs by 50° from the mean of all the 10 remaining Early Permian poles. This pole is clearly an outlier from the main group. Therefore, the results derived from the Galula coalfield adjacent to the Rukwa rift have not been included in the analysis.

There appears to be a rapid polar shift of about 25° between the Early and the Late Permian, but lack of good data in this time interval means that the details cannot be determined. During the Mesozoic the mean pole positions appear to circulate around a position at about 65N, 250E although the means listed in Table 6.12 suggest there is some fine structure involved. During the Triassic the mean pole position is located near 60N, then during the Early/Middle Jurassic it is more northerly (at 74N) than its subsequent position during the Late Jurassic

TABLE 6.12
Phanerozoic APWP for Madagascar[a]

Geological period/epoch	Age range (Ma)	Mean (Ma)	*m*	*N*	Mean pole position Lat N	Long E	A_{95} (°)
Late Cretaceous 1	68-80	75	3	30	63.5	219.6	4.1
Late Cretaceous 2	88-91	90	3	33	69.1	240.1	4.9
Late Triassic/Early Jurassic	178-208	205	1	8	74.0	277.1	6.3
Late Permian/Early Triassic	241-256	250	3	7	65.8	291.6	6.1
Late Carboniferous/Early Permian	256-323	290	2	5	47.9	264.1	8.1

[a]As summarized by Andriamirado (1971) and McElhinny *et al.* (1976). Symbols are as in Table 6.5.

to Early Cretaceous (near 55N). During the Late Cretaceous its position returns to near 70N.

The data from Madagascar are listed in Table 6.12 with the corresponding APWP illustrated in Fig. 6.20b. The Late Cretaceous results from volcanics along the east coast are due to Andriamirado (1971) and the remaining data are those summarized by McElhinny *et al.* (1976). At the time of writing no new data had been reported for more than twenty years. Only 12 pole positions representing 83 sites have been determined for Madagascar.

The major cratons of South America are the *Amazonia* and *Rio de la Plata* cratons (Fig. 6.19b). The *São Francisco* craton to the east was formerly a western extension of the Congo craton of Africa. In the northeast there is a small remnant of the West Africa craton shown in Fig. 6.19a. The western margin of South America contains displaced terranes of various ages, the most important of which (in the context of this book) is the *Arequipa Massif.* This is thought to be a former part of Laurentia (North Britain terrane) that became detached when it collided with the western margin of South America near the Precambrian–Cambrian boundary (Dalziel, 1997; see §7.4.3).

The first paleomagnetic measurements in South America were reported by Creer (1958), and his later studies and those of his students and associates are summarized in Creer (1970). Workers in Argentina and Brazil set up their own laboratories and a later summary is given by Vilas (1981). Much work has been carried out in Chile and along the western margin of South America. Most of this work relates to various displaced terranes and is not therefore considered in the context of constructing an APWP for South America. Currently, there are 65 acceptable results derived from 1061 sites as listed in Table 6.13.

Most of these results are concentrated in the time interval Late Carboniferous to Late Cretaceous. The large number of sites for the Early Cretaceous come from the many studies that have been carried out on the Serra Geral Formation in Brazil and Uruguay. There appear to have been some difficulties in obtaining good data for pre-Late Carboniferous times. Furthermore, especially for the Ordovician and Cambrian, ages are only known within the broad limits of the

TABLE 6.13
Phanerozoic APWP for South America[a]

Geological period/epoch	Age range (Ma)	Mean (Ma)	m	N	Mean pole position Lat N	Long E	A_{95} (°)
Late Cretaceous	65-97	80	7	68	84.6	196.9	7.2
Early Cretaceous	97-146	120	13	652	84.7	253.6	2.5
Middle/Late Jurassic	146-185	165	4	80	85.4	41.6	3.6
Early Jurassic	185-208	200	2	10	70.7	74.0	9.4
Late Triassic	225-235	230	5	69	79.4	1.7	5.4
Early/Middle Triassic	235-245	240	4	27	83.7	126.1	2.8
Late Permian	245-256	250	4	36	72.6	71.8	9.5
Early Permian	256-290	275	10	58	77.7	137.8	5.8
Late Carboniferous	290-323	305	9	28	55.2	168.6	3.6
Early Carboniferous	323-363	340	2	17	31.5	136.7	4.0
Middle Devonian	377-391	385	1	1	30.0	133.0	-
Ordovician	443-495	470	1	3	-11.0	153.3	-
Cambrian	495-54	520	3	12	-47.7	191.1	22.6

[a]Symbols and method of averaging are as given in Table 6.5.

whole period. Consequently, the few results summarized in Table 6.13 for times prior to the Late Carboniferous cannot be considered reliable as is reflected by the large circles of 95% confidence. Therefore, these early Paleozoic mean pole positions should be viewed with some caution.

The corresponding APWP for South America is illustrated in Fig. 6.21. It

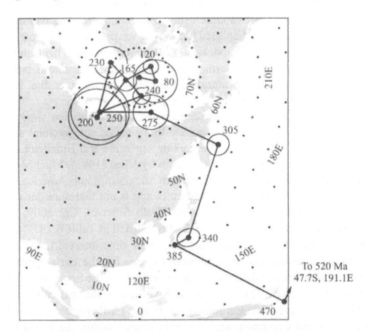

Fig. 6.21. Phanerozoic APWP for South America using the mean pole positions given in Table 6.13. Each mean and its circle of 95% confidence is labeled with the mean age in millions of years.

shows that the mean pole position remained close to the present pole for most of the Mesozoic and Permian. However, the pole path as it moves in these high-latitude positions is quite distinctive and clear.

East Gondwana

Simplified geological outlines of the three constituent parts of East Gondwana (Australia, Antarctica, and India) are given in Fig. 6.22. The western two-thirds of Australia is made up of three major Precambrian cratons, the *Yilgarn* and

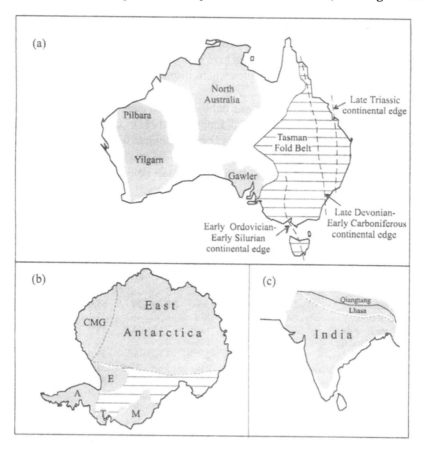

Fig. 6.22. Simplified tectonic maps of the constituent parts of East Gondwana. (a) Australia showing the three major cratons (shaded) and the Tasman Fold Belt. Three successive positions of the eastern continental edge during the Paleozoic and Mesozoic are indicated. (b) Antarctica with its main craton of East Antarctica and the Coats Land–Maudheim–Grunehogna (CMG) Province. Four terranes are labeled as follows: A, Antarctic Peninsula; E, Ellsworth Mountains–Whitmore Mountains; M, Marie Byrd Land; T, Thurston Island–Eights Coast. (c) India showing the main India craton with the northern terranes of Lhasa and Qiangtang.

Pilbara in the west, *North Australia* in the north, and the *Gawler* craton in the south (Fig. 6.22a). To the east the Tasman Fold Belt consists of many displaced terranes that accreted to the Australian continent through the Paleozoic and Mesozoic. At the Precambrian/Cambrian boundary the western edge of the Tasman Fold Belt was the eastern edge of the Australian continent.

The eastern continental edges at three successive ages are shown in Fig. 6.22a for the Late Ordovician/Early Silurian, Late Devonian/Early Carboniferous, and the Late Triassic. From a paleomagnetic point of view, therefore, it is important that data from the Tasman Fold Belt be derived from sites at which displacements have ceased before they can be considered to be representative of the main Australian continent. Powell *et al.* (1990) discussed these aspects in detail and their reasoning is followed here in deciding whether specific data from the Tasman Fold Belt are from sites where displacements can be considered to have ceased.

The first APWP for Australia was published by Irving and Green (1958) and updated by Irving *et al.* (1963) and Embleton (1981). The APWP during the early and middle Paleozoic has always been the subject of much debate. As first discussed by McElhinny and Embleton (1974), the problem is one of deciding whether data in this time interval from the Tasman Fold Belt are relevant to the

TABLE 6.14

Phanerozoic APWP for Australia[a]

Geological period/epoch	Age range (Ma)	Mean (Ma)	m	N	Mean pole position Lat N	Long E	A_{95} (°)
Eocene	35-56	45	7	131	69.1	295.7	5.2
Paleocene	56-65	60	3	36	56.1	302.3	11.9
Late Cretaceous	65-97	85	2	24	55.6	318.6	3.1
Early Cretaceous	97-146	120	6	46	47.2	334.8	9.6
Middle/Late Jurassic	146-178	160	7	51	48.2	352.6	8.4
Early/Middle Jurassic	178-208	195	4	29	48.7	355.5	4.0
Late Triassic	208-235	220	3	8	33.0	0.3	11.4
Late Permian/Early Triassic	245-256	250	5	51	33.4	326.1	11.8
Early Permian	256-290	275	5	5	34.3	292.8	23.0
Late Carboniferous A	290-315	305	4	106	51.4	321.0	2.6
Late Carboniferous B	315-323	320	2	9	85.7	331.7	22.1
Early Carboniferous	323-363	345	1	10	37.6	232.6	-
Late Devonian	363-377	370	5	46	58.0	211.3	8.0
Middle Devonian	377-391	385	3	15	73.2	159.1	11.0
Early Devonian	391-417	405	1	10	74.3	42.7	-
Late Ordovician/Silurian	417-458	440	1	6	-15.7	242.7	-
Early/Middle Ordovician	458-495	475	6	23	15.5	216.7	15.0
Middle/Late Cambrian	495-518	510	7	63	31.1	201.6	6.0
Early/Middle Cambrian	518-530	525	3	4	-15.6	207.6	15.3
Early Cambrian	530-545	535	3	32	19.3	175.9	25.3

[a]Symbols and method of averaging are as given in Table 6.5.

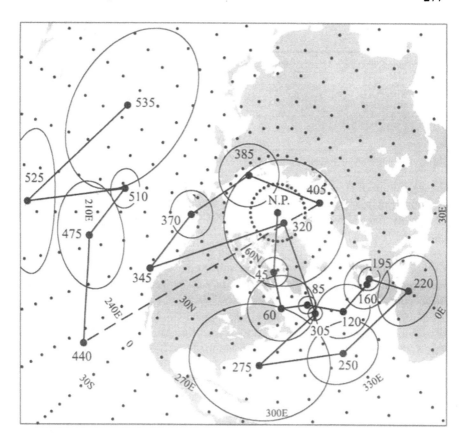

Fig. 6.23. Phanerozoic APWP for Australia using the mean pole positions given in Table 6.14. Each mean and its circle of 95% confidence is labeled with the mean age in millions of years.

whole of Australia.

Following the careful analysis by Powell *et al.* (1990) of the tectonics of the Tasman Fold Belt and hence of the relevance of the paleomagnetic data to Australia as a whole, Li *et al.* (1990) and Schmidt *et al.* (1990) produced an APWP making use of appropriate data from eastern Australia. It is now clear that certain Devonian results from eastern Australia may be considered as valid for the whole of continental Australia because there is excellent agreement between these results and those of similar age from central Australia (Chen *et al.*, 1995).

Mean pole positions for Australia for the Phanerozoic are listed in Table 6.14 and the resulting APWP is illustrated in Fig. 6.23. There are 78 pole positions that have been analyzed representing 618 sites. Australia now has the best determined APWP for the early and middle Paleozoic, although many more data are still required. Rapid apparent polar wander during the Tertiary arises from

TABLE 6.15

Phanerozoic APWP for Antarctica[a]

Geological period/epoch	Age range (Ma)	Mean (Ma)	m	N	Mean pole position Lat N	Long E	A_{95} (°)
Early Cretaceous	97-146	130	2	2	52.8	352.8	49.7
Middle/Late Jurassic	146-178	155	9	141	51.4	36.9	4.9
Early Jurassic	178-208	195	4	59	48.2	44.9	5.9
Early Ordovician	458-495	475	5	57	3.7	201.9	8.6
Cambrian	495-545	520	1	5	-1.5	208.5	-

[a]Symbols and method of averaging are as given in Table 6.5.

the separation of Australia from Antarctica and its northward drift to the present position. For many years it was thought that the pole position had remained essentially fixed through the Mesozoic (McElhinny, 1973a), but more detailed studies have shown that there is fine structure (Schmidt, 1977; Embleton, 1981).

There is rapid polar movement with respect to Australia during the Carboniferous, followed by distinctive movement during the Devonian, which has been carefully delineated by Chen *et al.* (1995). A large polar shift of about 90° occurs between the Late Ordovician/Silurian and Early Devonian. However, each of these poles is based on a single result and so the shift is not well constrained. This part of the APWP is therefore shown as a dashed line in Fig. 6.23. Although three mean pole positions have been calculated through the Cambrian, the associated errors in the case of the Early and Middle Cambrian means are large. Many of these data are more than 25 years old and the dating of the rocks is often not known as well as authors have suggested.

Antarctica consists of the main *East Antarctica* craton, whose extent is mostly inferred from coastal exposures (Fig. 6.22b). However, the *Coats Land– Maudheim–Grunehogna* (CMG) Province in the western part of this craton was formerly an eastern part of the Kalahari craton of Africa prior to the formation of

TABLE 6.16

Phanerozoic APWP for India[a]

Geological period/epoch	Age range (Ma)	Mean (Ma)	m	N	Mean pole position Lat N	Long E	A_{95} (°)
Eocene	35-56	45	6	28	61.6	210.6	8.0
Paleocene	56-65	60	4	8	50.1	287.1	17.0
Late Cretaceous	65-80	70	9	222	36.7	281.2	3.4
Mid Cretaceous	80-120	110	7	116	9.0	297.7	3.5
Early/Middle Jurassic	157-208	180	3	13	3.5	309.4	5.2
Late Permian/Early Triassic	245-256	250	6	39	-3.8	305.8	5.0
Early Permian	256-290	275	3	15	-13.7	312.4	13.9
Late Carboniferous	290-323	305	3	11	-28.3	300.3	12.5
Early/Middle Cambrian	505-545	525	4	10	26.7	217.9	8.6
Late Vendian/Early Cambrian	530-560	545	4	17	45.9	212.6	9.8

[a]Symbols and method of averaging are as given in Table 6.5.

Gondwana in Early Cambrian times (Groenewald *et al.*, 1991, 1995; Grunow *et al.*, 1996), as is discussed in §7.4.3. Four displaced terranes are recognized as the *Antarctic Peninsula*, the *Ellsworth Mountains–Whitmore Mountains* terrane, the *Marie Byrd Land* terrane and the *Thurston Island–Eights Coast* terrane. Extensive measurements have been made on the Ferrar dolerites of Middle Jurassic age, first reported by Turnbull (1959) and Bull *et al.* (1962). Unfortunately these widespread intrusions have remagnetized most of the underlying Mesozoic and late Paleozoic rocks. Despite several attempts to obtain data from this time interval, there has been no success. Therefore, there is a large gap between data from the Jurassic and those from several early Paleozoic rock units (see also Embleton, 1981) as indicated by the dashed line in the APWP shown in Fig. 24a. Phanerozoic paleomagnetic data for East Antarctica are summarized as mean pole positions in Table 6.15.

The Indian subcontinent comprises the main *India* craton as shown in Fig. 6.22c. North of the Himalayas are two displaced terranes named the *Lhasa* terrane and the *Qiangtang* terrane. These terranes accreted to Asia during the Tertiary prior to the arrival of India, as will be discussed further in §7.3.5. The first paleomagnetic measurements from India were reported by Irving (1956) on a few samples of the widespread Deccan Traps of age 67 Ma. It was at once

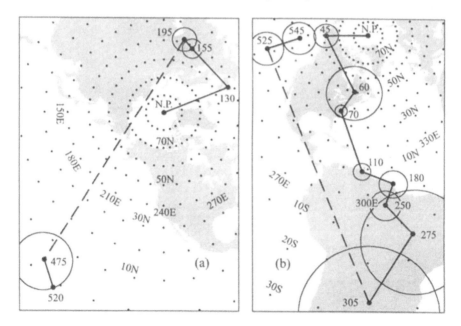

Fig. 6.24. Phanerozoic APWP for (a) Antarctica using the mean pole positions given in Table 6.14, and (b) India using the mean pole positions given in Table 6.15 respectively. Each mean and its circle of 95% confidence is labeled with the mean age in millions of years.

realized that India must have drifted north a considerable distance during the Tertiary. Many measurements have been made on the Deccan Traps with successive reviews given by McElhinny (1968), Wensink (1973), and Vandamme *et al.* (1991). The other widely studied formation is the Rajmahal Traps of eastern India of age 116 Ma, results from which were first reported by Clegg *et al.* (1958). Unfortunately, all the rocks of age between the Cambrian and Late Carboniferous that have been sampled are from the Himalayan region and there is the strong possibility that these rocks have suffered tectonic displacements associated with the collision of India with Asia. Klootwijk and Radhakrishnamurty (1981) reviewed all the data from the Indian subcontinent and included these data in a complex APWP. Following Van der Voo (1993), the more conservative approach is followed here and the data from the Himalayas have been omitted. The mean paleomagnetic pole positions during the Phanerozoic are given in Table 6.16 and the corresponding APWP is illustrated in Fig. 6.24b.

The data from each of the Gondwana continents have strengths and weaknesses, and most of the APWPs are incomplete in some respect or another. These data will be considered again in the context of the reconstruction of Gondwana in §7.2.3, in which a composite APWP is presented based on all the data.

Chapter Seven

Paleomagnetism and Plate Tectonics

7.1 Plate Motions and Paleomagnetic Poles

7.1.1 Combining Euler and Paleomagnetic Poles

In plate tectonic theory the relative motion between two plates is simply described by means of a rotation about an Euler pole as was shown in §5.1.1. The Euler pole defining the relative rotation between two plates is fixed to these two plates and describes only the instantaneous motion that takes place (McKenzie and Parker, 1967). For three contiguous plates, *A*, *B*, and *C*, it is not possible for all three to rotate simultaneously about their instantaneous relative rotation axes (McKenzie and Morgan, 1969). However, it is possible for two of the relative rotations (e.g., *A–B* and *A–C*) to remain simple and for each to be described by rotation about a single Euler pole, whereas the relative motion *B–C* is changing continuously in a complex manner (Le Pichon, 1968). Bullard *et al.* (1965) have shown how to reconstruct the original relative positions of two continents by finite rotations about suitably chosen Euler poles, but no physical significance can be attached to these rotations; they are nothing more than construction entities. They do, however, provide an important constraint because they represent an integral of the motion over some time. If there is no sea floor or magnetic record between two plates *A* and *B*, the understanding of the relative motion between them would require that the timing of the motion between, for example, *A* and *C* and *B* and *C* be accurately known because the order of the rotations is important. For an excellent text on how to perform such reconstructions, see Cox and Hart (1986).

The above concepts relate to the relative motions between plates and demand full information about the relative movements. A record of the relative movements is typically available from the sea floor for movements younger than about 160 Ma. In paleomagnetism, however, the available datum is the position of a paleomagnetic pole relative to a continental block. For ease of description "continental block" here implies the associated plate as well. A sequence of paleomagnetic poles provides an APWP as discussed in §6.4. A paleomagnetic pole provides information about the latitude and the azimuthal orientation of a continental block but provides no longitudinal information. This is reasonably obvious because if a continent moves along a line of latitude and maintains its azimuthal orientation, an observer on the continent will not perceive any apparent motion of the pole. The problem then is how to use a paleomagnetic pole to determine the location of the continental block and how to assess the relative motion between continental blocks.

First it must be recognized that a large amount of apparent polar wander does not necessarily equate to large movement of the continental block. Furthermore, the absence of apparent polar motion does not necessarily mean that the continental block was stationary. For example, if a continental block remains at the same location on the equator but rotates, an observer on the block would perceive a large amount of apparent polar wander along a circle 90° away. Conversely, for the same rotation but with the continent located at the north pole, there would be no apparent polar wander. As already noted, a continental block can move a long way along a line of latitude and, provided it retains its azimuthal orientation, there will be no apparent polar movement.

The obvious way to restore a paleomagnetic pole at latitude λ_p and longitude ϕ_p, $P(\lambda_p, \phi_p)$, to the north pole, G, is simply to move it along the great circle joining P and G. This is equivalent to rotating through an angle $-(90-\lambda)$ about a pole C on the equator at longitude $(\phi+90)$ as shown in Fig. 7.1. Note that the rotation here is clockwise when viewed from outside the Earth and the rotation angle is therefore negative (see §5.1.1). It is unlikely that this was the actual motion of the paleomagnetic pole so C is merely a construction pole that represents the integrated motion. The continental block is restored to its appropriate longitude and azimuthal orientation by rotating it through the same angle about the construction pole C. As shown in Fig. 7.1, if the paleomagnetic pole was at an angular distance p from the continental block a, then after the rotation to a' the continental block is at the correct colatitude p and has the correct azimuthal orientation. Note, however, that the longitude remains indeterminable and can only be defined by reference to some arbitrarily chosen point on the plate.

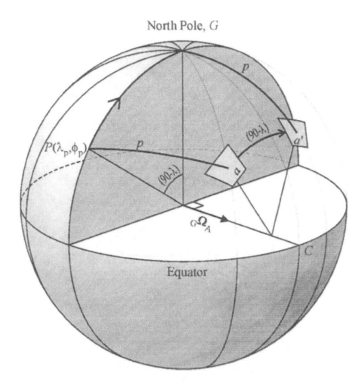

Fig. 7.1. The paleomagnetic north pole, P, is at latitude λ_p and east longitude ϕ_p. The construction pole C is on the equator at longitude $(\phi+90)$. Rotation of the pole P and its associated continental block a through an angle $-(90-\lambda)$ about C will rotate P to the geographic pole and will restore the continental block to its previous latitude and azimuthal orientation at a'.

Let $_G\omega_A$ be the angular velocity vector describing the real motion of plate A relative to the geographic pole G and let $\hat{z}$ be a unit vector along the rotational axis. As with any vector, $_G\omega_A$ can be split into two components, one along the rotational axis and the other, $_G\Omega_A$, passing through the equator (Fig. 7.1). The component along the rotational axis, $(_G\omega_A \cdot \hat{z})\hat{z}$, contains the longitudinal information and is indeterminable. Thus, in paleomagnetism the only available information is in $_G\Omega_A$, given formally by

$$_G\Omega_A = {_G\omega_A} - (_G\omega_A \cdot \hat{z})\hat{z} \ . \tag{7.1.1}$$

Similarly, if the angular velocity $_A\omega_B$ of another plate B relative to the plate A is known then the only measurable component $_G\Omega_B$ of the angular velocity of B relative to the geographic pole is given by

$$_G\Omega_B = {_G\Omega_A} + {_A\omega_B} - (_A\omega_B \cdot \hat{z})\hat{z} \tag{7.1.2}$$

and it also passes through the equator. Consequently, when reconstructing past positions using paleomagnetic information, rotation about a construction pole on the equator is most commonly used.

Three examples of the application of (7.1.2) are illustrated in Fig. 7.2. Continent a is taken to be fixed so $_G\omega_A = 0$, which automatically implies that $_G\Omega_A = 0$ for each of the situations depicted.

In the first case (Fig. 7.2a) the Euler pole E for the rotation between plates A and B is at the north pole so continent b moves along a line of latitude. This

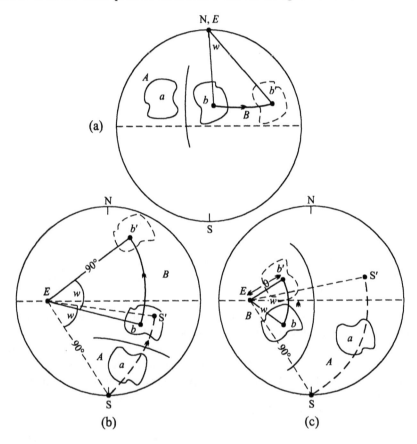

Fig. 7.2. The motions between plates and the resulting movement of the paleomagnetic pole. Continent a on plate A is regarded as fixed relative to the north (N) or south (S) pole. The Euler pole E describes the relative motion between plate B and plate A. Continent b on plate B moves through an angle w to position b'. In (a) E is at the north pole (N) so that no polar motion is observed, whereas in (b) and (c) E is on the equator and so coincides with the pole of rotation for the polar motion. In (b) continent b lies 90° away from E so that the polar motion SS' has the same apparent length as the continental motion bb'. In (c) continent b lies near E so that the polar motion SS' is greatly amplified compared with the continental motion bb'. From McElhinny (1973a).

means that $_G\omega_B$ is entirely along the spin axis (and therefore is paleomagnetically indeterminable). Consequently $_G\Omega_B = 0$, consistent with there being no apparent polar wander observed for plate B. Thus, from (7.1.2)

$$_A\omega_B = (_A\omega_B \cdot \hat{z})\hat{z} , \qquad (7.1.3)$$

which means that $_A\omega_B$ can have any value and is not constrained in any way by the paleomagnetic information. Hence, the motion between the two plates can continue at any rate without the change in relative positions being recorded paleomagnetically. An example of this type of situation can be imagined with respect to South America. The Triassic paleomagnetic pole for South America (§6.5.4, Table 6.13) lies close to the present geographic pole and remains close to it up to the present time. Yet it is known that South America separated from Africa during that time but this separation has been largely along lines of present latitude. No change in paleomagnetic pole position need necessarily be observed.

In the two other examples (Figs 7.2b and 7.2c), the Euler pole for the rotation between plates A and B lies on the equator. There is therefore no component of $_A\omega_B$ along the spin axis so that the last term in (7.1.2) is zero. Remembering that $_G\Omega_A = 0$ because plate A is taken to be fixed, (7.1.2) gives

$$_G\Omega_B = {_A\omega_B} . \qquad (7.1.4)$$

Consequently, the *angular* motion of plate B with respect to the paleomagnetic pole is the same as the *angular* motion of plate B relative to plate A. However, this does not mean that the speed (or distance moved) of plate B relative to plate A is necessarily the same as the apparent speed (or distance moved) of plate B with respect to the paleomagnetic pole. The relationship depends on the angular distance θ between, for example, the center of continent b and the axis of the rotation vector. The velocity $_A\mathbf{v}_B$ between the center of the continent on plate B and plate A is given by (5.1.1) in §5.1.1 as

$$_A\mathbf{v}_B = r{_A}\omega_B \sin\theta , \qquad (7.1.5)$$

where r is the radius of the Earth. Bearing in mind that $_G\Omega_B$ is always on the equator, the paleomagnetically discernible velocity component $_G\mathbf{V}_B$ between the center of the continent and the paleomagnetic pole is given by

$$_G\mathbf{V}_B = r{_G}\Omega_B . \qquad (7.1.6)$$

Combining (7.1.4), (7.1.5) and (7.1.6) gives

$$_G\mathbf{V}_B = \operatorname{cosec}\theta {_A}\mathbf{v}_B . \qquad (7.1.7)$$

Since the magnitude of cosec θ lies between 1 and ∞, the apparent motion of the paleomagnetic pole relative to the continent on plate B (b in Fig. 7.2) is either equal to or much larger than the actual motion of plate B relative to plate A.

In Fig. 7.2b, the Euler pole lies 90° away from the continent (cosec θ = 1) so that $_G V_B = _A V_B$ and the apparent polar motion with respect to the continent b on plate B reflects its motion with respect to A. This is essentially the situation involving the separation of Australia and Antarctica during the Tertiary. The position of Antarctica changes little with respect to the geographic pole during this time, whereas with respect to Australia the paleomagnetic pole (the south pole) moves essentially north–south (§6.5.4).

In Fig. 7.2c the continent on plate B lies near the Euler pole of rotation between A and B (θ small and cosec θ large), so the apparent polar motion is greatly amplified. This is because the motion of continent b now has a large rotation component and a small translation component. Hence, the continent need only move through a small angular distance for the apparent motion of the paleomagnetic pole to be large. This is essentially the situation involved in the rotation of the Iberian Peninsula with respect to Europe (§7.3.4). Note that this apparent polar wander does not even require that sea floor be created between A and B, the motion could be along a transform fault between them.

Despite the fact that, in general, real situations are unlikely to be one of the extremes shown in Fig. 7.2, the examples do illustrate the problems involved when using apparent polar wander to draw conclusions about the past motions between plates. Furthermore, in the real situation it is typically the case that only a part of the original plate (the portion of continental crust) is now discernible, so that the position of the Euler pole describing the relative motion is not known. Consequently, θ is not known and this exacerbates the problems.

The paleomagnetically determinable motion $_G \Omega_A$ of a plate A relative to the geographic pole G is only a component of the full motion $_G \omega_A$. Therefore, the movement of a plate as estimated from paleomagnetism is only a minimum movement. The paleolongitude cannot be constrained from paleomagnetism so the estimated movement between successive pole positions is only the latitudinal change and any change in azimuthal orientation. If the paleomagnetic pole position, (λ_p, ϕ_p), for a plate is known for any epoch, then the paleolatitude, λ, of any site, (λ_s, ϕ_s), on that plate can be calculated from (1.2.9) as

$$\lambda = \sin^{-1}[\sin\lambda_s \sin\lambda_p + \cos\lambda_s \cos\lambda_p \cos(\phi_p - \phi_s)] , \qquad (7.1.8)$$

so the latitudinal changes can be easily calculated. Estimates of the minimum movement of some continental plates (Africa, Baltica, Laurentia, and Siberia) through time since the Archean have been carried out by Ullrich and Van der Voo (1981) using this method. The center of mass was used as the reference point for each continent. These results showed that plate speeds in the past, although going through peaks and troughs, have at times easily exceeded those for present-day continents. At the present time plates containing a significant amount of continental crust ($>2 \times 10^7$ km^2) are known to move at speeds approaching those of mainly oceanic plates (60–100 km Myr^{-1}). The speeds determined by Ullrich and Van der Voo (1981) for Africa, Baltica, and Laurentia were at times in the range 100–150 km Myr^{-1}. This observation has been further emphasized by Meert *et al.* (1993), who showed that the most recent paleomagnetic results from Laurentia and Gondwana indicated that in the past large continental plates have moved with speeds in excess of 160 km Myr^{-1}.

Gordon *et al.* (1979) pointed out that the magnitude and direction of the latitudinal component of the velocity vector depend on the location of the site with respect to the absolute angular velocity vector. This results in differing tangential components of this vector through the plate. They developed a new method for determining minimum rms velocity that takes this factor into account. The method requires an integration over the plate for each pair of poles in order to compute each minimum velocity. Estimates of minimum velocities by this method range from less than a few percent to more than 20% greater than those calculated by the latitudinal method from (7.1.8). If these calculations are made for the center of mass of each plate, then the problem of varying velocities for each site in a plate is eliminated. A comparison of the values for several plates obtained by the latitudinal method as carried out by Ullrich and Van der Voo (1981) and Meert *et al.* (1993) is then quite valid.

7.1.2 Making Reconstructions from Paleomagnetism

It is possible in favorable circumstances to determine the past relative positions of two continents, if they were previously part of the same plate. For the period of time during which they occupied the same plate, these blocks should have the same APWP. Matching the two paths enables the previous relative positions of the blocks to be determined uniquely. The method is illustrated in Fig. 7.3 following Graham *et al.* (1964) and is based on the proposal by Irving (1958), although of course not originally stated in plate tectonic terms.

In practice it is rare to have a whole series of points on both APWPs for carrying out the match. Usually only two or three points are available. Also, it is important not to try to match the pole paths of two blocks that were in relative motion because this will obviously lead to erroneous conclusions. Just because two paths appear to be similar does not necessarily mean that a unique or correct solution for their former relative positions can be found. If at any time the paleomagnetic pole lay near the pole of relative motion between the blocks, an undetected component of motion can result, and the similarity in paths is coincidental. The matching of two paths should therefore be undertaken with some additional information, such as geological comparisons between the blocks or the matching of coast lines.

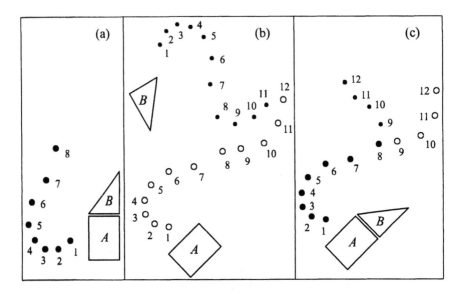

Fig. 7.3. Reconstruction of the relative positions of two continents, *A* and *B*, which were previously part of the same plate. (a) Originally when *A* and *B* occupied the same plate, the APWPs for both *A* and *B* were the same and successive pole positions from times 1 to 8 are shown. (b) After time 8, *A* and *B* split apart, each "carrying" its APWP with it, until the present-day situation at time 12 is reached. (c) To reconstruct the previous relative positions, *A* and *B* are moved until the initial portions of the paths coincide. The time at which the paths diverge then also dates the time at which they separated. From Graham *et al.* (1964).

The method illustrated in Fig. 7.3 applies to the case of an expanding ocean, such as the Atlantic and Indian Oceans of today. The inverse of this method can be applied in the case of a contracted ocean and is a means of demonstrating the existence of a collision zone between former subcontinents, such as is the case in many parts of Asia today. The present-day situation corresponds to that shown in Fig. 7.3c, where time 1 refers to the present. Going back in time through to time 8, the paths then diverge showing that the two regions *A* and *B* did not occupy the same plate prior to time 8, which therefore dates the time of collision.

Kono and Uchimura (1994) formalized the method shown in Fig. 7.3 using geophysical inverse theory. Reconstruction models are evaluated by the combined effects of misfit and penalty. Misfits in the reconstruction are allowed within the statistical uncertainties attributed to the paleomagnetic data. The penalty function minimizes the movement of the blocks over the surface of the Earth. Kono and Uchimura (1994) applied this method to the reconstruction of the North Atlantic using paleomagnetic data from Europe and North America. Unfortunately, they used an old data set from McElhinny (1973a) and the apparent discrepancy between the best fit obtained by this method and that of Bullard *et al.* (1965) most likely arises from this source.

7.2 Phanerozoic Supercontinents

7.2.1 Laurussia

Irving (1956) and Runcorn (1956) noted that the APWPs for Laurentia and Europe were of similar shape, but the APWP for Europe was to the east of that for Laurentia. This suggested they were once joined together and that the Atlantic Ocean opened in Jurassic times. Comparisons of these APWPs have been carried out in many publications since then (Irving, 1964; McElhinny, 1973a; Van der Voo, 1990b, 1993). Magnetic anomalies in the North Atlantic suggest that Laurentia and Europe separated at 175 Ma (see Table 5.2), or in early Middle Jurassic times. Therefore a comparison of the APWPs for Laurentia (Table 6.5 and Fig. 6.12) and Europe (Table 6.6 and Fig. 6.14) is made for times earlier than 175 Ma. However, Baltica only became a part of Europe after about 425 Ma (earliest Silurian), so for times older than 425 Ma the APWP for the North Britain terrane, the remnant of "Laurentian Europe", is used to represent the APWP for Europe.

The APWPs for Laurentia and Europe are compared in Fig. 7.4 before and after closure of the Atlantic Ocean using the rotation parameters of Bullard *et al.* (1965) derived from the best fit of the opposing coastlines. This involves rotation of Europe to Laurentia through an angle of -38.0° (clockwise rotation; see §5.1.1 for Euler rotation conventions) about an Euler pole at 88.5N, 27.7E. Several

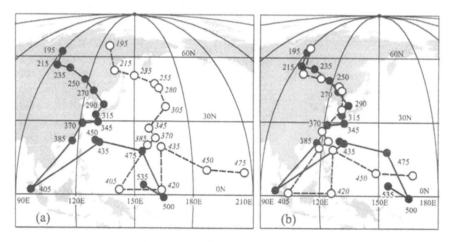

Fig. 7.4. Comparison of the APWPs of Laurentia (solid circles and solid line) and Europe (open circles and dashed line). Ages at each mean pole position are indicated in millions of years with those for Europe in italics. Laurentia and Europe (a) in their present positions and (b) after closing the Atlantic Ocean according to the rotation parameters of Bullard *et al.* (1965).

other possible rotation parameters have subsequently been proposed by Herron *et al.* (1974), Le Pichon *et al.* (1977), Sclater *et al.* (1977), Srivastava and Tapscott (1986), Savostin *et al.* (1986), and Rowley and Lottes (1988). The main difference between the rotation parameters of these subsequent models is that the angle of rotation varies between -22.5 and -30.2°. Each of these has been tested against the available paleomagnetic data by Van der Voo (1990b). The paleomagnetic data clearly supported the Bullard *et al.* (1965) reconstruction parameters. Therefore, these parameters have been used in the analysis shown in Fig. 7.4b.

After closure of the Atlantic Ocean there is excellent agreement between the APWPs at least until the Early Devonian poles for 405 Ma (interval 391-417 Ma) which lie near the equator at about 100E. For Europe, however, the Late Silurian poles at 420 Ma (interval 417-423 Ma) still lie near the equator but then lie just to the east of 125E. This corresponds to the mean of a group of 11 poles (155 sites) and there are no equivalent poles from North America. The paths then converge again in the Early/Middle Silurian at 435 Ma (interval 423-443 Ma). During the Ordovician there is only general agreement between the results from the two continents with both paths showing trends to more easterly longitudes.

Laurentia and Europe are the two continents with most paleomagnetic data.

TABLE 7.1

Reference Combined Phanerozoic APWP for Laurussia after Rotation of Paleomagnetic Poles for Europe to Laurentia using the Rotation Parameters of Bullard *et al.* (1965)[a]

Geological period	Age range (Ma)	Mean (Ma)	m	N	Mean pole position Lat N	Long E	A_{95} (°)
Early Jurassic 2	175-190	180	11	256	64.2	96.1	6.0
Early Jurassic 1	190-210	200	26	468	62.9	91.2	2.7
Late Triassic	210-230	220	28	541	55.2	97.5	1.9
Late Permian/Early Triassic	230-250	240	48	453	49.3	117.0	2.3
Mid Permian	250-270	260	42	541	46.6	123.6	1.8
Early Permian	270-290	280	50	457	41.6	128.0	1.8
Late Carboniferous 2	290-310	300	33	421	36.3	130.5	3.2
Late Carboniferous 1	310-330	320	22	240	33.4	126.8	4.4
Early Carboniferous	330-350	340	18	153	25.2	122.4	5.8
Late Devonian/Early Carboniferous	350-370	360	20	136	23.9	125.0	3.9
Middle Devonian	370-390	380	19	113	21.8	119.2	5.1
Early Devonian	390-410	400	31	384	1.4	101.5	3.0
Late Silurian/Early Devonian	410-425	420	14	199	3.0	125.7	4.6
Early/Middle Silurian	425-450	440	18	173	19.2	126.9	3.9
Middle/Late Ordovician	450-470	460	9	85	13.6	148.5	7.6
Early Ordovician	470-490	480	5	40	13.8	162.5	9.0
Late Cambrian	495-505	500	8	59	-1.9	165.0	8.6
Early Cambrian	525-545	535	7	92	4.3	154.7	10.1

[a]Symbols and method of averaging are as given in Table 6.5.

Their data can be combined after rotation of Europe to Laurentia for times older than 175 Ma. The resulting APWP in North American co-ordinates is then a reference path for *Laurussia* with which other paleomagnetic data can be compared. The data for the two continents are not uniformly distributed in either case, but, in determining a reference path for the combined data set, they complement each other. For pre-425 Ma, only the data from North Britain have been combined with the data for Laurentia to produce the composite APWP.

The mean poles for this reference path for Laurussia are listed at 20-Myr intervals in Table 7.1. The path has been calculated using 409 groups of data from 4811 sites and represents a large body of paleomagnetic data, more than twice that for the Gondwana continents (§7.2.3).

7.2.2 Paleo-Asia

Asia is a composite continent made up of accreted terranes that merged with Europe in the late Paleozoic/early Mesozoic. The situation is therefore the inverse of that discussed in §7.2.1, where two currently separated continents were once joined together. Figure 7.5 compares the APWPs of Siberia and Europe. For pre-425-Ma times Europe as such did not exist, so the comparison for pre-425 Ma has to be either with North Britain or with Baltica (see Fig. 6.14). Because Siberia subsequently collided with Baltica, it is appropriate to use the APWP for Baltica for that time. The paths are widely separated during the early Paleozoic and merge during the Early Triassic as indicated by the close proximity of the mean pole positions at 235 Ma. However, it should be noted that the APWP for Siberia is not well constrained during the late Mesozoic and Cenozoic, as it is for Europe. Therefore, it is convenient to define a reference APWP for Eurasia by combining the data for Europe and Siberia for Mesozoic

TABLE 7.2
Reference Combined Phanerozoic APWP for Eurasia for the Mesozoic and Cenozoic.

Geological period	Age range (Ma)	Mean (Ma)	m	N	Mean pole position Lat N	Long E	A_{95} (°)
Eocene	35-56	45	13	965	82.2	170.2	2.0
Paleocene	56-65	60	12	1086	77.4	169.4	3.0
Late Cretaceous	65-97	80	24	86	74.8	176.3	4.5
Early Cretaceous	97-146	120	25	197	74.6	201.8	4.3
Late Jurassic	146-157	150	8	81	74.8	149.9	3.7
Middle Jurassic	157-178	165	4	20	64.4	123.6	7.1
Early Jurassic	178-208	195	10	125	66.1	128.0	8.6
Late Triassic	208-225	215	7	65	50.8	133.1	8.2
Middle Triassic	225-235	230	10	120	52.0	147.4	5.6
Early Triassic	235-245	240	30	689	48.9	141.3	3.8

[a]Symbols and method of averaging are as given in Table 6.5.

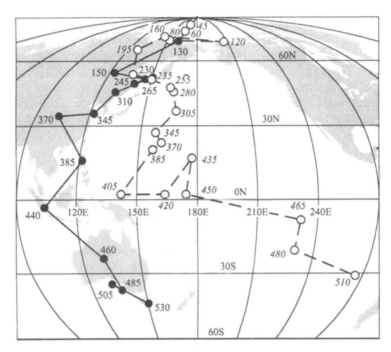

Fig. 7.5. Comparison of the APWP for Siberia (solid line and circles) with that of Europe (dashed line and open circles) during the Phanerozoic. Ages for each mean pole position are indicated in millions of years (Europe given in italics). For Europe pre-425 Ma, results are from Baltica only.

and Cenozoic times following the amalgamation of Siberia with Europe. This reference path for the Mesozoic and Cenozoic is listed in Table 7.2.

To the south of Siberia, the terranes of North China, South China, and Tarim merged with Siberia and the Kazakhstan subcontinent (see Fig. 6.15). Some details of the times of collision can be obtained from a comparison of the APWPs for the three blocks of China with that of Eurasia using the reference APWP given in Table 7.2 for the Mesozoic and Cenozoic and the APWP for Siberia for the Paleozoic as shown in Fig. 7.6. The APWPs for these four terranes are widely separated during all of the Paleozoic and start to merge during the Jurassic. The Early Cretaceous (120-Ma) poles for all four blocks are not significantly different; therefore, as a large continent Asia is a relatively recent phenomenon. For times younger than 120 Ma the joint paths are represented by that shown for Eurasia.

The paleolatitudes of Tarim and Eurasia have been similar since the Carboniferous, the main differences between the two data sets being in declination for younger times. This appears to conform with geological observations that Tarim first collided with Eurasia (Junggar) during the Early

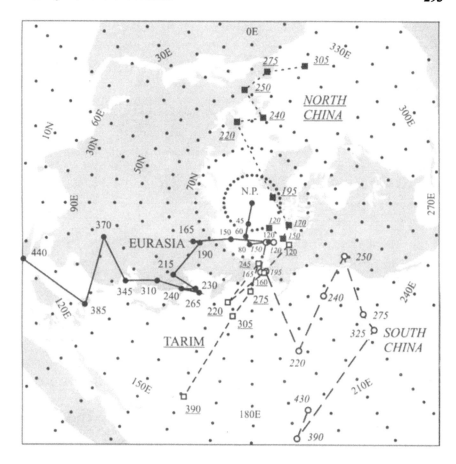

Fig. 7.6. APWPs for Eurasia (Siberia for >240 Ma)(solid circles), North China (solid squares), South China (open circles) and Tarim (open squares). Ages in millions of years are indicated against each mean pole position in the style corresponding to the terrane names for each APWP (normal, italics, underlined, or some combination). From 120 Ma to the present the APWPs all follow the path for Eurasia.

Permian along the Tianshan Mountains. Following the initial collision, there has been relative rotation between Tarim and Eurasia. However, the 160-Ma poles for Tarim and South China are virtually identical, so it appears that these two blocks probably merged at that time.

To assess the time of collision of South China with North China and their amalgamation with Eurasia, the details of the paths for the Jurassic and Early Cretaceous for these blocks are shown in Fig. 7.7. Both the Early Jurassic (195 Ma) and Middle Jurassic (165 Ma) poles for North and South China are significantly different because their circles of 95% confidence do not overlap.

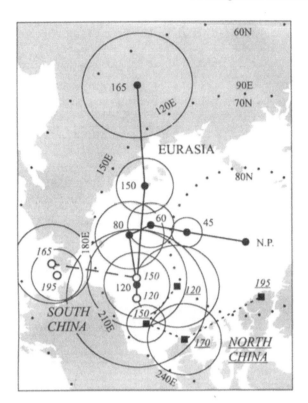

Fig. 7.7. Comparison of the APWP for Eurasia (165–45 Ma) with those for North China and South China during the Jurassic and Early Cretaceous (195–120 Ma). Matching symbols are as in Fig. 7.6. Circles of 95% confidence are shown around each mean pole except for North China at 195 Ma, which is based on a single result. Note that the poles for North and South China at 150 Ma are not significantly different but are clearly significantly different from the Eurasian pole at 150 Ma. The poles for all three blocks agree at 120 Ma.

Only by the Late Jurassic (150 Ma) are their poles not significantly different. Therefore it appears that the final amalgamation of the North and South China blocks occurred at about 150 Ma.

Gilder and Courtillot (1997) deduced a similar age of ~159 Ma (equivalent to the Middle/Late Jurassic boundary) for the timing of the amalgamation of North and South China. However, the paleolatitudes of these two blocks have been similar since the Early Triassic, so they may have been close together since that time. The final suture between these blocks is then one of rotation about a hinge at their eastern ends in a scissor-like fashion (Zhao and Coe, 1987; Zhao *et al.*, 1996). This is consistent with the geological observation that marine sediments of Triassic age are well developed along the northwestern margin of the South

China block but thin eastwards and are represented by continental facies on the northeastern margin (Yang *et al.*, 1986).

Although the North and South China poles merge at 150 Ma, they are still significantly different from the Eurasian 150-Ma pole (Fig. 7.7). Therefore, the final amalgamation of the combined North and South China (and Tarim) blocks with Eurasia is later than this and probably was completed by the earliest Cretaceous with the closure of the Amurian Sea. The poles for these blocks are in agreement with each other at 120 Ma (see also Gilder and Courtillot, 1997). This is consistent with the continuous observation of marine fossils between Cambrian and Late Jurassic times in the Mongol–Okhotsk orogenic belt (Kosygin and Parfenov, 1981).

Where were these blocks of Asia formerly located? McElhinny *et al.* (1981) first showed that all these blocks were located in the equatorial region in Permian times. They may have been a part of an equatorial archipelago during the late Paleozoic (Zhao *et al.*, 1996). They were certainly not a part of eastern Gondwana since Australia and India were in low latitudes in Permo-Carboniferous times (§7.2.3).

7.2.3 Gondwana

The breakup of the supercontinent of Gondwana commenced slightly earlier than 160 Ma with the separation of Antarctica from Africa (Table 5.2). The separation of South America from Africa, however, started at about 130 Ma. Gondwana was probably not finally assembled until about 530 Ma following the Brasiliano (600–530 Ma) and Pan-African (≈550 Ma) orogenies as summarized by Powell *et al.* (1993) and Meert and Van der Voo (1997). Discussion of the formation of Gondwana from the breakup of its ancestor Rodinia is given in §7.4.3. The purpose here is to establish a reference APWP for Gondwana for the Phanerozoic using the data for each of its constituent parts as set out in §6.5.4.

The paleomagnetic poles for the Gondwana continents have been rotated using the reconstruction parameters of Lottes and Rowley (1990). Their reconstruction takes account of the internal deformation of Africa arising from the formation of an Early Cretaceous extensional rift system from the Benue Trough into central and northern Africa (see Fig. 6.19a). The rotation parameters that apply to data from Africa and Arabia are given in §6.5.4, following Lottes and Rowley (1990). Northwest (NW) Africa is the reference region and all rotations internal to Africa are made to this region. The rotation parameters of Lottes and Rowley (1990) for reconstructing Gondwana also use NW Africa as the reference region (see §5.1.1 for explanation of Euler rotation conventions) and are as follows.

(i) Africa treated internally as three plates reconstructed to NW Africa (see §6.5.4).

(ii)   South America to NW Africa; Euler pole at 53N, 325E, +51.0° (counter-clockwise).
(iii) India to northwest Africa; Euler pole at 26.7N, 37.3E, -69.4° (clockwise).
(iv)   Australia to NW Africa; Euler pole at 28.1S, 66.8W, +52.1° (counter-clockwise)
(v)   Antarctica to NW Africa; Euler pole at 12.4S, 33.8W, +53.3° (counter-clockwise).
(vi)   Madagascar to NW Africa; Euler pole at 14.9S, 82.4W, +15.7° (counter-clockwise).

Table 7.3 gives the reference APWP for Gondwana for ages between 180 and 530 Ma. For times between 130 and 170 Ma, only data for Africa and South America have been combined. From 140 to 325 Ma, the data have been combined so as to give mean pole positions at 20-Myr intervals. It should be noted that the assignment of poles to a particular 20-Myr interval is occasionally somewhat arbitrary because the ages of many of the poles are not known with such precision. For earlier times the mean pole positions are determined according to the availability of data.

The resulting *south pole* APWP for Gondwana is illustrated in Fig. 7.8 in NW Africa co-ordinates. The south pole APWP has traditionally been plotted for Gondwana, because Gondwana moved across the south pole throughout much of

TABLE 7.3

Reference Combined Phanerozoic APWP for Gondwana after Rotation of Paleomagnetic Poles to Northwest Africa Using the Rotation Parameters of Lottes and Rowley (1990)[a]

Geological period	Age range (Ma)	Mean (Ma)	m	N	Mean pole position Lat N	Long E	A_{95} (°)
Late Jurassic/Early Cretaceous[b]	130-150	140[b]	9	591	56.1	262.4	2.5
Late Jurassic[b]	150-170	160[b]	9	117	63.2	254.5	6.0
Middle Jurassic	170-190	180	22	256	69.4	251.5	4.0
Early Jurassic	190-210	200	15	138	68.0	247.0	4.2
Late Triassic	210-230	220	11	118	65.8	259.7	5.1
Late Permian/Early Triassic	230-250	240	20	122	50.9	259.8	6.1
Mid Permian	250-270	260	14	78	44.8	253.7	8.3
Early Permian	270-290	280	14	75	39.6	241.5	7.9
Late Carboniferous 2	290-310	300	13	149	38.3	241.2	4.2
Late Carboniferous 1	310-325	320	10	50	27.7	234.9	5.1
Early Carboniferous	325-363	345	8	60	15.0	213.2	18.1
Late Devonian	363-377	370	4	24	4.3	189.6	6.5
Middle Devonian	377-391	385	8	47	24.8	184.7	7.8
Early Devonian	391-417	405	2	22	51.0	187.9	15.0
Early Silurian	428-443	435	1	6	-53.5	236.6	-
Late Ordovician	443-465	455	5	18	-35.6	172.2	15.0
Early/Middle Ordovician	465-495	480	9	86	-35.2	185.3	8.1
Middle/Late Cambrian	495-518	505	15	82	-11.5	176.2	4.9
Early Cambrian	518-545	530	9	37	-33.9	169.7	6.9

[a]Symbols and method of averaging are as given in Table 6.5.

[b]Africa and South America only.

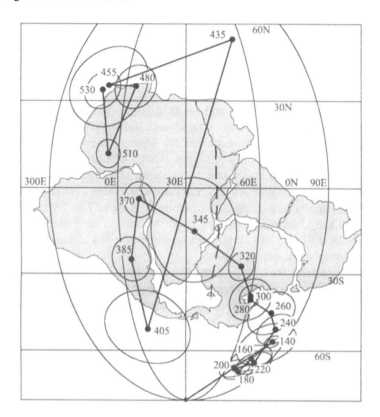

Fig. 7.8. Combined reference south pole APWP for Gondwana in NW Africa co-ordinates using the mean pole positions listed in Table 7.3. Mean ages at each mean pole are given in millions of years. The dashed line indicates the possible suture between East and West Gondwana.

its history and the relationship of the pole path to the occurrence of various glacial deposits is of some interest. The mean pole positions listed in Table 7.3 are the usual north pole positions; the corresponding south poles are determined by inverting the sign of the latitude and subtracting 180° from the longitude.

During the Ordovician the south pole is situated in northwest Africa consistent with the well-known occurrence of glacial deposits of this age. A single pole from the Mereenie Sandstone of central Australia (Li *et al.*, 1991) of possible Early Silurian age is the youngest of the poles situated in the north Africa region but lies 47° to the east of the Ordovician poles. During the Silurian the pole then makes a rapid shift through an angle of more than 90° from north Africa to a position off southern Africa. Van der Voo (1994) suggested that this rapid polar shift (~90° in what appears to be ~30 Myr) may be due to true polar wander (TPW), as will be discussed in §7.5.1. The pole then moves up towards the equator during the Devonian and Early Carboniferous then to the east during the

Late Carboniferous and Permian consistent with the occurrence of glacial deposits in India and Australia at that time.

7.2.4 Pangea

The reconstruction of the North and South Atlantic Oceans by Bullard *et al.* (1965), when combined with that of Smith and Hallam (1970) for Gondwana, created the first formal reconstruction of the supercontinent of *Pangea*, which existed just prior to the opening of the North Atlantic Ocean at 175 Ma (Table 5.2). Therefore, to investigate the existence of Pangea before 175 Ma, a comparison of the reference APWPs for Laurussia (Table 7.1) and Gondwana (Table 7.3) is appropriate. Such a comparison is made in Fig. 7.9 for the poles between 180 and 320 Ma after rotating North America to northwest Africa about an Euler pole at 61.3N, 343.2E through an angle of +79.5° (counterclockwise), following Lottes and Rowley (1990).

The reference poles for Laurussia and Gondwana are not significantly different at 180 and 200 Ma, just prior to the opening of the Atlantic Ocean. However, for earlier times the Gondwana poles all plot systematically eastwards of those from Laurussia. Although the overall lengths of the two paths between 180 and 320 Ma are almost identical, the angular changes in pole positions for successive 20-Myr intervals are quite different. It must first be determined whether the difference arises from a data problem or from a reconstruction problem. The difficulty of assigning poles from the Gondwana continents into the 20-Myr time groupings has already been noted in §7.2.3. Restricting the Gondwana poles to West Gondwana (Africa and South America) substantially reduces the eastward trend of the Gondwana path, as shown in Fig. 7.9. Thus, it appears that the poles from the individual Gondwana continents do not, when reconstructed, match each other as well as they might. The larger circles of 95% confidence for the mean Gondwana poles compared with those from Laurussia reflect this fact.

Two factors may be combining to create the problem illustrated in Fig. 7.9. The first is age assignment for each pole and the second is a reconstruction problem. A change in age assignments would move the positions of the mean poles north or south and therefore account for differences in angular change between pole positions; the easterly trend of the Gondwana path would remain. However, it does appear as though the poles have not been reconstructed tightly enough, so perhaps the reconstruction of East Gondwana with West Gondwana is not tight enough. For example, Gondwana reconstructions do not fully account for the extension that occurs on continental margins prior to sea-floor spreading (e.g., see Tikku and Cande, 1999).

The factors described above have led to many discussions regarding the reconstruction of Pangea between Late Carboniferous and Early Jurassic, as has been well summarized by Smith and Livermore (1991). Taken at their face value

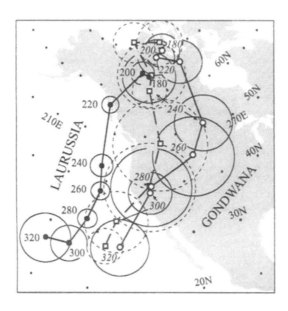

Fig. 7.9. Comparison of the reference poles for Laurussia (Table 7.2, solid circles) and Gondwana (Table 7.3, open circles) in the co-ordinates of northwest Africa after rotation of North America to northwest Africa. Ages are shown in millions of years (italics for Gondwana poles). The APWP with data restricted to West Gondwana (open squares) is shown as a dashed line. Circles of 95% confidence are shown for each of the mean poles (dashed for West Gondwana only).

the difference in pole paths requires that Gondwana be rotated with respect to Laurussia in some way so as to bring the two paths into coincidence. Morel and Irving (1981) referred to the conventional reconstruction of Pangea just prior to the opening of the Atlantic Ocean as Pangea A1 (Fig. 7.10a). Van der Voo and French (1974) proposed a reconstruction that brought the older parts of the curves into coincidence (Fig. 7.10b), named by Morel and Irving (1981) as Pangea A2. This involves rotating Laurussia about an Euler pole located in the Sahara at 19.3N, 0.7W through an angle of about +20° (counterclockwise) relative to its position in Pangea A1. Here the cratonic edge of South America is fitted closely to the Gulf coast of North America by tightly closing the Gulf of Mexico. Although parts of Mexico would then overlap with South America, this is a geologically plausible scheme considering the existence of displaced terranes in this part of Mexico.

In order to bring the poles for the Early Permian (280 Ma) into coincidence, Irving (1977) and Morel and Irving (1981) proposed Pangea B (Fig. 7.10c). Here Gondwana is rotated clockwise so that the northwest coast of South America is placed against the Atlantic margin of North America and Africa lies to the south of eastern Europe and southwestern Asia. Hallam (1983) generalized this approach by naming the paleomagnetic reconstructions for 240, 280, and

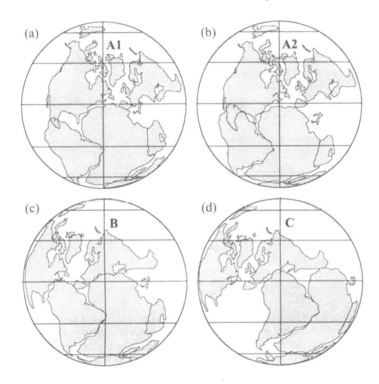

Fig. 7.10. Various reconstructions of Pangea. (a) Pangea A1, the conventional reconstruction just prior to the opening of the Atlantic Ocean (Bullard *et al.*, 1965). (b) Pangea A2, with the cratonic edge of South America fitted closely to the Gulf coast of North America (Van der Voo and French, 1974). (c) Pangea B, with the northwest coast of South America fitted to the Atlantic coast of North America (Morel and Irving, 1981). (d) Pangea C, with the northwest coast of South America fitted against southeast Europe (Smith *et al.*, 1981). After Smith and Livermore (1991), with permission from Elsevier Science.

320 Ma proposed by Smith *et al.* (1981) as Pangea C (Fig. 7.10d). Alternative paleomagnetic reconstructions made by fitting linear segments of the Laurussia and Gondwana paths have been termed Pangea D by Smith and Livermore (1991). They are similar to Pangea A2 and require rotation of Laurussia about an Euler pole located in central Africa at 6S, 19E through an angle of about +30° (counterclockwise) relative to its position in Pangea A1.

It is generally agreed that Pangea A1 existed just prior to the opening of the Atlantic Ocean at 175 Ma. The Pangea B and C configurations must therefore have evolved to the A1 configuration, presumably during the (Permo-) Triassic. The transition between Pangea B and A1 requires a dextral strike slip (referred to as the "Tethys Twist" – see Van der Voo, 1993) of 3500 km with respect to Laurussia at rates of up to 100 km Myr^{-1}. The transition from Pangea C to A1 would require an even larger displacement (see Fig. 7.10). There is no plausible geological evidence that would account for a megashear of such magnitude. However, some recent analyses of paleomagnetic data by Muttoni *et al.* (1996) and Torcq *et al.* (1997) still favor Pangea B. Pangea B and C are discounted here as implausible, and both Smith and Livermore (1991) and Van der Voo (1993) agree that the solution that is both geologically plausible and paleomagnetically acceptable is that of Pangea A2. Final resolution of this problem requires the restudy of many of the Permian and Triassic poles of Gondwana. In addition, the paleomagnetic data strongly suggest that the reconstruction of Gondwana needs careful reassessment.

7.2.5 Paleogeography: 300 Ma to the present

Pangea formed in the Late Carboniferous (~300 Ma) when the Rheic Ocean between Gondwana and Laurussia closed. At that time the south pole lay in Antarctica (Fig. 7.8) and most of Gondwana was covered by an ice sheet. Asia as a continent did not exist and Siberia, Kazakhstan, and North and South China lay in equatorial latitudes. Because of the controversy surrounding the evolution of Pangea during the Permo-Carboniferous (§7.2.4), a discussion of global paleogeography will be restricted to Late Permian and younger times at this point. An overview of Earth history since 1000 Ma is given in §7.4.3, which includes a representation of the paleogeographic information here in a cladistic-style diagram (see Fig. 7.24).

Paleomagnetic data for the major blocks on the surface of the Earth can be combined with the information derived from marine magnetic anomalies summarized in §5.3.1 (see Fig. 5.19) to produce paleogeographic maps through time. Such a set of maps has been compiled by Scotese (1997) and is shown in Fig. 7.11 for times since the Late Permian (255 Ma). The maps include topographic and sea level information and are set in their correct paleolatitude according to paleomagnetic data. Paleogeographic maps for the Paleozoic based on combined paleomagnetic and geological information can be found in Scotese and McKerrow (1990) and are updated by Scotese (1997). A much more complete set of maps than those presented here can be viewed on the internet site *http://www.scotese.com*.

In the Late Permian (255 Ma), Pangea dominated global paleogeography (Fig. 7.11a). North and South China and Indochina were located in low latitudes, whereas Gondwana occupied high latitudes. Turkey, Iran, and the terranes of Tibet were the components of the Cimmerian continent postulated by Sengör *et al.* (1984). As a protocontinent located in Gondwana, Cimmeria was situated just north of India and Arabia, whereas the Paleotethys Ocean lay to the north. At 255 Ma Cimmeria had just separated from Gondwana resulting in the initial opening of the Tethys Ocean. Siberia and Kazakhstan were situated in high latitudes but lay close to Europe.

By the Middle Triassic (237 Ma) Siberia and Kazakhstan had amalgamated with Europe (Fig. 7.11b) to form the Ural Mountains. The Tethys Ocean was well developed and the map shows Cimmeria occupying a position between Gondwana and Eurasia. However, it should be noted that the latest paleomagnetic data from Iran (Besse *et al.*, 1998) suggest that Iran had already merged with Eurasia at that time (see §7.3.5, Fig. 7.16). South China and Indochina lay in equatorial latitudes and North China had by then moved into intermediate northerly latitudes. In Early Jurassic time (195 Ma) the Tethys Ocean was fully developed (Fig. 7.11c). North and South China had not yet amalgamated but lay close to the southern margin of Eurasia. The Amurian Seaway separated North China from Siberia.

Pangea started to break up about 175 Ma when Africa separated from North America. The breakup of Gondwana commenced at about 160 Ma when Antarctica (together with Greater India) separated from Africa (see Table 5.2). At about 130 Ma, North America separated from Iberia and South America commenced its separation from Africa. Note that East Gondwana (Greater India, Madagascar, Antarctica, and Australia) first drifted southwards on separation from Africa. Then Greater India and Madagascar separated from Antarctica at about 130 Ma and drifted northwards. By Mid-Cretaceous time (94 Ma), the major blocks that make up present-day Eurasia (with the exception of India) had amalgamated and the development of narrow North and South Atlantic Oceans had occurred (Fig. 7.11d).

By the latest Cretaceous (69 Ma), Greater India had separated from Madagascar and was drifting rapidly northwards in the central Indian Ocean (Fig. 7.11e). The Tethys Ocean had by then virtually disappeared. Although the separation between Australia and Antarctica commenced at about 95 Ma (see Table 5.2), at 69 Ma they still lay close to one another because the rate of separation was initially very slow. The Labrador Sea between Greenland and North America opened at about 85 Ma and was still only a narrow seaway at 69 Ma. The opening of the South Atlantic continued and South America and Africa were well separated at this time (Fig. 7.11e). The North Atlantic between Europe and Greenland did not open until about 60 Ma (see Table 5.2).

By the Eocene (50 Ma) Greater India was about to collide with Asia, resulting in the formation of the Himalayas and underthrusting of the Indian plate. The separation between Antarctica and Australia was well under way (Fig. 7.11f) and the Indian Ocean was well developed. The North Atlantic Ocean between Europe and Greenland had started to open. Meanwhile, opening of the Labrador Sea continued.

Fig. 7.11(a–c). Global paleogeographic maps for (a) Late Permian (255 Ma), (b) Middle Triassic (237 Ma), and (c) Early Jurassic (195 Ma). Oceanic trenches are shown in red. From Scotese (1997), see also *http://www.scotese.com*.

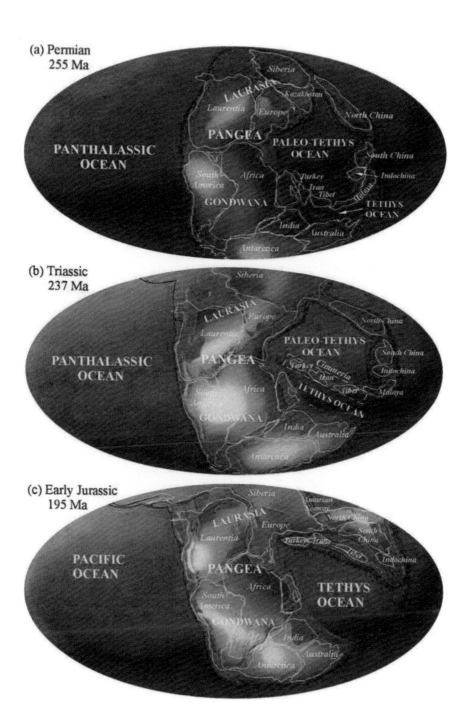

(a) Permian
255 Ma

Siberia

LAURASIA

Kazakhstan

Laurentia

Europe

North China

PANGEA

PALEO-TETHYS
OCEAN

PANTHALASSIC
OCEAN

South China

South
America

Africa

Turkey
Iran
Tibet

Indochina

Malaya

GONDWANA

TETHYS
OCEAN

India

Australia

Antarctica

(b) Triassic
237 Ma

Siberia

LAURASIA

Europe

North China

Laurentia

PALEO-TETHYS
OCEAN

PANTHALASSIC
OCEAN

PANGEA

South China

Turkey

Cimmeria

Iran

Indochina

South
America

Africa

Tibet

Malaya

TETHYS OCEAN

GONDWANA

India

Australia

Antarctica

(c) Early Jurassic
195 Ma

Siberia

Amurian
Seaway

North China

LAURASIA

Europe

South
China

PACIFIC
OCEAN

Laurentia

Turkey Iran

PANGEA

Tibet

Indochina

TETHYS
OCEAN

South
America

Africa

GONDWANA

India

Australia

Antarctica

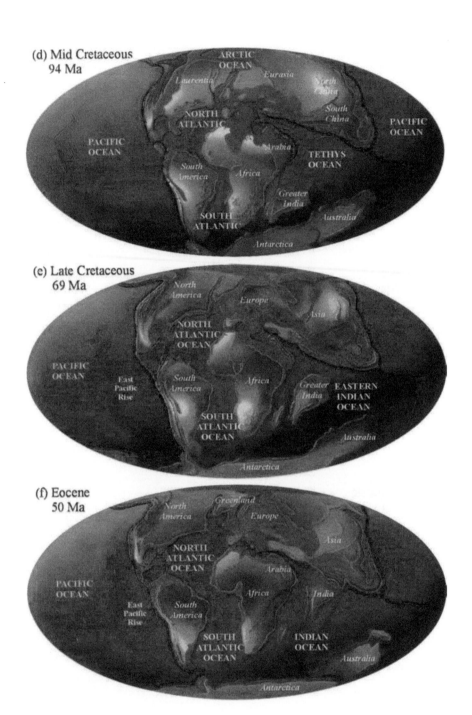

(d) Mid Cretaceous
94 Ma

ARCTIC OCEAN
Laurentia
Eurasia
North China
South China
PACIFIC OCEAN
NORTH ATLANTIC
PACIFIC OCEAN
Arabia
TETHYS OCEAN
South America
Africa
Greater India
Australia
SOUTH ATLANTIC
Antarctica

(e) Late Cretaceous
69 Ma

North America
Europe
Asia
NORTH ATLANTIC OCEAN
PACIFIC OCEAN
East Pacific Rise
South America
Africa
Greater India
EASTERN INDIAN OCEAN
SOUTH ATLANTIC OCEAN
Australia
Antarctica

(f) Eocene
50 Ma

North America
Greenland
Europe
Asia
NORTH ATLANTIC OCEAN
Arabia
India
PACIFIC OCEAN
East Pacific Rise
South America
Africa
SOUTH ATLANTIC OCEAN
INDIAN OCEAN
Australia
Antarctica

7.3 Displaced Terranes

7.3.1 Western North America

In the early 1970s it was noted that the paleomagnetic record from rocks along the western margin of North America showed consistent and systematic discordance with results of the same age from the stable Laurentian craton (Irving and Yole, 1972; Packer and Stone, 1972). Beck (1976, 1980) and Irving (1979) reviewed all available paleomagnetic results from the western margin and then compared them with an established set of reference poles from the cratonic interior (e.g., Irving, 1977). Only a few pre-Miocene directions were found to be concordant with those from the craton. Most rock units exhibited more easterly paleomagnetic declinations and/or shallower inclinations than would have been expected from the reference poles. It was concluded that the likely cause of this systematic difference was northward translation and/or clockwise rotation of exotic terranes within or alongside the westernmost Cordillera.

These observations made it clear that most of the western edge of North America is made up of terranes (commonly referred to as *displaced terranes*) that have been displaced from their original positions and have been transported, rotated, and accreted to the Laurentian craton. Northward transport and clockwise rotation have been the prime elements in shaping the western Cordillera. For example, south of Cape Mendocino (40°N) and north of Vancouver Island (50°N) the main displacement has been northward transport of these displaced terranes, whereas between these locations in the Pacific northwest only clockwise rotations are found. This is consistent with the steady underthrusting of the coastal Pacific northwest by the Farallon plate for most of the Cenozoic, contrasting with the transform activity to the north or south (Beck, 1980). Batholith belts, such as in the Peninsular Range batholith of southern California and the Coast Plutonic Complex of British Columbia, also appear to have been involved in this northward transport and clockwise rotation. Although undetected tectonic tilts could also explain the results from batholiths, Beck (1980) argues that postmagnetization tilts are probably small.

Some of the displaced terranes north of Vancouver Island were shown in Fig. 6.11. The most important of these in southern Alaska from a paleomagnetic viewpoint are the Alexander, Peninsular, and Wrangell terranes, the latter being named Wrangellia by Jones *et al.* (1977). Based on paleomagnetic data, Monger

---

Fig. 7.11(d–f). Global paleogeographic maps for (d) Mid-Cretaceous (94 Ma), (e) Latest Cretaceous (94 Ma), and (f) Eocene (50 Ma). Oceanic trenches are shown in red. From Scotese (1997) (see also *http://www.scotese.com*).

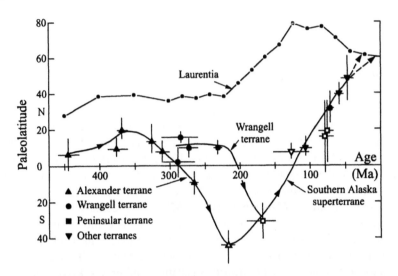

Fig. 7.12. Paleolatitude variation of the terranes of southern Alaska compared with those that would have been observed had they been attached to Laurentia since 450 Ma. Solid symbols indicate known polarity (and hence hemisphere). Open symbols indicate equivocal polarity. After Panuska and Stone (1985).

and Irving (1980) suggested a Jurassic or Cretaceous age for the accretion of the inboard Stikine terrane. Estimates of the age of suturing of the Alexander, Peninsular and Wrangell terranes with each other (but not necessarily with Laurentia) are between 140 and 150 Ma based on geological considerations (Wallace *et al.*, 1989). Summaries of paleomagnetic investigations of terranes in British Columbia and southern Alaska have been given by Hillhouse and Grommé (1984), Symons and Litalien (1984), Irving *et al.* (1985), Coe *et al.* (1985), and Panuska and Stone (1985).

The first APWP for the Alexander terrane was constructed by Van der Voo *et al.* (1980). Panuska and Stone (1985) summarized the paleomagnetic data for the Alexander, Peninsular, Wrangell, and associated terranes in southern Alaska and deduced the paleolatitude changes shown in Fig. 7.12. The proposed southern hemisphere position of the Alexander terrane in the Triassic has been the subject of some discussion. Because the polarity of the paleomagnetic results is not known, it is possible for the observed paleolatitude to be either north or south. The latitudinal motion shown in Fig. 7.12 is based on the simplest APWP path and least latitudinal motion. The Alexander, Peninsular, and Wrangell terranes, as well as other terranes, amalgamated to form the Southern Alaska superterrane in the Middle to Late Jurassic when located in low to middle southerly latitudes. The Southern Alaska superterrane accreted to Laurentia sometime during the Cenozoic.

Coe *et al.* (1985) concluded that the Peninsular and Wrangell terranes collided with North America in the vicinity of present-day Cape Mendicino in Mid- to Late Cretaceous time, about 20° south of their present position with respect to the Laurentian craton. They were then moved northwestward along strike-slip faults parallel to the ancient margin by oblique subduction of the Kula plate and arrived at interior Alaska some time before 52 Ma. Irving *et al.* (1985) examined the data from the Wrangell terrane and concluded that it arrived at the continental margin at the present latitude of Baja California at about 100 Ma. It was then intruded by the Coast Plutonic Complex and subsequently moved tangentially along the coast driven by oblique convergence of the Farallon and Kula plates. Engebretson *et al.* (1985) produced a model of the evolution of the Pacific plate since 140 Ma, as was described in §5.5.2 and illustrated in Fig. 5.25. Debiche *et al.* (1987) used that model to relate the timing of the accretion of displaced terranes along the western margin of North America to the evolution of the Pacific plate.

The possible trajectories taken by these displaced terranes across the present-day Pacific depends on the timing and the position at which they docked along the western margin. Two possible scenarios are shown in Fig. 7.13 for docking times at 60 and 90 Ma using five different docking positions down the entire western margin. For each trajectory the starting time is the age of the oldest sea floor now adjacent to the continental margin. For docking times at 30 Ma, the trajectories are short because the Pacific–Farallon spreading system is close to the margin. For a docking time at 60 Ma (Fig. 7.13a), the possible terrane trajectories indicate northward transport by 60° or more since 125 Ma. For a

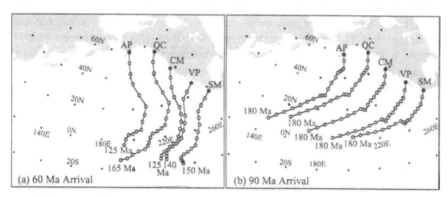

Fig. 7.13. Trajectories of oceanic plates in fixed North American co-ordinates based on the plate model of Engebretson *et al.* (1985) shown in Fig. 5.25. Trajectories are determined at 5-Myr intervals by backward modeling for arrival times of (a) 60 Ma and (b) 90 Ma at five locations: AP, Alaskan Peninsula (56°N); QC, Queen Charlotte Islands (53°N); CM, Cape Mendocino (40°N); VP, Vizcaino Peninsula (27°N); SM, Southern Mexico (18°N). Eurasia is reconstructed to its position in North American co-ordinates for the two arrival times. After Debiche *et al.* (1987), with permission from The Geological Society of America.

docking time at 90 Ma (Fig. 7.13b), and also at 120 Ma, the trajectories indicate easterly transport by as much as 60° of longitude since 180 Ma.

As a terrane rides passively on an oceanic plate it also undergoes rotation relative to North America arising purely from the geometry of its motion (Debiche *et al.*, 1987). For example, rocks with age 140 Ma, docking at the latitude of the Queen Charlotte Islands at 60 Ma (Fig. 7.13a), will have rotated 60° clockwise from the expected direction simply as a result of transport across the Farallon and Kula plates. Therefore, the clockwise rotations so evident in the paleomagnetic record of the displaced terranes along the western margin arise not from tectonic rotation in the shear zone along the continental margin, but purely as a result of their trajectory across the Pacific. These rotations have been referred to as *trajectory rotations* by Debiche *et al.* The net amount of tectonic rotation is actually zero.

Debiche *et al.* (1987) also modeled the coastwise translation of terranes following their docking on the western margin. Such translation arises either from transform motion or from oblique convergence between North America and the oceanic plates lying to the west. Applying their model to the Wrangell terrane, they find that the following scenario is compatible with paleomagnetic data. The Alaskan part of the Wrangell terrane rode with the Farallon plate until docking near the present-day latitude of Vancouver Island at ~90 Ma and was then driven tangentially along the margin until reaching its present latitude at ~56 Ma. This is not unlike the suggestion made by Coe *et al.* (1985) from their analysis of paleomagnetic data. The Vancouver Island portion of the Wrangell terrane rode with the Farallon plate until docking near the present-day latitude of Baja California at ~90 Ma and was then driven tangentially along the margin until it reached its present position at ~55 Ma. This agrees well with the proposal of Irving *et al.* (1985) based on paleomagnetic data. However, Harbert (1990) argues that models involving coastwise translation of the southern Alaska terranes do not fit the most reliable paleomagnetic data. His interpretation is that these terranes rode on the Kula plate and sutured to North America at about 50 Ma without significant coastwise translation.

7.3.2 The East and West Avalon Terranes

The West Avalon terrane of North America shown in Fig. 6.11 is named after its type locality on the Avalon Peninsula of eastern Newfoundland. It also includes a coastal strip in eastern New Brunswick and Nova Scotia and the Boston Basin in Massachusetts. Paleomagnetic results have been summarized by Van der Voo (1993). The East Avalon terrane includes the southern parts of England, Wales, and Ireland (see Fig. 6.13). Paleomagnetic data from East Avalon have been reviewed by Trench and Torsvik (1991), Van der Voo (1993), and Torsvik *et al.* (1993) to produce an APWP for this terrane. The West Avalon terrane has many characteristics in common with Armorica, including aspects of the coastal landscape in Newfoundland and Brittany and geological and geophysical similarities (Van der Voo, 1982).

In Fig. 7.14 the paleolatitudes determined from the East and West Avalon terranes during the Paleozoic are compared with those expected if these terranes were a part of Laurentia or Gondwana. The predicted paleolatitudes have been calculated from the mean poles listed in Tables 7.1 and 7.3, respectively, using a present-day reference point at 45N, 65W in West Avalon and 54N, 4W in East Avalon. The shaded regions represent the fact that there is an error associated with each of these curves. The observed paleolatitudes for Ordovician and younger times for the West Avalon terrane show excellent agreement with those predicted from Gondwana (Fig. 7.14a). The high paleolatitudes expected for the northern part of Gondwana during the Ordovician are also seen in the data for West Avalon. During the Early to Middle Cambrian, however, there is no such agreement. However, Gondwana very likely only formed in the Early Cambrian, so this disagreement has little significance.

For the East Avalon terrane, the results for Ordovician times show both intermediate and high paleolatitudes (Fig. 7.14b). However, there is a tendency for high paleolatitudes during both the Cambrian and Ordovician, a result suggesting more of an affinity with Gondwana than with Laurentia. The results from both the West and East Avalon terranes suggest they cannot have been a part of Laurentia during the early and middle Paleozoic. However, the northern

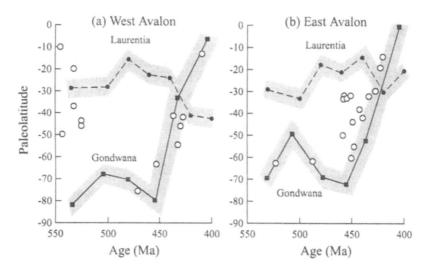

Fig. 7.14. Paleolatitudes (open circles) observed from rocks in (a) the West Avalon terrane for a present-day location at 45N, 65W and (b) the East Avalon terrane for a present-day location at 54N, 4W. Paleolatitudes predicted using the results from Laurentia (solid circles) and Gondwana (solid squares) are shown for comparison.

part of Gondwana was located in high latitudes during the Ordovician, with the south pole located in northwest Africa (see Fig. 7.8). Therefore, it is generally presumed that these terranes originally occupied a position in the region of northwest Africa (see Fig. 7.16 below). Van der Voo (1993) has given a comprehensive review of the data from these two terranes and their geological setting.

7.3.3 Armorica

Armorica refers to the collection of Paleozoic blocks presently located within Hercynian Europe as was discussed in §6.5.2 and illustrated in Fig. 6.13. Although these blocks may have been separate units within the Iapetus Ocean prior to their Hercynian collision with Baltica, their Precambrian and Cambrian geology shows more similarities than differences. There is also some paleomagnetic evidence for considering these as a single Paleozoic plate (Van der Voo, 1982, 1988).

The rock sequences of the Armorican, Bohemian and Iberian Massifs all exhibit a pronounced late Precambrian to Cambrian unconformity associated with the Cadomian orogeny that is of the same age as the Pan-African orogeny of Gondwana. Hagstrum *et al.* (1980) and Perigo *et al.* (1983) analyzed the paleomagnetic data from Armorica (mainly from the Armorican Massif) for the time interval 650–500 Ma. Their comparisons of the APWPs during this interval for Armorica and for what was then considered to be Gondwana showed remarkable similarities. Both APWPs suggested rapid apparent polar wander during this time interval and appeared to have the same shape. These similarities extended to results from the East Avalon terrane. Although these comparisons appear impressive, it should be noted that the Pan-African orogeny (~550 Ma) is now generally regarded as the time at which the various cratons of Gondwana amalgamated. Therefore, it is not clear that a "Gondwana" APWP compiled for times older than 550 Ma has any meaning.

Armorica was certainly amalgamated with Baltica at ~300 Ma but the question remains as to where it came from; Gondwana is a prime candidate for its origin. Thus, in Fig. 7.15 the paleolatitudes observed from Armorica for the time interval 550–300 Ma have been compared with those expected from Baltica and Gondwana during the same interval. These have been calculated from the mean poles listed in Tables 6.6 and 7.3, respectively, using a present-day reference point at 47N, 0E. The general agreement with the expected Gondwana high paleolatitudes during the Cambrian and Ordovician is excellent. The paleomagnetic data for Gondwana and Baltica show similar predicted paleolatitudes for Armorica for post-Ordovician times, so the precise timing of the amalgamation of Armorica and Baltica is unclear.

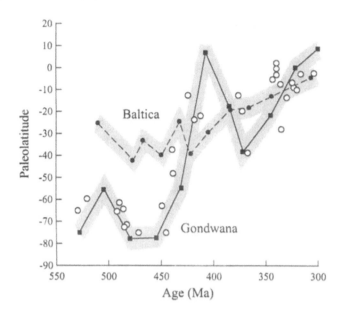

Fig. 7.15. Paleolatitudes (open circles) observed from rocks in Armorica for a present-day location at 47N, 0E. Paleolatitudes predicted using the results from Baltica (solid circles) and Gondwana (solid squares) are shown for comparison.

During Ordovician times Armorica and East and West Avalon all exhibited the high paleolatitudes to be expected for locations close to northwest Africa (Figs. 7.14 and 7.15). On the basis of the paleomagnetic data, Torsvik *et al.* (1996) proposed a reconstruction of these terranes adjacent to northwest Gondwana as shown in Fig. 7.16 for Early Ordovician times (490 Ma). The Iapetus Ocean between Gondwana and Laurentia had reached its maximum extent in Late Cambrian and Early Ordovician (Fig. 7.16). Torsvik *et al.* (1996) propose that Armorica and the East and West Avalon terranes rifted away from Gondwana in early Middle Ordovician (Llanvirn) times. By the close of the Ordovician (440 Ma), East and West Avalon had drifted into low latitudes close to Baltica and Laurentia. In the Middle Silurian (430 Ma), Baltica, East and West Avalon, and Laurentia had amalgamated. Finally, by the Late Silurian, Armorica was sutured to Baltica, but there are few paleomagnetic constraints on this timing (Torsvik *et al.*, 1996).

A detailed paleogeographic scenario for the Ordovician evolution of the Iapetus Ocean has been proposed by Mac Niocaill *et al.* (1997). They suggest there was an extensive peri-Laurentian arc that collided with Laurentia in the Early and Middle Ordovician and produced the Taconian orogenic pulse. By the Middle Ordovician a peri-Avalonian arc had developed on the southern margin of the Iapetus Ocean. Paleomagnetic data from the Robert's Arm, Sommerford,

Fig. 7.16. Reconstruction of the positions of the Armorica (Ar), East Avalon (EA), and West Avalon (WA) terranes in relation to Gondwana for the Early Ordovician (490 Ma) after Torsvik *et al.* (1996), with permission from Elsevier Science.

and Chanceport groups of central Newfoundland indicate the presence of a third arc system, the Exploits arc, situated in the middle of the Iapetus Ocean. Convergence of Baltica and Avalon with Laurentia continued through the Middle and Late Ordovician, and final closure of the Iapetus Ocean seems to have been completed in the Silurian.

7.3.4 The Western Mediterranean

Some of the earliest paleomagnetic studies in Europe were made in the Mediterranean region in the early 1960s. In the western Mediterranean the most significant discovery from these studies was the observation that the paleomagnetic declinations from late Paleozoic and younger rocks of Iberia, Corsica–Sardinia, and Italy deviated counterclockwise from those expected from stable Europe. Therefore, it was proposed that each of these regions had suffered counterclockwise rotation with respect to Europe. In the eastern Mediterranean region the situation is much more complex and there have been many studies made in Greece, Turkey, Bulgaria, and the Carpathians. An excellent review of the current situation with respect to paleomagnetic studies in this region has been given by Van der Voo (1993). Therefore, only the results from Iberia and Italy will be mentioned here as examples of terranes where rotations have been the important feature determined from paleomagnetism.

Iberia

Clegg *et al.* (1957) reported the first paleomagnetic results from Iberia that indicated such rotations. These were from Triassic rocks in northern Spain. Because of difficulties with structural control outside the Iberian Meseta (see Fig. 6.13), it is convenient to restrict any analysis to data from the Meseta itself. Paleolatitudes from the Iberian Meseta tend to agree with those predicted either from Europe or from Gondwana (Van der Voo, 1993). It is the paleomagnetic declinations that differ, as illustrated in Fig. 7.17a. The declinations predicted from stable Europe have been calculated using the mean poles from Tables 6.6 (Europe pre-Triassic) and 7.2 (Eurasia for Triassic and younger) using a present-day reference point at 40N, 4W. Those from Gondwana have been calculated using the data from Tables 6.11 (Africa for <140 Ma) and 7.3 (Gondwana).

The declinations observed from Iberia show no systematic agreement with those expected from either Europe or Gondwana (Fig. 7.17a). The declinations change by an angle of about 35° from those predicted from Gondwana during the Jurassic and Early Cretaceous to those expected from Europe during the Late Cretaceous and Tertiary. This corresponds with the opening of the Bay of Biscay through a similar angle. Kinematic analyses based on sea-floor spreading, sedimentary basin evolution, and structural geology favor the view that Iberia was probably an independent plate (Malod, 1989; Srivastava *et al.*, 1990), even though it may have been associated with Gondwana during the Jurassic and Early Cretaceous.

Italy (Adria)

In the Italian Peninsula some of the regions along the Adriatic coast, in the southern Alps and in northeast Sicily, are thought to be underlain by major detachment zones and have been referred to collectively as *Adria* (Channell *et al.*, 1979). However, it should be noted that they may not necessarily have together formed a single rigid block. As with the results from the Iberian Meseta, the paleolatitudes determined from Adria show correspondence with those expected both from stable Europe and from Gondwana (Van der Voo, 1993). The observed paleomagnetic declinations (Fig. 7.17b) are compared with those predicted from stable Europe and Gondwana using the same mean poles as in Fig. 7.17a for a present-day reference point at 42N, 12.5E. The observed declinations show strong affinity with those predicted from Gondwana since the Late Carboniferous. More detailed work in Umbria suggests that there are differences between the declinations observed in North and South Umbria that

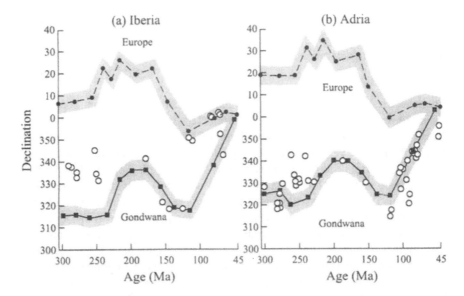

Fig. 7.17. Paleomagnetic declinations (open circles) observed from (a) the Iberian Meseta for a present-day location at 40N, 4W, and (b) Adria for a present-day location at 42N, 12.5E. The declinations, predicted using the results from Europe (solid circles) and Gondwana (solid squares), are shown for comparison.

are related to thrust sheet rotations (Channell *et al.*, 1978, 1984; VandenBerg *et al.*, 1978; Hirt and Lowrie, 1988; Jackson, 1990).

On the basis of the paleomagnetic agreement with Gondwana, Channell *et al.* (1979) proposed that Adria was a fixed promontory of Africa. As a consequence, the kinematics of the collision of Africa with Europe are directly related to the relative movements between them. However, if there was rotation of the Umbrian (and other) basement with respect to Africa and the southern Alps, then Italy consisted of many small blocks that were not coherent (VandenBerg, 1983; Lowrie, 1986). In that case there was no Adriatic promontory or certainly not a permanent one. This issue is at present unresolved, but the correspondence between the paleomagnetic data of Adria and Gondwana in Fig. 7.17b appears on the face of it to support the idea of the fixed Adriatic promontory. For a more comprehensive review of paleomagnetic data from the Mediterranean region, see Van der Voo (1993).

7.3.5 South and East Asia

Iran
Paleomagnetic data from Iran have been acquired over the past two decades (Wensink, 1979, 1982, 1983; Soffel and Förster, 1981; Soffel *et al.*, 1996; Besse

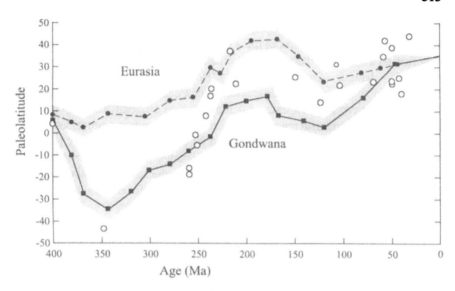

Fig. 7.18. Paleolatitudes (open circles) observed from Iran for a present-day location at 35N, 60E. The paleolatitudes predicted using the results from Eurasia (solid circles) and Gondwana (solid squares) are shown for comparison.

et al., 1998). Van der Voo (1993) and Besse *et al.* (1998) provided substantive reviews of all the data and their tectonic implications. The paleolatitudes observed from Devonian and younger rocks from Iran are illustrated in Fig. 7.18. They are compared with those expected from Eurasia and Gondwana during the same time interval. These reference paleolatitudes have been calculated using the data listed in Tables 6.6 and 7.2 for Eurasia and those in Tables 6.11 and 7.4 for Gondwana using a present-day reference point at 35N, 60E.

During the Devonian and Permian the paleolatitude of Iran tracks that expected from Gondwana rather than that from Eurasia. During the Paleozoic it appears that Iran was situated in Gondwana adjacent to Arabia, close to its present relative position. During the Triassic the paleolatitude changes rapidly from that predicted for Gondwana to that predicted for Eurasia (Fig. 7.18). Therefore, at the end of the Permian Iran must have rifted away from Gondwana and drifted rapidly northwards during the Early Triassic. Iran then converged with Turan, which was then the southern margin of Eurasia, during the Middle Triassic (Besse *et al.*, 1998).

Lhasa and Qiangtang Terranes
Paleomagnetic data for the Lhasa and Qiangtang terranes of Tibet are confined to the Cretaceous and Tertiary and have been reviewed by Van der Voo (1993) and Chen *et al.* (1993). Agreement between the Cretaceous paleomagnetic

directions for these terranes suggests they have been a single unit at least since then. Cretaceous paleolatitudes of these terranes are consistent with the view that they formed a stable southern margin of Eurasia with a general eastward trend at about 10°N latitude (Chen *et al.*, 1993). However, the various poles do not coincide and the change in paleodeclinations from west to east suggests that these terranes have suffered internal deformation on different scales up to 1000 km. These declination differences indicate that the suture was more or less linear prior to collision with Eurasia and acquired its curvature subsequently (Chen *et al.*, 1993). On the basis of paleomagnetic data, such a form of oroclinal bending was previously proposed, with larger amplitude, for the southern boundary of the Lhasa terrane as a result of the collision of India (Klootwijk *et al.*, 1985).

During the Tertiary it has been observed from these terranes, and the Kunlun and Qaidam terranes to the north, that the paleomagnetic inclinations are about 20° shallower than would be expected from the reference APWP for Eurasia. Westphal (1993) proposed that this implies the presence of some magnetic field anomaly over this part of Eurasia causing a large departure from the geocentric axial dipole field direction. However, Cogné *et al.* (1999) have proposed that this large-scale inconsistency can be satisfactorily explained by a combination of factors, including the nonrigid behavior of the Eurasia plate during the Tertiary and intracontinental crustal shortening resulting from the collision of India with Asia.

Kolyma and Sikhote Alin
In the far east of Asia the Kolyma block (see Fig. 6.15) is made up of several terranes, the most important being the Omolon Massif and the Chukotka terrane. Southeast of Siberia the Sikhote Alin terrane lies on the Asian mainland adjacent to the island of Sakhalin. Khramov and Ustritksy (1990) summarized the paleomagnetic data from these terranes. In Fig. 7.19 the paleolatitudes observed for the Omolon and Sikhote Alin terranes since the Permian are compared with those expected using the mean poles for Eurasia (Table 7.2) using present-day reference points at 64N, 159E and 45N, 135E respectively. The Omolon Massif was located far south of Eurasia during the Permian and Triassic and then amalgamated with Eurasia during the Cretaceous (Fig. 7.19a). Closely associated with the Omolon Massif is the Chukotka terrane. However, the paleolatitudes observed from this terrane show good agreement with those predicted from North America throughout the Mesozoic and Cenozoic. Khramov and Ustritsky (1990) therefore propose that the Chukotka terrane may have been part of a single Chukotka–Arctic Alaska block that existed in Mesozoic time.

The Sikhote Alin terrane was also located far south of Eurasia during the Permo-Triassic but must have been closely associated with Asia by the Late Cretaceous (Fig. 7.19b). However, the final amalgamation with Asia may only have occurred sometime during the Tertiary.

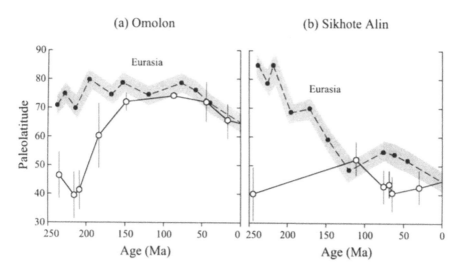

Fig. 7.19. Observed mean paleolatitudes (open circles) and their 95% confidence limits from Khramov and Ustritsky (1990). (a) Omolon for a present-day location at 64N, 159E and (b) Sikhote Alin for a present-day location at 45N, 135E. The paleolatitudes predicted using the results from Eurasia (solid circles) are shown for comparison.

7.4 Rodinia and the Precambrian

7.4.1 Rodinia

Geological similarities between the late Proterozoic margins of western North America and eastern Australia (Bell and Jefferson, 1987) led to the hypothesis of the existence of a late Proterozoic supercontinent (Moores, 1991; Dalziel, 1991; Hoffman, 1991b). The proposed reconstruction of this supercontinent is based on geological evidence linking truncated Grenville-aged mobile belts (~1.3–1.1 Ga) on the margins of Laurentia, East Gondwana, Amazonia, and Baltica. The western margin of Laurentia faces East Antarctica in the so-called SWEAT (southwest U.S.A.–East Antarctica) connection (Moores, 1991). This supercontinent has been named *Rodinia* (McMenamin and McMenamin, 1990) from the Russian word "rodit" meaning "to beget" or "to grow". Rodinia begat all subsequent continents and the continental shelves were the cradle of the earliest shelled animals.

Fig. 7.20. Reconstruction of the Neoproterozoic supercontinent of Rodinia. Grenville-age (~1.3–1.1 Ga) mobile belts are shaded. KA, Kalahari craton, MA, Madagascar. Reproduced with permission from Hoffman (1991b). © American Association for the Advancement of Science.

Over the past decade the reconstruction, evolution, and breakup of Rodinia have become major topics in global tectonics and paleomagnetic data will clearly play a major role in their resolution. The original reconstruction of Hoffman (1991b) also includes Siberia and the separated Kalahari, Congo, and West Africa cratons (Fig. 7.20), although these links are poorly constrained. Based on stratigraphic correlations and tectonic analysis, Li *et al.* (1995) proposed, in a reconstruction of Rodinia, that the Yangtze block of South China could have been a continental fragment caught between the northeast part of the Australian craton and the northwest part of Laurentia.

The assembly of Rodinia is assumed to have been of Grenville age (~1300–1100 Ma). The timing of its breakup is not known with certainty, but it is generally thought to have been in existence for at least 300 Myr. Therefore, any paleomagnetic test of possible reconstructions of Rodinia involves an analysis of paleomagnetic data in the time range 1100–700 Ma.

7.4.2 Paleomagnetism and Rodinia

The key to the reconstruction of Rodinia is the fit between the eastern margin of Australia and the western margin of Laurentia as they were constituted at about 1100 Ma. At that time Rodinia included East Gondwana (Australia, Antarctica, and India) as a unit (Fig. 7.20). Therefore, a viable paleomagnetic test of the Australia–Laurentia connection can be made by comparing poles from East Gondwana with Laurentia. In a seminal paper, Powell *et al.* (1993) carried out the first such test using poles between 1050 Ma and the Cambrian (Fig. 7.21). There are few data for East Gondwana in this time interval, but the agreement between paleomagnetic poles at 1055 Ma and again at 725 Ma for Laurentia and East Gondwana in a reconstructed Rodinia confirmed the Australia–Laurentia connection. The optimum paleomagnetic fit shown in Fig. 7.21 allows for a gap between northwest Laurentia and northeast Australia. Li *et al.* (1995) subsequently proposed that South China occupied this gap. Australia (and East Gondwana) is reconstructed to Laurentia by rotation about an Euler pole at 35.5N, 132.2E through an angle of +125.7° (counterclockwise). After 725 Ma the paleomagnetic poles for East Gondwana and Laurentia diverge, so Powell *et al.* (1993) proposed that Rodinia broke up after that time.

The only significant paleomagnetic data set for the time interval 1100–725 Ma is that for Laurentia. The data for this time interval have been mainly obtained from the region adjacent to the Grenville Front, and as a consequence have been the subject of much debate. Paleomagnetic data in the time range 1100–1000 Ma define a well-determined segment of the APWP, known as the Keweenawan track. Data for the subsequent time interval, 1000–800 Ma, define a segment of the APWP known as the Grenville loop. In the region of the Grenville Front the magnetization of the rocks has been overprinted during post-Grenville uplift and cooling, which occurred ~1000–900 Ma (Hyodo *et al.*, 1986). To make use of such data, it is necessary to calibrate the ages at which the observed magnetizations were acquired. Using $^{40}Ar/^{39}Ar$ thermochronometry Berger *et al.* (1979) have shown that cooling from ~600 to ~350°C occurred at ~980–820 Ma and took place over 160 million years. This was the first attempt to relate isotopic closure temperatures with magnetic blocking temperatures, as was outlined in §6.4.3, and enabled the necessary age calibration of the Grenville paleomagnetic data.

McWilliams and Dunlop (1978) pointed out that the Grenville paleomagnetic poles trace out a simple path when plotted in order of increasing distance from the Grenville Front. There was also a parallel pattern of younging in K–Ar and other radiometric ages away from the front. Constanzo–Alvarez and Dunlop (1998) made a regional paleomagnetic study of lithotectonic domains in the Central Gneiss Belt of the Grenville Province in Ontario along ten north–south and east–west traverses up to 200 km in length. They found that the younging

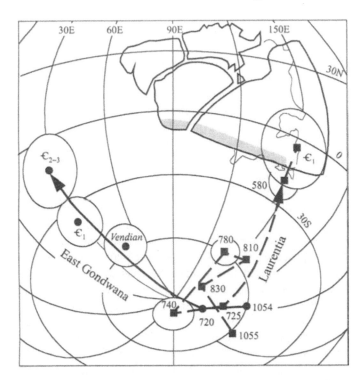

Fig. 7.21. Comparison of paleomagnetic poles from East Gondwana and Laurentia from 1055 Ma to the Cambrian in a Rodinia reconstruction as described in the text, using the present-day co-ordinates of India. Mean poles (with 95% circles of confidence) are shown as solid circles for East Gondwana and solid squares for Laurentia. The shaded region is the Grenville Front and its extension into Antarctica. After Powell *et al.* (1993), with permission from The Geological Society of America.

patterns proposed by McWilliams and Dunlop (1978) based on K–Ar thermochron maps were unreliable because of hydrothermal alteration, which causes both chemical remagnetization and anomalously old K–Ar ages near the Grenville Front.

The Constanzo–Alvarez and Dunlop (1998) study confirmed that the Grenville loop is a clockwise loop when plotted in the Pacific, as shown in Fig. 7.22 for the time interval 1000–800 Ma. In this case the poles in the APWP are presumed to be north poles. The time calibration of the APWP was based on joint paleomagnetic and $^{40}Ar/^{39}Ar$ thermochronometric studies. The ages of paleopoles corresponding to high and low blocking temperature partial TRMs were determined from $^{40}Ar/^{39}Ar$ plateau ages for hornblende and biotite by matching closure temperatures for Ar diffusion in these minerals to the measured blocking temperature ranges of the partial TRMs (ages at ~980, ~900, and ~820 Ma in Table 7.4).

TABLE 7.4
Mean Pole Positions for Laurentia in the Time Interval 1100–725 Ma[a]

Age range (Ma)	Method of dating	Mean age (Ma)	m	N	Mean pole position		A_{95} (°)
					Lat N	Long E	
1090-1100	U-Pb, Rb-Sr	1100	19	281	38.7	181.9	4.1
1070-1090	U-Pb, Rb-Sr	1080	6	73	25.5	180.2	5.3
1030-1050	Rb-Sr	1040	5	53	7.8	172.7	9.1
1010-1030	Rb-Sr	1020	4	52	-15.7	184.5	7.3
~1000		~1000	6	40	-38.0	167.9	8.4
~980	^{40}Ar/^{39}Ar hornblende	~980	7	85	-35.8	144.0	3.1
~940		~940	6	43	-24.1	133.6	3.5
~900	^{40}Ar/^{39}Ar biotite	~900	11	90	-12.6	147.4	4.0
~880		~880	11	112	-6.1	161.2	3.7
~850		~850	6	47	11.5	153.7	6.3
~820	^{40}Ar/^{39}Ar biotite	~820	5	47	23.6	163.9	3.6
~800	Stratigraphy	~800	4	12	26.7	150.2	7.3
770-790	U-Pb	780	4	36	-3.4	138.0	3.3
715-735	U-Pb, Stratigraphy	725	3	70	5.2	162.7	5.9

[a]Symbols and method of averaging are as given in Table 6.5.

The data for Laurentia in the whole time range 1100–725 Ma are summarized in Table 7.4 and illustrated in Fig. 7.22. For the time covered by the Grenville loop (1000–820 Ma) the data as compiled by Constanzo–Alvarez and Dunlop (1998) have been grouped to calculate successive mean pole positions. Results from the Yukon and the MacKenzie Mountains of the Northwest Territories, Canada (Park, 1981, 1984, 1997; Park and Aitken, 1986a,b; Park and Jefferson, 1991) have been accommodated within this scheme. The final two pole positions based on the Tzezotene Sills and the Franklin Intrusions have U–Pb zircon ages of 778 and 723 Ma, respectively.

The paleomagnetic poles now available from East Gondwana are also plotted in Fig. 7.22. In addition to the poles at 1054±14 Ma and 720±10 Ma used by Powell *et al.* (1993) in Fig. 7.21, three new poles are plotted: the Mt Isa Intrusives dated at 1116±12 Ma (Tanaka and Idnurm, 1994), the Walsh tillite at 750 Ma (Li, 1999) from Australia, and the Wajrakarur kimberlite at 1079 Ma from India (Miller and Hargraves, 1994). These results further support the Australia–Laurentia connection in the Rodinia reconstruction.

Paleomagnetic data from the constituent parts of Rodinia are not as extensive as those for Laurentia in the key time interval 1100–700 Ma. The major problem is the quality of age determinations, although use of the U–Pb zircon and ^{40}Ar/^{39}Ar methods is providing better constraints than used to be available. Each of the various cratons tends to have a few pole positions limited to short age segments so that it is difficult to justify drawing a separate APWP for each craton. Therefore, it is convenient to use the Laurentian APWP as a reference path for the analysis of paleomagnetic data from the constituent parts of Rodinia.

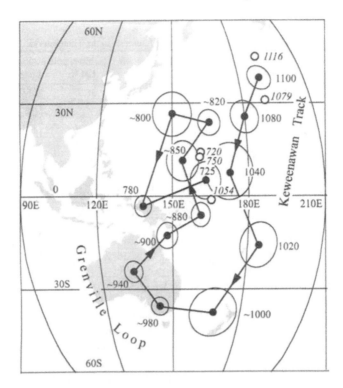

Fig. 7.22. APWP for Laurentia in the time interval 1100–725 Ma from the mean pole positions (solid circles) listed in Table 7.4. The Keweenawan track merges into the clockwise Grenville loop with ages in millions of years indicated against each mean. Circles of 95% confidence are shown around the means. Poles from East Gondwana (open circles, ages in italics) are plotted in their positions after rotation into the Rodinia configuration as discussed in the text.

Elming *et al.* (1993) proposed that the north pole APWP for Baltica in the time interval 1000–850 Ma performed a counterclockwise loop, named the Sveconorwegian loop. This is in exactly the opposite sense to that of the Grenville loop and so produced a major conflict because Baltica is also considered to have been a part of Rodinia in that time interval. Weil *et al.* (1998) therefore made a critical examination of the Laurentian data and produced a set of poles that suggested the Grenville poles performed a matching counterclockwise loop. However, new data from Baltica now show that the equivalent loop indeed moves in a clockwise direction (Pisarevsky and Bylund, 1998; Bylund and Pisarevsky, 1999) as has always been presumed for the Grenville loop. For Siberia, Smethurst *et al.* (1998) reviewed all the available data to produce an APWP extending from 1100 to 250 Ma. They conclude that although the APWPs for Siberia, Baltica, and Laurentia differ, they imply broadly similar paleolatitudinal drifts for the three continents.

In the Kalahari craton there was widespread igneous activity at 1105 Ma, corresponding with the age of the Umkondo dolerites of Zimbabwe (Hanson *et al.*, 1998). Widespread paleomagnetic studies have been made of this igneous province including the Umkondo dolerites (McElhinny and Opdyke, 1964) and lavas (McElhinny, 1966), the Timbavati gabbros of northeast South Africa (Hargraves *et al.*, 1994), and the post-Waterberg diabases of Botswana and South Africa (Jones and McElhinny, 1966). In a reconstruction of Gondwana the Coats Land–Maudheim–Grunehogna (CMG) Province of the western part of East Antarctica (see Figs. 6.22b and 7.8) is presumed to be an eastern extension of the Kalahari craton (Groenewald *et al.*, 1991, 1995; Grunow *et al.*, 1996). Equivalents of the Umkondo dolerites and lavas are found in Dronning Maud Land (Hanson *et al.*, 1998). Gose *et al.* (1997) carried out paleomagnetic studies on 1112 Ma nunataks in Coats Land confirming that the CMG Province was most likely linked to the Kalahari craton at that time.

Idnurm and Giddings (1995) compared paleomagnetic data for Australia and Laurentia in the time interval 1700–1600 Ma and found that the APWPs match each other in the Rodinia configuration. However, it should be emphasized that this match refers only to the Australia–Laurentia connection, which is the core of the Rodinia supercontinent. The Grenville-age mobile belts, which are presumed to mark sutures relating to the formation of Rodinia, are not involved in this connection. So it remains a possibility that the Australia–Laurentia connection existed for at least 1000 Myr prior to its breakup in the late Precambrian.

7.4.3 Earth History: 1000 Ma to the Present

Dalziel (1992, 1997) and Dalziel *et al.* (1994) discussed the possible sequence of events that occurred between the breakup of Rodinia and the formation of Gondwana. They proposed that the present eastern margin of Laurentia collided with the western margin of South America during this time interval to form a short-lived supercontinent near the Precambrian–Cambrian boundary (545 Ma). Powell (1995) named this supercontinent *Pannotia* [from the Greek *pan* ("all") and *notios* ("southern")] because it lay almost entirely in the southern hemisphere.

Powell *et al.* (1995) have shown that in the Pannotia reconstruction (Fig. 7.23) the paleomagnetic south poles for Laurentia and Gondwana all lie in the region of west Africa at 545 Ma (see Fig. 7.8). Following Dalziel (1997), in Fig. 7.23 the paleolatitudes drawn are based upon the south pole occupying such a position. An interesting implication of this reconstruction is that the North Britain terrane, when reconstructed to Laurentia, fits next to the Arequipa Massif of southern Peru (see also Fig. 6.19b). This Arequipa–Scotland connection suggests that it was a piece of North Britain left behind in southern Peru when Laurentia broke away and continued on its northward journey (Dalziel, 1997).

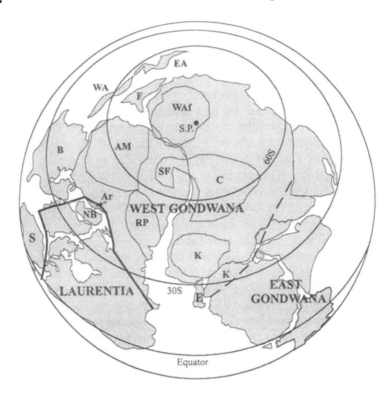

Fig. 7.23. Geologically constrained reconstruction of the Pannotia supercontinent that may have existed ephemerally near the Precambrian–Cambrian boundary (545 Ma). S.P. is the paleomagnetic south pole. Various cratons and terranes are labeled as follows: AM, Amazonia, Ar, Arequipa Massif, B, Baltica, C, Congo, E, Ellsworth–Whitmore Mountains, EA, East Avalon, F, Florida, K, Kalahari, NB,North Britain, RP, Rio de la Plata, S, Siberia, SF, São Francisco, WA, West Avalon, WAf, West Africa. The dashed line indicates the suture between East and West Gondwana (see Fig. 7.12). After Dalziel (1997), with permission from The Geological Society of America.

Torsvik *et al.* (1996) have argued against the juxtaposition of Laurentia and Gondwana on paleomagnetic and faunal grounds. On the other hand, Meert and Van der Voo (1997) regard the existence of Pannotia as an open question that may ultimately be resolved when more paleomagnetic, geochronologic, and geologic data are available.

It is now generally agreed that the Argentine Precordillera, in the foothills of the Argentine Andes, is an exotic fragment derived from Laurentia (Astini *et al.*, 1995; Thomas and Astini, 1996; Dalziel *et al.*, 1996). Rapalini and Astini (1998) carried out a paleomagnetic study of Early Cambrian rocks from the Argentine Precordillera. The results are not consistent with the APWP for Gondwana at that time but are consistent with the Laurentian path if the Argentine Precordillera is positioned as the conjugate margin of the Ouachita embayment in southeast

Laurentia. This confirms that the Precordillera is a displaced terrane, which was derived from Laurentia during the Cambrian and was probably accreted to Gondwana in the Middle Ordovician.

The evolution of the Earth's crust following the breakup of Rodinia thus involved Laurentia breaking away from Rodinia and traveling clockwise around the western margin of South America as West Gondwana was being formed. Laurentia collided with the Andean margin of South America to form the short-lived supercontinent of Pannotia and then continued on its northward journey until it lay north of Gondwana. The Iapetus Ocean closed as the Avalon terranes collided with Laurentia, and together they merged with Baltica to form Laurussia. Armorica then collided with Laurussia and finally the Rheic Ocean between Gondwana and Laurussia closed to create the supercontinent of Pangea at about 300 Ma. The history of the Earth's crust from 1000 Ma to the present is summarized in the cladistic-style diagram of Fig. 7.24. It combines the pre-Pangea history described above with the post-Pangea history illustrated in the paleogeographic maps of Fig. 7.11. This diagram emphasizes that the history of the Earth's crust involves the continual amalgamation and dispersal of supercontinents over time scales of hundreds of millions of years.

7.4.4 Precambrian Cratons

In the 1960s and 1970s there was considerable interest in Archean and Proterozoic paleomagnetism, and studies were made in all the major cratons of the present continents. The general method of presentation of these studies was to allow the paleomagnetic poles to fall within an APWP swathe of ~20° width. These swathes were then drawn so as to incorporate all the pole positions by making suitable bends and curves in the swathe where necessary. As a result of this approach Piper *et al.* (1973) proposed that the principal cratons of Africa and South America were approximately in their present relative positions and orientations as early as 2200 Ma. This implied that the younger orogenic belts separating these cratons did not arise from the convergence of separate plates but favored a dominance of intracrustal processes in Proterozoic times.

The approach outlined above was favored by many workers at that time, including Irving and McGlynn (1976) and Roy and Lapointe (1976) for Laurentia, McElhinny and Embleton (1976) for Australia, and Piper (1980) and Pesonen and Neuvonen (1981) for Baltica. Burke *et al.* (1977) took an opposing view pointing out that the paleomagnetic record could just as easily be interpreted in terms of the opening and closing of ocean basins in terms of plate tectonics. McElhinny and McWilliams (1977) criticized this approach claiming that the "fixed cratons" model was the simplest interpretation of the paleomagnetic data. However in the previous sections it has been shown that the

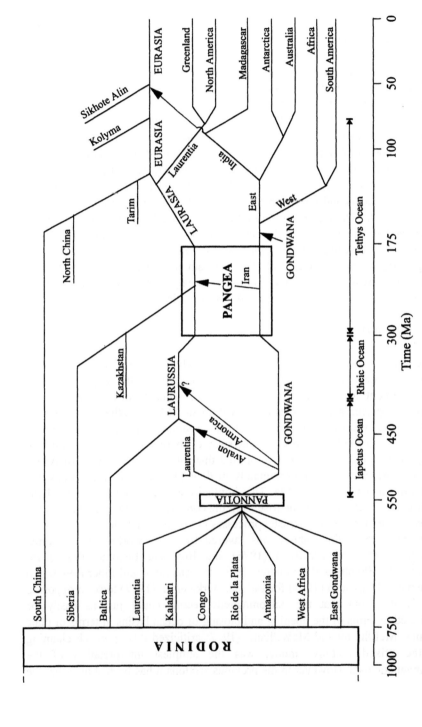

Fig. 7.24. 1000 million years of Earth history. From the Rodinia supercontinent through its breakup and the subsequent formation and breakup of Pannotia, Gondwana, Laurussia and Pangea. The time scale at the base is approximate and not uniform.

various cratons of West Gondwana were separated and were a part of the Rodinia supercontinent. It is now widely presumed that the Pan-African mobile belts of age of ~550 Ma between these cratons represent the suture zones associated with the formation of Gondwana. Therefore, the interpretation of the paleomagnetic data argued by Burke *et al.* (1977) is more realistic.

The rather simplistic approach adopted in the 1970s and 1980s by most paleomagnetists for analyzing Precambrian paleomagnetic data is clearly incorrect. The methodology used was flawed. With limited data for each of the cratons covering times of hundreds to thousands of million years, it is always possible to construct APWPs that appear to be coherent by making suitable loops and bends in wide swathes to include any new data as they arise. There is the increased difficulty when dealing with Precambrian rocks in knowing accurate ages associated with paleomagnetic poles. Furthermore, it can always be argued that isotopic ages and magnetic ages are not the same, thus providing a convenient additional variable to explain almost anything. Use of this rather broad brush approach led Piper (1976, 1982, 1987) to propose the existence of a single supercontinent covering the time interval 2800–800 Ma.

The approach currently being used by those researching the Rodinia supercontinent is much more appropriate than that outlined above. There must be well-determined paleomagnetic data coupled with precise age determinations and these must be complemented with detailed geological syntheses. Thus, Precambrian paleomagnetism has commenced a new era in which one slowly works one's way backwards in time and where possible, obtaining the crucial information that resolves each problem as it arises. It seems clear now that plate tectonics has certainly operated through much of geological time. For example, the Grenville-age belts, which were one of the keys in the original reconstruction of Rodinia, are presumed to be collision belts relating to the formation of Rodinia. Therefore, the complexity involved in the application of paleomagnetism to global tectonics in pre-Rodinia times will involve a new and exciting dimension.

7.5 Non-Plate Tectonic Hypotheses

7.5.1 True Polar Wander

In paleomagnetism it is important to distinguish between the concepts of polar wandering and continental drift. All measurements in paleomagnetism are made with respect to the geographic pole as the frame of reference. Therefore, in plate tectonics the concept of polar wander may appear to be meaningless because all one is saying is that the reference frame itself is in motion. *True polar wander*, TPW, on the other hand, has been defined as the displacement of the entire Earth with respect to its spin axis (Goldreich and Toomre, 1969; Gordon, 1987). In paleomagnetic terms TPW is more generally defined as a displacement of the entire Earth or of an outer shell with respect to the axis of rotation.

The shape of the Earth is approximately an oblate spheroid with flattening ~1/300. The oblateness is due to the equatorial bulge, a hydrodynamic effect that does not contribute to long-period TPW. The remaining variations in the surface gravity equipotential (the nonhydrostatic geoid) are related to the internal mass anomalies that control long-period TPW (Goldreich and Toomre, 1969; Richards *et al.*, 1997; Steinberger and O'Connell, 1997; Evans, 1998). Polar instability occurs when the maximum and intermediate moments of inertia of the nonhydrostatic geoid are roughly equal, that is for a quasi-sphere or prolate spheroid. In these cases small adjustments in internal mass anomalies can generate large TPW. The prolate spheroid situation would result in TPW confined generally to a single great circle orthogonal to the nonhydrostatic geoid's prolate axis. The quasi-spherical case would result in chaotic TPW paths.

Three reference frames are meaningful in a discussion of TPW (Marcano *et al.*, 1999): the Earth's spin axis, the mantle hotspot reference frame, and the mean-lithosphere framework. In paleomagnetic terms the Earth's spin axis is equated with the paleomagnetic axis. In the hotspot reference frame the basic assumption is that hotspot traces reflect plate movements over the stationary deep mantle. If the hotspot reference frame is valid, and there is now considerable doubt that this is so (§5.5.1), then TPW is equated with the displacement of a plate's mean paleopole while keeping the hotspots fixed (Livermore *et al.*, 1983, 1984; Andrews, 1985; Besse and Courtillot, 1991). By correcting the positions of the plates and their paleopoles for these movements, the displacement of the spin axis can be calculated with respect to the mantle hotspots. In the mean-lithosphere framework there may be some net motion common to the motion of each of the plates. TPW is then the movement of the rotation axis in this framework and is directly related to the concepts of the net angular momentum of the lithosphere or the net torque on the lithosphere (Simpson, 1975; Solomon *et al.*, 1977). Thus, the mean-lithosphere framework relies on precise knowledge of the relative motions between all plates.

In the mean-lithosphere framework TPW can be defined as the vector sum of the horizontal displacements of all points on the Earth's surface. McKenzie (1972) expressed this formally. If the angular velocity of the pole P relative to a plate A is $_A\mathbf{\Omega}_P$, the velocity and direction of polar motion $_A\mathbf{v}_P$ is

$$_A\mathbf{v}_P = r(_A\mathbf{\Omega}_P \times \hat{\mathbf{z}}), \tag{7.5.1}$$

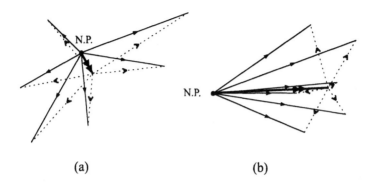

(a)   (b)

Fig. 7.25. True polar wander as a concept in the mean-lithosphere framework. The thin solid lines with arrows represent the vectors $_n V_P$ (7.5.3) of the north pole (N.P.) for each of six plates imagined to cover the Earth's surface. The thicker line with two arrows shows the vector V_m of (7.5.4) and the dashed lines $(_n V_P - V_m)$. True polar wander is a useful concept when V_m is large as in (b) but not as in (a). After Mckenzie (1972), with permission from the McGraw-Hill Companies.

where $\hat{z}$ is a unit vector along the spin axis and r is the radius of the Earth. Because TPW is a mean value for the whole of the Earth's surface, it is necessary to weight the velocity vector by f_A, which is the fraction of the Earth's surface occupied by plate A. This defines a new vector $_A V_P$ as

$$_A V_P = f_A \cdot {_A}\Omega_P . \tag{7.5.2}$$

TPW only becomes a useful concept if the rate of relative motion of all plates relative to the pole is much reduced by a particular choice of TPW direction and velocity. This can be expressed as

$$\sum_{n=1}^{N} |_n V_P| >> \sum_{n=1}^{N} |(_n V_P - V_m)|, \tag{7.5.3}$$

where

$$V_m = \frac{1}{N} \sum_{n=1}^{N} {_n}V_P \tag{7.5.4}$$

and represents the TPW vector. Two situations are illustrated in Fig. 7.25, where the thin solid lines with arrows are the vectors $_1 V_P, ..., _N V_P$ drawn from the north pole to represent the *apparent* polar wander directions of all N plates which cover the Earth's surface. The condition of (7.5.3) is satisfied by the situation shown in Fig. 7.25b but not by that shown in Fig. 7.25a.

   McElhinny (1973b) used the above method to calculate the amount of TPW for the past 50 Myr and showed that TPW was not significant over that time scale. Jurdy and Van der Voo (1974, 1975) used a different method that employs a technique for separating TPW from purely relative plate motions. Their results

TABLE 7.5
Magnitude and Direction of Polar Motion with Respect to Six Major Plates over the Past 50 Myr[a]

Plate	Pole (~ 50 Ma)	f	Ω_p (° Myr^{-1})	V_p (° Myr^{-1})	Pole of rotation
Africa	82.3N, 175.4E	0.16	0.15	0.025	0N, 85.4E
Antarctica	81.9, 158.2	0.11	0.16	0.018	0N, 68.2E
Eurasia	79.7N, 169.6E	0.12	0.21	0.025	0N, 79.6E
Indo-Australia	67.9N, 286.1E	0.14	0.44	0.062	0N, 196.1E
N+S America	77.7N, 164.1E	0.20	0.25	0.049	0N, 74.1E
Pacific	77.6N, 3.6E	0.22	0.25	0.055	0N, 273.6E

[a]Symbols are as in (7.5.2).

also showed no significant TPW for the Tertiary and that the cumulative amount of TPW since the Early Cretaceous is also insignificant.

The calculations made by McElhinny (1973b) are repeated here and are summarized in Table 7.5. The summary data for the Eocene and Paleocene for the Pacific plate have been used as given in Table 5.3. There are no reliable data for the early Tertiary of South America, so North and South America are considered as a single plate using the data in Table 6.5. Originally, Le Pichon (1968) regarded North and South America as a single plate, and analysis of Pacific Ocean magnetic anomalies suggests that negligible movement occurred between North and South America during the Tertiary (see Fig. 5.25). For Eurasia, the data from Europe (see Table 6.6) have been combined with those from South China (see Table 6.9). Data for Africa are given in Table 6.14. The break between the India and Australia plates is only a recent one (Wiens *et al.*, 1985). Therefore, they are considered to be a single plate throughout the Tertiary and the data from Tables 6.14 and 6.16 have been combined. The position of the

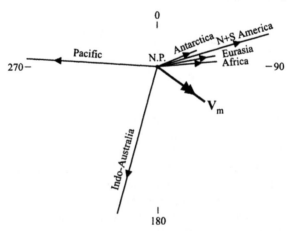

Fig. 7.26. Estimates of the vectors $_nV_p$ for the six major plates covering the Earth's surface during the past 50 Myr as given in Table 7.5. The situation corresponds with that illustrated in Fig. 7.25a.

Antarctic plate is found by subtracting the relative motion between Antarctica and Australia (from sea-floor spreading) from the Australian data.

The magnitude and poles of rotation of the six vectors $_n V_P$ calculated in Table 7.5 are illustrated in Fig. 7.26. The resultant vector V_m has pole of rotation at 0N, 126E and magnitude $2.5°$ corresponding to motion of the whole lithosphere of $2.5°$ along the $36°$ meridian. Within the error limits this is not discernible from zero and confirms the previous result of McElhinny (1973b).

For earlier times there has been much speculation regarding the possibility that TPW can been seen in the paleomagnetic APWPs for the major continents. Van der Voo (1994) observed that the APWPs for the Late Ordovician–Late Devonian interval for Laurentia, Baltica and Gondwana have nearly identical looping shapes (see §7.2) that can be brought into superposition. Also the less well-known paths for Siberia (Fig. 6.7) and South China (see Fig. 6.9a) reveal similar lengths. Kirschvink *et al.* (1997) suggested that Vendian to Cambrian data for all the major continents show anomalously fast rotations and latitudinal drift that might be attributed to TPW, although this is disputed by Torsvik *et al.* (1998). Mound *et al.* (1999) suggested that TPW involving movement of the rotation pole with respect to the Earth by $90°$ in a matter of 10 Myr would produce dramatic changes in sea-level of up to 200 m. Therefore a possible test for such TPW could lie in accurate estimates of sea-level changes for the time interval involved.

During the Permo-Triassic (295–205 Ma) the APWP for Pangea is about $35°$ in length (§7.2.4). This means that Pangea rotated through an angle of $35°$ with respect to the rotation axis about an Euler pole located on the equator. Marcano *et al.* (1999) argued that the rest of the world (mainly composed of the Panthalassic Ocean) rotated about the same Euler pole in the same sense as Pangea, so that TPW occurred during the Permo-Triassic at a rate of about $0.4°$ Myr^{-1}. On the basis of the above observations, Evans (1998) speculates that the geoid, and hence TPW, may be a legacy of supercontinental breakup, which can persist even after the next supercontinent has begun to form.

7.5.2 An Expanding Earth?

Despite the success of plate tectonics in explaining the evolution of the Earth's crust, there are still many who, following Carey (1958, 1976), believe that a more reasonable hypothesis is that the Earth has expanded with time. It is supposed that the continents form a scum on the surface of the Earth and maintain their primitive dimensions as the Earth expands underneath. Egyed (1960) was the first to point out that paleomagnetic data may be used to test the hypothesis of Earth expansion. In his *paleomeridian method* (Fig. 7.27a), if two widely separated paleomagnetic sampling sites are available for the same

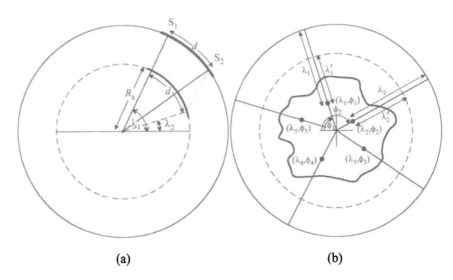

(a)   (b)

Fig. 7.27. Methods for determining the Earth's paleoradius. (a) The paleomeridian method of Egyed (1960). Cross-section through the Earth with two sampling sites S_1 and S_2 lying on a paleomeridian separated by distance d. If their paleolatitudes are λ_1 and λ_2, the ancient radius R_a is determined from (7.1.13). (b) The minimum dispersion method of Ward (1963). The positions of rock units (λ_n, ϕ_n) are transformed to a new co-ordinate system using the center of the continent as the pole (center of diagram). As the radius of the Earth is varied the latitudes of the rock units (λ_n) will change to some new value (λ_n') but the longitudes (ϕ_n) will remain unchanged. For any trial paleoradius, new values (λ_n') can be determined and the dispersion of paleomagnetic poles recalculated. After McElhinny (1973a, 1978).

geological epoch and they lie on the same paleomeridian, then the paleoradius, R_a, may be calculated from

$$R_a = \frac{d}{\lambda_1 - \lambda_2} \, , \tag{7.5.5}$$

where d is the present separation of the sampling sites and λ_1 and λ_2 are their paleolatitudes (in radians) calculated from the usual dipole equation (§1.2.3). It should be noted that the determination of paleolatitude from the dipole equation (1.2.3) is independent of the Earth's radius. Usually, it is the ratio of the ancient radius, R_a, to the present radius, R_p, that is calculated from

$$\frac{R_a}{R_p} = \frac{\theta_d}{\lambda_1 - \lambda_2} \, , \tag{7.5.6}$$

where θ_d is the great circle angular distance (in the same units as the paleolatitudes) between the sampling sites, corresponding to d in (7.5.5). Despite the simplicity of the paleomeridian method, it is rare to find two sampling sites situated exactly on a paleomeridian so in practice its application is limited.

To overcome this difficulty Egyed (1961) suggested a more general triangulation method that allowed sampling sites to be arbitrarily located. This method was later improved by Van Andel and Hospers (1968). Ward (1963) took a different approach in his *minimum dispersion method* (Fig. 7.27b) that is particularly appropriate to the spherical environment of paleomagnetic data. The dispersion of paleomagnetic poles is investigated using the Fisher (1953) method for dispersion on a sphere (§3.2.2) for different values of the Earth's radius. The radius for which the dispersion of the poles is a minimum (precision parameter estimate k is a maximum) is regarded as the best estimate of the paleoradius.

The basic data set consists of the co-ordinates (λ_m, ϕ_m) of n rock units each having paleodeclination and inclination (D_m, I_m). The Fisher (1953) mean position (λ_0, ϕ_0) of all rock units is calculated and the basic data are transformed to a new co-ordinate system with (λ_0, ϕ_0) as the pole to yield new values $(\lambda_n, \phi_n, D_n, I_n)$ in this system. In such a system, the latitude of the rock units (λ_n) will change to some new value (λ_n') as the radius of the Earth is changed (with the continent keeping the same physical dimensions) but the longitudes (ϕ_n) will remain unchanged (Fig. 7.27b). For each value of radius new values (λ_n') may be calculated and the resulting dispersion of the paleomagnetic poles determined.

Figure 7.28 shows examples of the determination of paleoradius using the two methods outlined above. It was noted in §6.5.4 that the paleomagnetic poles for Africa during the Mesozoic remained in much the same position except for some minor variations around a position at about 65N, 250E (see Table 6.11). Following McElhinny and Brock (1975), the Mesozoic data for Africa can be subdivided into a set from northern Africa (29 poles from 306 sites) and one from southern Africa (17 poles from 200 sites). The mean locations of these two subsets lie close to a paleomeridian with mean declinations 340.8 and 338.7° (see Fig. 7.28a). The present angular distance between the two mean sampling localities is 50.3° and their paleolatitude difference is 47.4 ± 4.3° corresponding to a paleoradius of 1.06 ± 0.09 times the present radius. This is not discernibly different from 1.00 and in any case would favor a small contraction rather than expansion over the past 150 Myr.

McElhinny *et al.* (1978) analyzed the available paleomagnetic data at that time using the minimum dispersion method and, in a variety of analyses, found that for the past 400 Myr the average paleoradius was 1.020 ± 0.028 times the present radius. Possible expansion during that time was therefore limited to only 0.8%. The substantial data set for the Permian of Europe was noted in the discussion of Table 6.10 summarizing the data from Europe. These 85 pole positions (from 843 sites) have been used in a new calculation of the Earth's paleoradius for the Permian using the minimum dispersion method as illustrated in Fig. 7.28b. The method for combining m groups of data from N sites as set out in §6.4.2 (McFadden and McElhinny, 1995) has been used for the calculations. Estimates of the Fisher precision parameter k (for $N = 843$ sites) for various

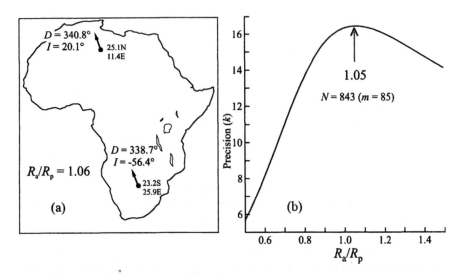

Fig. 7.28. Determinations of the Earth's paleoradius. (a) Use of the paleomeridian method by combining Mesozoic data for northern and southern Africa. The mean directions for the two regions have declinations that differ by only 2° (they should be the same if the sites are on the same paleomeridian). (b) The minimum dispersion method applied to Permian paleomagnetic data from Europe. The precision k is that calculated using the method for combining m groups of data from N sites set out in §6.4.2.

ratios R_a/R_p give a peak at 1.05. This is not discernibly different from unity and rules out any Earth expansion over the past 250 Myr, as was deduced by McElhinny *et al.* (1978).

Carey (1976) argued that the no-expansion result using paleomagnetic data is merely a consequence of Ward's (1963) method. For example, a continent cannot simultaneously preserve total surface area, intersite distances, and intersite angles while adjusting to an expanding Earth. Carey (1976) states that because all the parameters assumed to be constant vary with the radius, at a rate which increases rapidly with the size of the section of the Earth under consideration, the minimum dispersion inevitably occurs with the least change from this base radius (R_p) when $R_a/R_p = 1$. Schmidt and Clark (1980) examined this argument in some detail and found that the errors introduced by allowing for more realistic behavior of the continents, such as the "orange-peel effect" and crustal extension, are smaller by an order of magnitude than the response of paleomagnetic data to simplified expansion models. Therefore, it appears that the minimum dispersion method of Ward (1963) for determining paleoradius is a valid one.

References

Acton, G.D., and Gordon, R.G. (1991). A 65 Ma palaeomagnetic pole for the Pacific plate from the skewness of magnetic anomalies 27r–31. *Geophys. J. Int.*, **106**, 407–420. [*219*]

Acton, G.D., and Gordon, R.G. (1994). Paleomagnetic tests of Pacific plate reconstructions and implications for motion between hotspots. *Science*, **263**, 1246–1254. [*219*]

Ade-Hall, J.M., Palmer, H.C., and Hubbard, T.P. (1971). The magnetic and opaque petrological response of basalts to regional hydrothermal alteration. *Geophys. J. R. Astron. Soc.*, **24**, 137–174. [*44*]

Alfvén, H. (1942). On the existence of electromagnetic–hydromagnetic waves. *Arkiv. f. Mat. Astron. Fysik*, **29B(2)**, (7 pp). [*11*]

Alfvén, H. (1950). *Cosmical Electrodynamics*. Oxford Univ. Press, New York, 237pp. [*11*]

Algeo, T.J. (1996). Geomagnetic polarity bias patterns through the Phanerozic. *J. Geophys. Res.*, **101**, 2785–2814. [*145, 181*]

Allan, T.D. (1969). Review of marine geomagnetism. *Earth Sci. Rev.*, **5**, 217–254. [*189*]

Alvarez, W., Kent, D.V., Primoli Silva, L., Schweikert, R.A., and Larson, R.L. (1980). Franciscan complex limestones deposited at 17° South paleolatitude. *Geol. Soc. Am. Bull.*, **91**, 487–484. [*225*]

Alvarez, W., Arthur, M.A., Fischer, A.G., Lowrie, W., Napoleone, G., Premoli Silva, I., and Roggenthen, W.M. (1977). Upper Cretaceous–Paleocene magnetic stratigraphy at Gubbio, Italy. V. Type section for the Late Cretaceous–Paleocene geomagnetic reversal time scale. *Geol. Soc. Am. Bull.*, **88**, 383–389. [*159*]

Andrews, J.A. (1985). True polar wander: An analysis of Cenozoic and Mesozoic paleomagnetic poles. *J. Geophys. Res.*, **89**, 7737–7750. [*326*]

Andriamirado, C.R.A. (1971). Recherches paléomagnetiques sur Madagascar: Résultats et interprétations dans le cadre de la dislocation de la partie orientale du Gondwana. PhD thesis, University of Strasbourg. [*273*]

Anonymous (1979). Magnetostratigraphic polarity units. A supplementary chapter of the International Subcommission on Stratigraphic Classification. *Int. Stratigraphic Guide Geology*, **7**, 578–583. [*154*]

Anson, G.L., and Kodama, K.P. (1987). Compaction-induced inclination shallowing of the post-depositional remanent magnetization in a synthetic sediment. *Geophys. J. R. Astron. Soc.*, **88**, 673–692. [*70*]

333

Arason, P., and Levi, S. (1997). Intrinsic bias in averaging paleomagnetic data. *J. Geomag. Geoelect.*, **49**, 721–726. [*99*]

Argyle, K.S., and Dunlop, D.J. (1990). Low-temperature and high-temperature hysteresis of small muiltidomain magnetites (215–540 nm). *J. Geophys. Res.*, **95**, 7069–7083. [*57*]

Arkani-Hamed, J. (1989). Thermoviscous remanent magnetization of oceanic lithosphere inferred from its thermal evolution. *J. Geophys. Res.*, **94**, 17421–17426. [*203*]

Armstrong, R.L. (1966). K–Ar dating of plutonic and volcanic rocks in orogenic belts. In *Potassium–Argon Dating* (O.A. Schaefer and J. Zahringer, Eds.), pp.117–133. Springer–Verlag, Berlin. [*249*]

Arthur, M.A., and Fischer, A.G. (1977). Upper Cretaceous–Paleocene magnetic stratigraphy at Gubbio, Italy. I. Lithostratigraphy and sedimentology. *Geol. Soc. Am. Bull.*, **88**, 367–371. [*159*]

As, J.A., and Zijderveld, J.D.A. (1958). Magnetic cleaning of rocks in palaeomagnetic research. *Geophys. J. R. Astron. Soc.*, **1**, 308–319. [*115*]

Astini, R.A., Benedetto, J.L., and Vaccari, N.E. (1995). The early Paleozoic evolution of the Argentine Precordillera as a Laurentian rifted, drifted, and collided terrane – A geodynamic model. *Geol. Soc. Am. Bull.*, **107**, 253–273. [*322*]

Athanasopoulos, J. Fuller, M., and Weeks, R. (1996). Preliminary analysis of an atlas of reversal records. *Surv. Geophys.*, **17**, 177–182. [*165*]

Atwater, T. (1970). Implications of plate tectonics for the Cenozoic tectonic evolution of western North America. *Geol. Soc. Am. Bull.*, **81**, 3513–3536. [*207*]

Atwater, T., and Menard, H.W. (1970). Magnetic lineations in the north–east Pacific. *Earth Planet. Sci. Lett.*, **7**, 445–450. [*207*]

Atwater, T., and Molnar, P. (1973). Relative motion of the Pacific plate and North America deduced from seafloor spreading in the Atlantic, Indian and South Pacific Oceans. In *Proceedings of the Conference of Tectonic Problems of the San Andreas Fault System, 13* (R.L. Kovach and A. Nur, Eds.), pp.136–148, Stanford Univ. Press, Stanford, CA. [*225*]

Bachtadse, V., and Briden, J.C. (1990). Palaeomagnetic constraints on the position of Gondwana during Ordovician to Devonian times. In *Palaeozoic Paleogeography and Biogeography* (W.S. McKerrow and C.R. Scotese, Eds.), *Geol. Soc. London Mem.*, **12**, 43–48. [*271, 272*]

Bachtadse, V., and Briden, J.C. (1991). Palaeomagnetism of Devonian ring complexes from the Bayuda Desert, Sudan – new constraints on the apparent polar wander path for Gondwanaland. *Geophys. J. Int.*, **104**, 635–646. [*271, 272*]

Bai, Y., Chen, G, Sun, Q., Li, Y.A., Dong, Y. and Sun, D. (1987). Late Paleozoic polar wander path for the Tarim platform and its tectonic significance. *Tectonophysics*, **139**, 145–153. [*268*]

Bailey, M.E., and Dunlop, D.J. (1983). Alternating field characteristics of pseudo-single-domain (2–14 μm) and multidomain magnetite. *Earth Planet. Sci. Lett.*, **63**, 335–352. [*133*]

Bailey, R.C., and Halls, H.C. (1984). Estimate of confidence in paleomagnetic directions derived from mixed remagnetization circle and direct observational data. *J. Geophys.*, **54**, 174–182. [*125*]

Baksi, A.K. (1993). A paleomagnetic polarity time scale for the period 0–17 Ma, based on $^{40}Ar/^{39}Ar$ plateau ages for selected field reversals. *Geophys. Res. Lett.*, **20**, 1607–1610. [*151*]

Banerjee, S.K. (1980). Magnetism of the oceanic crust: Evidence from ophiolite complexes. *J. Geophys. Res.*, **85**, 3557–3566. [*192*]

Barbetti, M. (1977). Measurements of recent geomagnetic secular variation in southeast Australia and the question of dipole wobble. *Earth Planet. Sci. Lett.*, **36**, 207–218. [*22, 23*]

Barraclough, D.R. (1974). Spherical harmonic analysis of the geomagnetic field for eight epochs between 1600 and 1910. *Geophys. J. R. Astron. Soc.*, **36**, 497–513. [*14, 23*]

Bartlett, M.S. (1937). Properties of sufficiency and statistical tests. *Proc. R. Soc. London A*, **160**,

268–282. [93]

Barton, C.E. (1989). Geomagnetic secular variation: Directions and intensity. In *Encylopedia of Solid Earth Geophysics* (Ed., D.E. James), pp.560–577. Van Nostrand Reinhold, New York. [13]

Barton, C.E., and McFadden, P.L. (1996). Inclination shallowing and preferred transitional VGP paths. *Earth Planet. Sci. Lett.*, **140**, 147–158. [170]

Beck, M.E. (1976). Discordant paleomagnetic poles as evidence of regional shear in the western cordillera of North America. *Am. J. Sci.*, **276**, 694–712. [303]

Beck, M.E. (1980). Paleomagnetic record of plate-margin tectonic processes along the western edge of North America. *J. Geophys. Res.*, **85**, 7115–7131. [303]

Bell, R., and Jefferson, C.W. (1987). An hypothesis for an Australia–Canadian connection in the late Proterozoic and the birth of the Pacific Ocean. In *Proceedings of Pacific Rim Congress 1987*, pp.39–50, Australian Institute of Mining and Metallurgy, Parkville, Victoria. [315]

Berger, G.W., and York, D. (1981). Geochronometry from ^{40}Ar/^{39}Ar dating experiments. *Geochim. Cosmochim. Acta*, **45**, 795–811. [250]

Berger, G.W., York, D., and Dunlop, D.J. (1979). Calibration of Grenvillian palaeopoles by ^{40}Ar/^{39}Ar dating. *Nature*, **277**, 46–48. [317]

Berggren, W.A., Kent, D.V., and Flynn, J.J. (1985a). Jurassic to Paleogene. Part 2. Paleogene geochronology and chronostratigraphy. In *The Chronology of the Geological Record* (Ed., N.J. Snelling), pp.141–195, Blackwell, Oxford. [150]

Berggren, W.A., Kent, D.V., and van Couvering, J.A. (1985b). The Neogene. Part 2. Neogene geochronology and chronostratigraphy. In *The Chronology of the Geological Record* (Ed., N.J. Snelling), pp.211–260, Blackwell, Oxford. [150]

Berggren, W.A., McKenzie, D.P., Sclater, J.G., and van Hinte, J.E. (1975). World-wide correlation of Mesozoic magnetic anomalies and its implications: Discussion and reply. *Geol. Soc. Am. Bull.*, **86**, 267–272. [210]

Besse, J., and Courtillot, V. (1991). Revised and synthetic apparent polar wander paths for the African, Eurasian, North American and Indian plates, and true polar wander since 200 Ma. *J. Geophys. Res.*, **96**, 4029–4050. [262, 326]

Besse, J., Torcq, F., Gallet, Y., Ricou, L.E., Krystyn, L., and Saidi, A. (1998). Late Permian to Late Triassic palaeomagnetic data from Iran: Constraints on the migration of the Iranian block through the Tethyan Ocean and initial destruction of Pangaea. *Geophys. J. Int.*, **135**, 77–92. [313]

Bingham, C, (1964). Distributions on the sphere and on the projective plane. PhD. Thesis, Yale University. [99]

Bingham, C. (1974). An antipodally symmetric distribution on the sphere. *Ann. Stat.*, **2**, 1201–1225. [99]

Blackett, P.M.S. (1952). A negative experiment relating to magnetism and the earth's rotation. *Philos. Trans. R. Soc. London*, **A245**, 309–370. [82]

Blackett, P.M.S. (1961). Comparison of ancient climates with the ancient latitudes deduced from rock magnetic measurements. *Proc. R. Soc. London*, **A263**, 1–30. [241, 243]

Blackett, P.M.S., Clegg, J.A., and Stubbs, P.H.S. (1960). An analysis of rock magnetic data. *Proc. R. Soc. London*, **A256**, 291–322. [15]

Blakely, R.J. (1974). Geomagnetic reversals and crustal spreading rates during the Miocene. *J. Geophys. Res.*, **79**, 2979–2985. [150]

Blakely, R.J. (1995). *Potential Theory in Gravity and Magnetic Applications*, Cambridge Univ. Press, Cambridge, UK, pp.441 [215]

Blakely, R.J., and Cox, A. (1972). Evidence for short geomagnetic polarity intervals in the early Cenozoic. *J. Geophys. Res.*, **77**, 7065–7072. [196, 201, 213]

Bleil, U., and Petersen, N. (1983). Variations in magnetization intensity and low-temperature oxidation of ocean floor basalts. *Nature*, **301**, 384–388. [*192, 193*]

Blow, R.A., and Hamilton, N. (1978). Effect of compaction on the acquisition of a detrital remanent magnetization in fine-grained sediments. *Geophys. J. R. Astron. Soc.*, **52**, 13–23. [*70*]

Bloxham J., and Gubbins, D. (1985). The secular variation of the Earth's magnetic field. *Nature*, **317**, 777–781. [*7, 14*]

Bloxham J., and Gubbins, D. (1986). Geomagnetic field analysis, IV, Testing the frozen-flux hypothesis, *Geophys. J. R. Astron. Soc.*, **84**, 139–152. [*7*]

Bloxham, J., and Jackson, A. (1992). Time-dependent mapping of the magnetic field at the core–mantle boundary. *J. Geophys. Res.*, **97**, 19537–19563. [*137*]

Bogue, S.W., and Merrill, R.T. (1992). The character of the field during geomagnetic reversals. *Annu. Rev. Earth Planet. Sci.* **20**, 181–219. [*172*]

Bogue, S.W., and Paul, H.A. (1993). Distinctive field behavior following geomagnetic reversals. *Geophys. Res. Lett.*, **20**, 2399–2402. [*173*]

Bonhommet, N., and Babkine, J. (1967). Sur la présence d'aimantations inversées dans la Chaîne des Puys. *C. R. Acad. Sci. Paris*, **B264**, 92–94. [*174*]

Borradaile, G.J. (1993). Strain and magnetic remanence. *J. Struct. Geol.*, **15**, 383–390. [*75*]

Bott, M.H.P. (1967). Solution of the linear inverse problem in magnetic interpretation and application to oceanic magnetic anomalies. *Geophys. J. R. Astron. Soc.*, **13**, 313–323. [*199, 200*]

Boulin, J. (1991). Structures in southwest Asia and evolution of the eastern Tethys. *Tectonophysics*, **196**, 211–268. [*262*]

Briden, J.C. (1965). Ancient secondary magnetization in rocks. *J. Geophys. Res.*, **70**, 5205–5221. [*72*]

Briden, J.C., and Duff, B.A. (1981). Pre-Carboniferous palaeomagnetism of Europe north of the Alpine Orogenic Belt. In *Paleoreconstructions of the Continents* (M.W. McElhinny and D.A. Valencio, Eds.), *Am. Geophys. Union Geodynam. Ser.*, **2**, 138–149. [*228, 259*]

Briden, J.C., and Irving, E. (1964). Paleolatitude spectra of sedimentary paleoclimatic indicators. In *Problems in Paleoclimatology* (Ed., A.E.M. Nairn), pp.199–250, Wiley–Interscience, New York. [*242, 243, 244*]

Briden, J.C., and Ward, M.A. (1966). Analysis of magnetic inclination in borecores. *Pure Appl. Geophys.*, **63**, 133–152. [*97, 221*]

Briden, J.C., Morris, W.A., and Piper, J.D.A. (1973). Palaeomagnetic studies in the British Caledonides – VI. Regional and global implications. *Geophys. J. R. Astron. Soc.*, **34**, 107–134. [*259*]

Brock, A. (1981). Paleomagnetism of Africa and Madagascar. In *Paleoreconstructions of the Continents* (M.W. McElhinny and D.A. Valencio, Eds.), *Am. Geophys. Union Geodynam. Ser.*, **2**, 65–76. [*270*]

Brunhes, B. (1906). Recherches sur le direction d'aimantation des roches volcaniques. *J. Phys.*, **5**, 705–724. [*14, 139*]

Bryan, P., and Gordon, R.G. (1986). Rotation of the Colorado Plateau: An analysis of paleomagnetic data. *Tectonics*, **5**, 661–667. [*253*]

Bryan, P., and Gordon, R.G. (1990). Rotation of the Colorado Plateau: An updated analysis of paleomagnetic poles. *Geophys. Res. Lett.*, **17**, 1501–1504. [*253*]

Buchan, K.L., and Hodych, J.P. (1989). Early Silurian paleopole for red beds and volcanics of the King George IV Lake area, Newfoundland. *Can. J. Earth Sci.*, **26**, 1904–1917. [*109*]

Bull, C., Irving, E., and Willis, I. (1962). Further palaeomagnetic results from South Victoria Land, Antarctica. *Geophys. J. R. Astron. Soc.*, **6**, 320–336. [*279*]

Bullard, E.C. (1949). The magnetic field within the earth. *Proc. R. Soc. London*, **A197**, 433–453.

[*9*]

Bullard, E.C., Everett, J.E., and Smith, A.G. (1965). A Symposium on Continental Drift IV. The fit of the continents around the Atlantic. *Philos. Trans. R. Soc. London*, **A258**, 41–51. [*254, 281, 288, 289, 290, 298, 300*]

Bullard, E.C., Freedman, C., Gellman, H., and Nixon, J. (1950). The westward drift of the earth's magnetic field. *Philos. Trans. R. Soc. London*, **A243**, 67–92. [*7, 13*]

Burke, K., Dewey, J.F., and Kidd, W.S.F. (1977). Precambrian paleomagnetic results compatible with contemporary operation of the Wilson cycle. *Tectonophysics*, **33**, 287–299. [*323*]

Burov, B.V., Mirgaliev, D.K., and Heller, F. (1996). The problems of paleomagnetic correlation of the Upper Permian deposits of the stratotype and marine formations of Tethys (in Russian). *Permskie Ottozheniya Respubliki Tatarstan, Kazan*, 93–96. [*163*]

Butler, R.F. (1992). *Paleomagnetism: Magnetic Domains to Geologic Terranes*. Blackwell, Boston, 319pp. [*72, 91*]

Butler, R.F., and Banerjee, S.K. (1975). Theoretical single domain size range in magnetite and titanomagnetite. *J. Geophys. Res.*, **80**, 4049–4058. [*58*]

Bylund, G., and Pisarevsky, S. (1999). Remagnetization in Mesoproterozoic dykes from the Protogine zone, southern Sweden and the Sveconorwegian loop. *Precambrian Res.* in press [*320*]

Cande, S.C. (1976). A palaeomagnetic pole from Late Cretaceous marine magnetic anomalies in the Pacific. *Geophys. J. R. Astron. Soc.*, **44**, 547–566. [*202*]

Cande, S.C. (1978). Anomalous behavior of the paleomagnetic field inferred from the skewness of anomalies 33 and 34. *Earth Planet. Sci. Lett.*, **40**, 275–286. [*203*]

Cande, S.C., and Kent, D.V. (1976). Constraints imposed by the shape of marine magnetic anomalies on the magnetic source. *J. Geophys. Res.*, **81**, 4157–4162. [*203*]

Cande, S.C., and Kent, D.V. (1992a). A new geomagnetic polarity time scale for the Late Cretaceous and Cenozoic. *J. Geophys. Res.*, **97**, 13917–13951. [*151, 206, 209*]

Cande, S.C., and Kent, D.V. (1992b). Ultrahigh resolution marine magnetic anomaly profiles: A record of continuous paleointensity variations? *J. Geophys. Res.*, **97**, 15075–15083. [*209, 212*]

Cande, S.C., and Kent, D.V. (1995). Revised calibration of the geomagnetic polarity timescale for the Late Cretaceous and Cenozoic. *J. Geophys. Res.*, **100**, 6093–6095. [*145, 150, 151, 152, 153, 162, 179, 206*]

Cande, S.C., and Kristoffersen, Y. (1977). Late Cretaceous magnetic anomalies in the North Atlantic. *Earth Planet. Sci. Lett.*, **35**, 215–224. [*150*]

Cande, S.C., and Mutter, J.C. (1982). A revised identification of the oldest sea-floor spreading anomalies between Australia and Antarctica. *Earth Planet. Sci. Lett.*, **58**, 151–160. [*206*]

Cande, S.C., Larson, R.L., and LaBrecque, J.L. (1978). Magnetic lineations in the Pacific Jurassic Quiet Zone. *Earth Planet. Sci. Lett.*, **41**, 434–440. [*207, 211*]

Cande, S.C., LaBrecque, J.L., Larson, R.L., Pitman III, W.C., Golovchenko, X., and Haxby, W.F. (1989). *Magnetic Lineations of the World's Ocean Basins*, Am. Assoc. Pet. Geol., Tulsa, OK, United States, pp.13, 1 sheet. [*207*]

Carey, S.W. (1958). A tectonic approach to continental drift. In *Continental Drift – A Symposium* (Ed., S.W. Carey), pp.177–355, Univ. of Tasmania, Hobart. [*329*]

Carey, S.W. (1976). *The Expanding Earth*. Elsevier, Amsterdam, 488pp. [*329, 332*]

Carlson, R.L. (1982). Cenozoic convergence along the California coast: A quantitative test of the hotspot approximation. *Geology*, **10**, 191–196. [*225*]

Chamalaun, F.H. (1964). Origin of the secondary magnetization of the Old Red Sandstone of the Anglo–Welsh cuvette. *J. Geophys. Res.*, **69**, 4327–4337. [*72*]

Champion, D.E. (1980). Holocene geomagnetic secular variation in the western United States: Implications for the global geomagnetic field. *Rep. Open file series US Geol. Surv.*, **80–824**,

314pp. [*23*]

Champion, D.E., Lanphere, M.A., and Kuntz, M.A. (1988). Evidence for a new geomagnetic reversal from lava flows in Idaho: Discussion of short polarity reversals in the Brunhes and Late Matuyama polarity chrons. *J.Geophys. Res.*, **93**, 11667–11680. [*174, 175, 181*]

Channell, J.E.T., and McCabe, C. (1994). Comparison of the magnetic hysteresis properties of unremagnetized and remagnetized limestones. *J. Geophys. Res.*, **99**, 4613–4623. [*68, 134, 135*]

Channell, J.E.T., D'Argenio, B., and Horvath, F. (1979). Adria, the African promontory in Mesozoic Mediterranean paleogeography. *Earth Sci. Rev.*, **15**, 213–292. [*311, 312*]

Channell, J.E.T., Lowrie, W., Medizza, F., and Alvarez, W. (1978). Paleomagnetism and tectonics in Umbria. *Earth Planet. Sci. Lett.*, **39**, 199–210. [*312*]

Channell, J.E.T., Lowrie, W., Pialli, P., and Venturi, F. (1984). Jurassic magnetic stratigraphy from Umbrian (Italian) land sections. *Earth Planet Sci. Lett.*, **68**, 309–325. [*312*]

Chapman, S., and Bartels, J. (1940, 1962). *Geomagnetism, Vols. 1 and 2* (1940); 2nd edition (1962). Oxford Univ. Press, Oxford, 1049pp. [*8*]

Chase, C.G. (1978). Plate kinematics: The Americas, East Africa, and the rest of the world. *Earth Planet. Sci. Lett.*, **37**, 355–368. [*187*]

Chen, Y., Cogné, J.P., Courtillot, V., Tapponnier, P., and Zhu, X.Y. (1993). Cretaceous paleomagnetic results from western Tibet and tectonic implications. *J. Geophys. Res.*, **98**, 17981–17999. [*313, 314*]

Chen, Z., Li, Z.X., and Powell, C.McA. (1995). Paleomagnetism of the Upper Devonian reef complexes, Canning Basin, Western Australia. *Tectonics*, **14**, 154–167. [*272, 277, 278*]

Cheng, X.Y., and Barton, C.E. (1991). Onset of aridity and dune-building in central Australia: Sedimentological and magnetostratigraphic evidence from Lake Amadeus. *Palaeogeogr. Palaeoclimatol. Palaeoecol.*, **84**, 55–73. [*138*]

Chevallier, R. (1925). L'aimantation des laves de l'Etna et l'orientation du champ terrestre en Sicile du XIIe and XVIIe siecle. *Ann. Phys.*, **4**, 5–162. [*14, 231*]

Christoffel, D.A., and Falconer, R.K.H. (1972). Marine magnetic mesaurements in the south–west Pacific Ocean and the identification of new tectonic features. In *Antarctic Oceanology II. The Australian–New Zealand Sector* (Ed., D.E. Hayes), pp.197–209, American Geophysical Union, Washington, DC, Antarctic Research Series. [*207*]

Clegg, J.A., Radhakrishnamurty, C., and Sahasrabudhe, P.W. (1958). Remanent magnetism of the Rajmahal traps of north–eastern India. *Nature*, **181**, 830–831. [*280*]

Clegg, J.A., Deutsch, E.R., Everitt, C.W.F., and Stubbs, P.H.S. (1957). Some recent palaeomagnetic measurements made at Imperial College. *Acad. Phys.*, **6**, 219–231. [*311*]

Clement, B. (1991). Geographical distribution of transitional VGPs: Evidence for non-zonal equatorial symmetry during the Matuyama–Brunhes geomagnetic reversal. *Earth Planet. Sci. Lett.*, **104**, 48–58. [*167*]

Clement, B., and Kent, D.V. (1991). A southern hemisphere record of the Matuyama–Brunhes polarity reversal. *Geophys. Res. Lett.*, **18**, 81–84. [*165*]

Coe, R.S. (1977). Source models to account for Lake Mungo palaeomagnetic excursion and their implications. *Nature*, **269**, 49–51. [*174*]

Coe, R.S., and Prévot, M. (1989). Evidence suggesting extremely rapid field variation during a geomagnetic reversal. *Earth Planet. Sci. Lett.*, **92**, 292–298. [*165*]

Coe, R.S., Prévot, M., and Camps, P. (1995). New evidence for extraordinarily rapid change of the geomagnetic field during a geomagnetic reversal. *Nature*, **374**, 687–692. [*165, 166, 173*]

Coe, R.S., Globerman, B.R., Plumley, P.W., and Thrupp, G.A. (1985). Paleomagnetic results from Alaska and their tectonic implications. In *Tectono-stratigraphic Terranes of the Circum-Pacific Region 1* (Ed., D.G. Howell), pp.85–108. Circum-Pacific Council for Energy and Mineral

Resources, Houston, Texas. [*304, 305, 306*]

Cogné, J-P. (1987). TRM deviations in anisotropic assemblages of multidomain magnetite. *Geophys. J. R. Astron. Soc.*, **91**, 1013–1023. [*76*]

Cogné, J-P., and Perroud, H. (1985). Strain removal applied to paleomagnetic directions in an orogenic belt: The Permian red slates of the Alpes Maritimes, France. *Earth Planet. Sci. Lett.*, **72**, 125–140. [*108*]

Cogné, J-P., and Perroud, H. (1987). Unstraining paleomagnetic vectors: The current state of debate. *EOS, Trans. Am. Geophys. Union*, **68**, 705 and 711–712. [*75, 101, 106, 107, 108*]

Cogné, J-P., Halim, N., Chen, Y., and Courtillot, V. (1999). Resolving the problem of low Tertiary magnetic inclinations in Asia: New insights from paleomagnetic data from the Qiangtang, Kunlun, and Qaidam blocks (Tibet, China). *J. Geophys. Res.*, **104** (in press). [*314*]

Cogné, J-P., Perroud, H., Texier, M.P., and Bonhommet, N. (1986). Strain reorientation of hematite and its bearing upon remanent magnetization. *Tectonics*, **5**, 753–767. [*108*]

Collinson, D.W. (1965). Origin of remanent magnetization and initial susceptibility of certain red sandstones. *Geophys. J. R. Astron. Soc.*, **9**, 203–217. [*118*]

Collinson, D.W. (1966). Magnetic properties of the Taiguati formation. *Geophys. J. R. Astron. Soc.*, **11**, 337–347. [*118*]

Collinson, D.W. (1983). *Methods in Rock Magnetism and Palaeomagnetism: Techniques and Instrumentation*. Chapman & Hall, London, 503pp. [*80, 82, 114*]

Collinson, D.W., Creer, K.M., and Runcorn, S.K. (Eds.) (1967). *Methods in Palaeomagnetism*. Elsevier, Amsterdam. [*80*]

Coney, P.J. (1972). Cordilleran tectonics and North America plate motion. *Am. J. Sci.*, **272**, 603–628. [*225*]

Constable, C.G. (1992). Link between geomagnetic reversal paths and secular variation of the field over the past 5 My, *Nature*, **358**, 230–233. [*170*]

Constable, C.G. (1999). On rates of occurrence of geomagnetic reversals. *Phys. Earth Planet. Int.*, In Press. [*180*]

Constable, C.G., and Parker, R.L. (1988). Statistics of the geomagnetic secular variation for the past 5 m.y. *J. Geophys. Res.*, **93**, 11569–11581. [*99*]

Constable, C.G., and Parker, R.L. (1991). Deconvolution of long-core paleomagnetic measurements, spline theray for the linear problem, *Geophys. J. Int.*, **104**, 453–468. [*84*]

Constable, C.G., and Tauxe L. (1990). The bootstrap for magnetic susceptibility tensors. *J. Geophys. Res.*, **95**, 8383–8395. [*100*]

Constanzo-Alvarez, V., and Dunlop, D.J. (1998). A regional paleomagnetic study of lithotectonic domains in the Central Gneiss Belt, Grenville Province, Ontario. *Earth Planet. Sci. Lett.*, **157**, 89–103. [*317, 319*]

Cook, P.J., and McElhinny, M.W. (1979). A re-evaluation of the spatial and temporal distribution of sedimentary phosphate deposits in the light of plate tectonics. *Econ. Geol.*, **74**, 315–330. [*245*]

Coupland, D.H., and Van der Voo, R. (1980). Long-term non-dipole components in the geomagnetic field during the last 130 Ma. *J. Geophys. Res.*, **85**, 3529–3548. [*236*]

Courtillot, V., and Besse, J. (1987). Magnetic field reversals, polar wander, and core–mantle coupling. *Science*, **237**, 1140–1147. [*138, 179*]

Cox, A. (1961). Anomalous remanent magnetization of basalt. *U.S. Geol.Surv. Bull.*, **1083–E**, 131–160. [*81*]

Cox, A. (1962). Analysis of the present field for comparison with paleomagnetic results. *J. Geomag. Geoelect.*, **13**, 101–112. [*99*]

Cox, A. (1968). Lengths of geomagnetic polarity intervals. *J. Geophys. Res.*, **73**, 3247–3260. [*25, 175, 176*]

Cox, A. (1969). Geomagnetic reversals. *Science,* 163, 237–245. [*145, 176*]

Cox, A. (1981). A stochastic approach towards understanding the frequency and polarity bias of geomagnetic reversals. *Phys. Earth Planet. Int.,* 24, 178–190. [*138*]

Cox, A. (1982). Magnetostratigraphic time scale. In *A Geologic Time Scale* (W.B. Harland, A.V. Cox, P.G. Llewellyn, C.A.G. Pickton, A.G. Smith, and R. Walters, Eds.), pp.63–84. Cambridge Univ. Press, Cambridge. [*180*]

Cox, A., and Dalrymple, G.B. (1967). Statistical analysis of geomagnetic reversal data and the precision of Potassium–Argon dating. *J. Geophys. Res.,* 72, 2603–2614. [*173*]

Cox, A., and Doell, R.R. (1960). Review of paleomagnetism. *Geol. Soc. Am. Bull.,* 71, 645–768. [*15, 22, 232*]

Cox, A., and Gordon R.G. (1984). Paleolatitudes determined from paleomagnetic data from vertical cores. *Rev. Geophys. Space Phys.,* 22, 47–72. [*97*]

Cox, A., and R.B. Hart (1986). *Plate tectonics: How it works.* Blackwell, Oxford, 392pp. [*282*]

Cox, A., Doell, R.R., and Dalrymple, G.B. (1963a). Geomagnetic polarity epochs and Pleistocene geochronometry. *Nature,* 198, 1049–1051. [*144*]

Cox, A., Doell, R.R., and Dalrymple, G.B. (1963b). Geomagnetic polarity epochs – Sierra Nevada II. *Science,* 142, 382–385. [*144*]

Cox, A., Doell, R.R., and Dalrymple, G.B. (1964a). Geomagnetic polarity epochs. *Science,* 143, 351–352. [*144*]

Cox, A., Doell, R.R., and Dalrymple, G.B. (1964b). Reversals of the earth's magnetic field. *Science,* 144, 1537–1543. [*144*]

Creer, K.M. (1958). Preliminary palaeomagnetic measurements from South America. *Ann. Geophys.,* 15, 373–390. [*273*]

Creer, K.M. (1959). A.C. demagnetization of unstable Triassic Keuper marls from S.W. England. *Geophys. J. R. Astron. Soc.,* 2, 261–275. [*115*]

Creer, K.M. (1970). Palaeomagnetic survey of South American rocks, Parts I–V. *Philos. Trans. R. Soc. London,* A267, 457–558. [*273*]

Creer, K.M., and Ispir, Y. (1970). An interpretation of the behaviour of the geomagnetic field during polarity transitions. *Phys. Earth Planet. Int.,* 2, 283–293. [*166*]

Creer, K.M., Irving, E., and Nairn, A.E.M. (1959). The palaeomagnetism of the Great Whin Sill. *Geophys. J. R. Astron. Soc.,* 2, 306–323. [*99*]

Creer, K.M., Irving, E., and Runcorn, S.K. (1954). The direction of the geomagnetic field in remote epochs in Great Britain. *Philos. Trans. R. Soc. London,* A250, 144–156. [*246, 258*]

Creer, K.M., Tucholka, P., and Barton, C.E. (Eds.), (1983). *Geomagnetism of Baked Clays and Sediments.* Elsevier, Amsterdam. [*22, 27*]

Dalziel, I.W.D. (1991). Pacific margins of Laurentia and East Antarctica–Australia as a conjugate rift pair: Evidence and implications for an Eocambrian supercontinent. *Geology,* 19, 598–601. [*315*]

Dalziel, I.W.D. (1992). On the organization of the American plates in the Neoproterozoic and the breakout of Laurentia. *GSA Today,* 2, 237–241. [*321*]

Dalziel, I.W.D. (1997). Neoproterozoic–Paleozoic geography and tectonics: Review, hypothesis and environmental speculation. *Geol. Soc. Am. Bull.,* 109, 16–42. [*273, 321, 322*]

Dalziel, I.W.D., Dalla Salda, L.H., and Gahagan, L.M. (1994). Paleozoic Laurentia–Gondwana interaction and the origin of the Appalachian–Andean mountain system. *Geol. Soc. Am. Bull.,* 106, 243–252. [*321, 322*]

Dalziel, I.W.D., Dalla Salda, L.H., Congolani, C., and Palmer, A.R. (1996). The Argentina Precordillera: A Laurentian terrane? *GSA Today,* 6, 16–18. [*322*]

David, P. (1904). Sur la stabilité de la direction d'aimantation dans quelques roches volcaniques. *C. R. Acad. Sci. Paris,* 138, 41–42. [*14, 139*]

Davis, P.M., and Evans, M.E. (1976). Interacting single-domain properties of magnetite intergrowths. *J. Geophys. Res.*, 81, 989–994. [*60*]

Day, R., Fuller, M.D., and Schmidt, V.A. (1977). Hysteresis properties of titanomagnetites: Grain size and composition dependence. *Phys. Earth Planet. Int.*, 13, 260–267. [*57, 133*]

Debiche, M.G., Cox, A., and Engebretson, D.C. (1987). The motion of allochthonous terranes across the North Pacific Basin. *Geol. Soc. Am. Special Paper*, 207, 49pp. [*305, 306*]

DeMets, C. (1995). Plate motions and crustal deformation. *Rev. Geophys. Suppl.*, 33, 365–369. [*187*]

DeMets, C., Gordon, R.G., Argus, D.F., and Stein, S. (1990). Current plate motions. *Geophys. J. Int.*, 101, 425–478. [*184, 187*]

DeMets, C., Gordon, R.G., Argus, D.F., and Stein, S. (1994). Effects of current revisions to the geomagnetic polarity time scale on estimates of current plate motions. *Geophys. Res. Lett.*, 21, 2191–2194. [*184, 187, 188, 197*]

Desiderius, C.P.M., Tamaki, K., and Sager, W.W. (1997). Paleomagnetism of the Joban seamount chain: Its origin and tectonic implications for the Pacific plate. *J. Geophys. Res.*, 102, 5145–5155. [*224*]

Dickson, G.O., Pitman III, W.C., and Heirtzler, J.R. (1968). Magnetic anomalies in the South Atlantic and ocean floor spreading. *J. Geophys. Res.*, 73, 2087–2100. [*205*]

Diehl, J. F., and Shive, P.N. (1979). Paleomagnetic studies of the Early Permian Ingleside Formation of northern Colorado. *Geophys. J. R. Astron. Soc.*, 56, 278–282. [*163*]

Diehl, J.F., and Shive, P.N. (1981). Paleomagnetic results from the Late Carboniferous/Early Permian Casper Formation: Implications for northern Appalachian tectonics. *Earth Planet. Sci. Lett.*, 54, 281–291. [*163*]

DiVenere, V.J., and Opdyke, N.D. (1990). Paleomagnetism of the Maringouin and Shepody Formations, New Brunswick: A Namurian magnetic stratigraphy. *Can. J. Earth Sci.*, 27, 803–810. [*163*]

DiVenere, V.J., and Opdyke, N.D. (1991). Magnetic polarity stratigraphy and Carboniferous paleopole positions from the Joggins section, Cumberland Basin, Nova Scotia. *J. Geophys. Res.*, 96, 4051–4064. [*163*]

Dodson, M.H. (1973). Closure temperature in cooling geochronological and petrological systems. *Contrib. Mineral. Petrol.*, 40, 259–274. [*249*]

Dodson, M.H., and McClelland Brown, E.A. (1980). Magnetic blocking temperatures of single-domain grains during slow cooling. *J. Geophys. Res.*, 85, 2625–2637. [*249*]

Dodson, M.H., and McClelland Brown, E.A. (1985). Isotopic and palaeomagnetic evidence for rates of cooling, uplift and erosion. In *The Chronology of the Geological Record* (Ed., N.J. Snelling), *Geol. Soc. London Mem.*, 10, 315–325. [*249, 250*]

Dodson, R.E., Dunn, J.R., Fuller, M.D., Williams, I., Ito, H., Schmidt, V.A., and Wu, Yee-Ming (1978). Palaeomagnetic record of a late Tertiary field reversal. *Geophys. J. R. Astron. Soc.*, 53, 373–412. [*165, 173*]

Doell, R.R., and Cox, A. (1967a). Measurement of the natural remanent magnetization at the outcrop. In *Methods in Palaeomagnetism* (D.W. Collinson, K.M. Creer and S.K. Runcorn, Eds.), pp.159–162, Elsevier, Amsterdam. [*81*]

Doell, R.R., and Cox, A. (1967b). Analysis of alternating field demagnetization equipment. In *Methods in Palaeomagnetism* (D.W. Collinson, K.M. Creer and S.K. Runcorn, Eds.), pp.241–253, Elsevier, Amsterdam. [*115*]

Drewry, G.E., Ramsay, A.T.S., and Smith, A.G. (1974). Climatically controlled sediments, the geomagnetic field, and trade wind belts in Phanerozoic time. *J. Geol.*, 82, 531–553. [*241, 243*]

Duncan, R.A. (1981). Hotspots in the southern oceans – an absolute frame of reference for motion of the Gondwana continents. *Tectonophysics*, 74, 29–42. [*223*]

Duncan, R.A., and Richards, M.A. (1991). Hotspots, mantle plumes, flood basalts, and true polar wander. *Rev. Geophys.*, **29**, 31–50. [223]

Dunlop, D.J. (1972). Magnetic mineralogy of heated and unheated red sediments by coercivity spectrum analysis. *Geophys. J. R. Astron. Soc.*, **27**, 37–55. [129]

Dunlop, D.J. (1973). Thermoremanent magnetization in submicroscopic magnetite. *J. Geophys. Res.*, **78**, 7602–7613. [63]

Dunlop, D.J. (1977). The hunting of the 'psark'. *J. Geomag. Geoelect.*, **29**, 293–318. [57]

Dunlop, D.J. (1979). On the use of Zijderveld vector diagrams in multi-component palaeomagnetic studies. *Phys. Earth Planet. Int.*, **20**, 12–24. [120]

Dunlop, D.J. (1981). The rock magnetism of fine particles. *Phys. Earth Planet. Int.*, **26**, 1–26. [57]

Dunlop, D.J. (1990). Developments in rock magnetism. *Rep. Prog. Phys.*, **53**, 707–792. [38, 44]

Dunlop, D.J. (1995). Magnetism in rocks. *J. Geophys. Res.*, **100**, 2161–2174. [57]

Dunlop, D.J., and Argyle, K.S. (1997). Thermoremanence, anhysteretic remanence and susceptibility of submicron magnetites: Nonlinear field dependence and variation with grain size. *J. Geophys. Res.*, **102**, 20199–20210. [63]

Dunlop, D.J., and Özdemir, O. (1997). *Rock Magnetism: Fundamentals and frontiers*, Cambridge Univ. Press, Cambridge, UK, 573pp. [38, 50, 51, 57, 58, 62, 65, 70, 76, 127, 139, 191, 192]

Dunlop, D.J., and West, G.F. (1969). An experimental evaluation of single domain theories. *Rev. Geophys.*, **7**, 709–757. [51]

Dunn, J.R., Fuller, M.D., Ito, H., and Schmidt, V.A. (1971). Paleomagnetic study of a reversal of the earth's magnetic field. *Science*, **172**, 840–845. [165, 173]

Dunning, G.R., and Hodych, J.P. (1990). U/Pb zircon and baddeleyite ages for the Palisades and Gettysburg sills of the northeastern United States: Implications for the age of the Triassic/Jurassic boundary. *Geology*, **18**, 795–798. [159]

Dyment, J., Cande, S.C., and Arkani-Hamed, J. (1994). Skewness of marine magnetic anomalies created between 85 and 40 Ma in the Indian Ocean. *J. Geophys. Res*, **99**, 24121–24134. [203]

Egbert, C.G. (1992). Sampling bias in VGP longitudes. *Geophys. Res. Lett.*, **19**, 2353–2356. [234]

Egyed, L. (1960). Some remarks on continental drift. *Geofis. Pura Appl.*, **45**, 115–116. [329, 330]

Egyed, L. (1961). Palaeomagnetism and the ancient radii of the Earth. *Nature*, **190**, 1097–1098. [331]

Elming, S.A., Pesonen, L.J., Leino, M.A.H., Khramov, A.N., Mikhailova, N.P., Krasnova, A.F., Mertanen, S., Bylund, G., and Terho, M. (1993). The drift of the Fennoscandian and Ukranian shields during the Precambrian: A palaeomagnetic analysis. *Tectonophysics*, **223**, 177–198. [320]

Elmore, R.D., and McCabe, C. (1991). The occurrence and origin of remagnetization in the sedimentary rocks of North America. *Rev. Geophys.*, **29**, Suppl. (IUGG report), 377–383. [67]

Elsasser, W.M. (1946). Induction effects in terrestrial magnetism. 1. Theory. *Phys. Rev.*, **69**, 106–116. [9]

Embleton, B.J.J. (1981). A review of the paleomagnetism of Australia and Antarctica. In *Paleoreconstructions of the Continents* (M.W. McElhinny and D.A. Valencio, Eds.), *Am. Geophys. Union Geodynam. Ser.*, **2**, 77–92. [276, 278, 279]

Embleton, B.J.J., McElhinny, M.W., Ma, X., Zhang, Z., and Li., Z.X. (1996). Permo-Triassic magnetostratigraphy in China: The type section near Taiyuan, Shanxi Province, North China. *Geophys. J. Int.*, **126**, 382–388. [163]

Engebretson, D.C., Cox, A., and Gordon, R.G. (1984). Relative motions between oceanic plates of the Pacific basin. *J. Geophys. Res.*, **89**, 10291–10310. [225]

Engebretson, D.C., Cox, A., and Gordon, R.G. (1985). Relative motions between oceanic and continental plates in the Pacific Basin. *Geol. Soc. Am. Special Paper*, 206, 59pp. [225, 226, 305]

Enkin, R.J., Yang, Z., Chen, Y., and Courtillot, V. (1992). Paleomagnetic constraints on the geodynamic history of the major blocks of China from the Permian to the present. *J. Geophys. Res.*, 97, 13953–13989. [*265*]

Evans, D.A. (1998). True polar wander, a supercontinent legacy. *Earth Planet. Sci. Lett.*, 157, 1–8. [*326, 329*]

Evans, D.A., Beukes, N.J., and Kirschvink, J.L. (1997). Low-latitude glaciation in the Palaeoproterozoic era. *Nature*, 386, 262–266. [*244*]

Evans, M.E. (1976). Test of the dipolar nature of the geomagnetic field throughout Phanerozoic time. *Nature*, 262, 676–677. [*236, 237, 238*]

Evans, M.E., and McElhinny, M.W. (1969). An investigation of the the origin of stable remanence in magnetite-bearing igneous rocks. *J. Geomag. Geoelect.*, 21, 757–773. [*58*]

Evans, M.E., and Wayman, M.L. (1970). An investigation of small magnetic particles by means of electron microscopy. *Earth Planet. Sci. Lett.*, 9, 365–370. [*60*]

Evans, M.E., and Wayman, M.L. (1974). An investigation of the role of ultrafine titanomagnetite intergrowths in palaeomagnetism. *Geophys. J. R. Astron. Soc.*, 36, 1–10. [*60*]

Everitt, C.W.F., and Clegg, J.A. (1962). A field test of paleomagnetic stability. *Geophys. J. R. Astron. Soc.*, 6, 312–319. [*100, 109*]

Fedosh, M.S., and Smoot, J.P. (1988). A cored stratigraphic section through the northern Newark basin, New Jersey. *U.S. Geol. Surv. Bull.*, 1776, 19–24. [*159*]

Fisher, N.I., Lewis, T., and Embleton, B.J.J. (1987). *Statistical Analysis of Spherical Data*. Cambridge Univ. Press, Cambridge, UK, 329pp. [*96, 100, 108*]

Fisher, N.I., Lewis, T., and Willcox, M.E. (1981). Tests of discordancy for samples from Fisher's distribution on the sphere. *Appl. Statist.*, 30, 230–237. [*92*]

Fisher, R.A. (1953). Dispersion on a sphere. *Proc. R. Soc. London*, A217, 295–305. [*87, 88, 89, 90, 91, 247, 331*]

Flanders, F.J. (1988). An alternating gradient force magnetometer. *J. Appl. Phys.*, 63, 3940–3945. [*133*]

Folgerhaiter, G. (1899). Sur les variations séculaires de l'inclinaison magnétique dans l'antiquité. *J. Phys. Ser. 3.*, 8, 5–16. [*14, 22*]

Foster, J.H. (1966). A paleomagnetic spinner magnetometer using a fluxgate gradiometer. *Earth Planet. Sci. Lett.*, 1, 463–466. [*82*]

Foster, J.H., and Opdyke, N.D. (1970). Upper Miocene to Recent magnetic stratigraphy in deep-sea sediments. *J. Geophys. Res.*, 75, 4465–4473. [*148, 149*]

Francheteau, J., Sclater, J.G., and Menard, H.W. (1970a). Pattern of relative motion from fracture zone and spreading rate data in the north–eastern Pacific. *Nature*, 226, 746–748. [*225*]

Francheteau, J., Harrison, C.G.A. Sclater, J.G., and Richards, M.L. (1970b). Magnetization of Pacific seamounts: A preliminary polar curve for the northeastern Pacfic. *J. Geophys. Res.*, 75, 2035–2061. [*216*]

Fraser-Smith, A.C. (1987). Centered and eccentric geomagnetic dipoles and their poles, 1600–1985. *Rev. Geophys.*, 25, 1–16. [*12, 13, 14, 23*]

Fuller, M. (1984). On the grain size dependence of the behaviour of fine magnetic particles in rocks. *Geophys. Surv.*, 7, 75–87. [*63*]

Fuller, M. (1987). Experimental methods in rock magnetism and paleomagnetism. *Methods Exp. Phys.*, 24, 303–471. [*82*]

Fuller, M.D., and Kobayashi, K. (1964). Identification of the magnetic phases carrying natural remanent magnetization in certain rocks. *J. Geophys. Res.*, 69, 4409–4413. [*135*]

Fuller, M.D., Williams I., and Hoffman, K.A. (1979). Paleomagnetic records of geomagnetic field reversals and the morphology of the transitional fields. *Rev. Geophys. Space Phys.*, 17, 179–203, 1979. [*167*]

Gallet, Y., and Hulot, G. (1997). Stationary and nonstationary behavior within the geomagnetic polarity time scale. *Geophys. Res. Lett.*, **24**, 1875–1878. [*179*]

Gallet, Y., and Pavlov, V. (1996). Magnetostratigraphy of the Moyero river section (north–western Siberia): Constraints on the geomagnetic reversal frequency during the early Palaeozoic. *Geophys. J. Int.*, **125**, 95–105. [*162*]

Gauss, C.F. (1839). Allgemeine Theorie des Erdmagnetismus. In *Resultate aus den Beobachtungen magnetischen Vereins im Jahre 1838*, pp.1–57. [Reprinted in *Werke*, **5**, 121–193, Gottingen, 1877; Translated by E. Sabine, in Taylor, R., *Scientific Memoirs Vol. 2*, R. & J. E. Taylor, London, 1841]. [*4, 7, 8*]

Gee, J., and Kent, D.V. (1994). Variations in Layer 2A thickness and the origin of the central anomaly magnetic high. *Geophys. Res. Lett.*, **21**, 297–301. [*193, 194*]

Gee, J., and Kent, D.V. (1997). Magnetization of axial lavas from the southern East Pacific Rise (14°–23°S): Geochemical controls on magnetic properties. *J. Geophys. Res.*, **102**, 24873–24886. [*195*]

Gee, J., Schneider, D.A., and Kent, D.V. (1996). Marine magnetic anomalies as recorders of geomagnetic intensity variations. *Earth Planet. Sci. Lett.*, **144**, 327–335. [*209*]

Gee, J., Staudigel, H., and Tauxe, L. (1988). Nonuniform magnetization of Jasper seamount. *J. Geophys. Res.*, **93**, 12159–12175. [*216*]

Gee, J., Staudigel, H., and Tauxe, L. (1989). Contribution of induced magnetization to magnetization of seamounts. *Nature*, **342**, 170–173. [*215, 216*]

Gilder, S., and Courtillot, V. (1997). Timing of the North-South China collision from new middle to late Mesozoic paleomagnetic data from the North China block. *J. Geophys. Res.*, **102**, 17713–17727. [*294, 295*]

Gilder, S.A., Coe, R.S., Wu, H., Kuang, G., Zhao, X., Wu, Q., and Tang, X. (1993). Cretaceous and Tertiary paleomagnetic results from Southeast China and their tectonic implications. *Earth Planet. Sci. Lett.*, **117**, 637–652. [*104, 105*]

Glatzmaier, G.A., Coe, R.S., Hongre, L., and Roberts, P.H. (1999). How the Earth's mantle controls the frequency of geomagnetic reversals. Submitted for publication. [*179*]

Glen, W. (1982). *The Road to Jaramillo. Critical Years of the Revolution in Earth Science*. Stanford Univ. Press, Stanford, California, 459pp. [*15, 138, 144*]

Globerman, B.R., and Irving, E. (1988). Mid-Cretaceous paleomagnetic field for North America: Restudy of 100 Ma intrusive rocks from Arkansas. *J. Geophys. Res.*, **93**, 11721–11733. [*256*]

Goldreich, P., and Toomre, A. (1969). Some remarks on polar wandering. *J. Geophys. Res.*, **74**, 2555–2567. [*326*]

Gordon, R.G. (1987). Polar wandering and paleomagnetism. *Annu. Rev. Earth Planet. Sci.*, **15**, 567–593. [*326*]

Gordon, R.G., and Cox, A. (1980a). Paleomagnetic test of the early Tertiary plate circuit between the Pacific basin plates and the Indian plate. *J. Geophys. Res.*, **85**, 6534–6546. [*216*]

Gordon, R.G., and Cox, A. (1980b). Calculating palaeomagnetic poles for oceanic plates. *Geophys. J. R. Astron. Soc.*, **63**, 619–640. [*217*]

Gordon, R.G., Cox, A., and O'Hare, S. (1984). Paleomagnetic Euler poles and the apparent polar wander path and absolute motion of North America since the Carboniferous. *Tectonics*, **3**, 499–537. [*248*]

Gordon, R.G., McWilliams, M.O., and Cox, A. (1979). Pre-Tertiary velocities of the continents: A lower bound from paleomagnetic data. *J. Geophys. Res.*, **84**, 5480–5486. [*287*]

Goree, W.S., and Fuller, M. (1976). Magnetometers using RF-driven squids and their applications in rock magnetism and paleomagnetism. *Rev. Geophys. Space Phys.*, **14**, 591–608. [*82*]

Gose, W.A., and Helsley, C.E. (1972). Paleomagnetism and rock magnetism of the Permian Cutler and Elephant Canyon Formations in Utah. *J. Geophys. Res.*, **77**, 1534–1548. [*163*]

Gose, W.A., Helper, M.A., Connelly, J.N., Hutson, S.E., and Dalziel, I.W.D. (1997). Paleomagnetic data and U–Pb isotope age determinations from Coats Land, Antarctica: Implications for late Proterozoic plate reconstructions. *J. Geophys. Res.*, **102**, 7887–7902. *[321]*

Gough, D.I. (1964). A spinner magnetometer. *J. Geophys. Res.*, **69**, 2455–2463. *[82]*

Gough, D.I., Opdyke, N.D., and McElhinny, M.W. (1964). The significance of paleomagnetic results from Africa. *J. Geophys. Res.*, **69**, 2509–2519. *[270]*

Gradstein, F.M., Agterberg, F.P., Ogg, J.G., Hardenbol, J., van Veen, P., Thierry, J., and Huang, Z. (1994). A Mesozoic time scale. *J. Geophys. Res.*, **99**, 24051–24074. *[151, 156, 211, 212]*

Graham, J.W. (1949). The stability and significance of magnetism in sedimentary rocks. *J. Geophys. Res.*, **54**, 131–167. *[100, 101, 108]*

Graham, K.W.T. (1961). The remagnetization of a surface outcrop by lightning currents, *Geophys. J. R. Astron. Soc.*, **6**, 149–161. *[81]*

Graham, K.W.T., Helsley, C.E., and Hales, A.L. (1964). Determination of the relative positions of continents from paleomagnetic data. *J. Geophys. Res.*, **69**, 3895–3900. *[287, 288]*

Greenewalt, D., and Taylor, P.T. (1974). Deep-tow magnetic measurements across the axial valley of the mid-Atlantic ridge. *J. Geophys. Res.*, **79**, 4401–4405. *[190]*

Griffiths, D.H., King, R.F., Rees, A.I., and Wright, A.E. (1960). The remanent magnetism of some recent varved sediments. *Proc. R. Soc. London*, **A256**, 359–383. *[70]*

Groenewald, P.B., Grantham, G.H., and Watkeys, M.K. (1991). Geological evidence for a Proterozoic to Mesozoic link between southeastern Africa and Dronning Maud Land, Antarctica. *J. Geol. Soc. London*, **148**, 1115–1123. *[279, 321]*

Groenewald, P.B., Moyes, A.B., Grantham, G.H., and Krynauw, J.R. (1995). East Antarctic crustal evolution: Geological constraints and modelling in western Dronning Maud Land. *Precambrian Res.*, **75**, 231–250. *[279, 321]*

Grow, J.A., and Atwater, T. (1970). Mid-Tertiary tectonic transition in the Aleutian arc. *Geol. Soc. Am. Bull.*, **81**, 3715–3722. *[225]*

Grunow, A., Hanson, R.E., and Wilson, T.J. (1996). Were aspects of Pan-African deformation linked to Iapetus opening? *Geology*, **24**, 1063–1066. *[279, 321]*

Gubbins, D. (1999). The distinction between geomagnetic excursions and reversals. *Geophys. J. Int.*, **137**, F1–F3. *[174]*

Gubbins, D., and Coe, R.S (1993). Longitudinally confined geomagnetic reversal paths from non-dipole transition fields. *Nature*, **362**, 51–53. *[167]*

Gubbins, D., and Kelly, P. (1993). Persistent patterns in the geomagnetic field over the past 2.5 Myr. *Nature*, **365**, 829–832. *[235]*

Gubbins, D., and Zhang, K. (1993). Symmetry properties of the dynamo equations for palaeomagnetism and geomagnetism. *Phys. Earth Planet. Int.*, **75**, 225–241. *[12]*

Gurevich, E.L. (1984). Paleomagnetism of the Ordovician deposits of the Moyero river sequence. In *Paleomagnetic Methods in Stratigraphy*, [in Russian] VNIGRI, St. Petersburg, pp.35–41. *[263]*

Gurevich, Y.L., and Slautsitais, I.P. (1985). A paleomagnetic section in the Upper Permian and Triassic deposits on Novaya Zemlya. *Int. Geol. Rev.*, **27**, 168–177. *[163]*

Gurnis, M. (1991). Continental flooding and mantle–lithosphere dynamics. In *Global Isostacy, Sea Level and Mantle Rheology* (R. Sabatini *et al.*, Eds.), pp.445–491. Kluwer, Dordrecht, The Netherlands. *[210]*

Haag, M., and Heller, F. (1991). Late Permian to Early Triassic magnetostratigraphy. *Earth Planet. Sci. Lett.*, **107**, 42–54. *[163]*

Haggerty, S.E. (1976). Oxidation of opaque mineral oxides in basalts, and Opaque mineral oxides in terrestrial igneous rocks. In *Oxide Minerals*, 1st edn (Ed., D. Rumble), pp.Hg1–177, Mineral Society of America, Washington, DC. *[42, 46]*

Hagstrum, J.T., Van der Voo, R., Auvray, B., and Bonhommet, N. (1980). Eocambrian–Cambrian palaeomagnetism of the Armorican Massif. *Geophys. J. R. Astron. Soc.*, **61**, 498–517. [*308*]

Hale, C.J. (1987). The intensity of the geomagnetic field at 3.5 Ga: Paleointensity results from the Komati Formation, Barberton Mountain Land, South Africa. *Earth Planet. Sci. Lett.*, **86**, 354–364. [*25*]

Halgedahl, S.L. (1998). Revisiting the Lowrie–Fuller test: Alternating field demagnetization characteristics of single-domain through multidomain glass-ceramic magnetite. *Earth Planet. Sci. Lett.*, **160**, 257–271. [*132*]

Halgedahl, S.L., and Fuller, M.D. (1980). Magnetic domain observations of nucleation processes in fine particles of intermediate titanomagnetite. *Nature*, **288**, 70–72. [*57, 62*]

Halgedahl, S.L., and Fuller, M.D. (1983). The dependence of magnetic domain structure upon magnetization state with emphasis on nucleation as a mechanism for pseudo-single-domain behavior. *J. Geophys. Res.*, **88**, 6505–6522. [*57, 62*]

Hall, J.M., Walls, C.C., Yang, J-S., Hall, S.L., and Bakor, A.R. (1991). The magnetization of the oceanic crust: Contribution to knowledge from the Troodos, Cyprus, ophiolite. *Can. J. Earth Sci.*, **28**, 1812–1826. [*192*]

Hallam, A. (1983). Supposed Permo-Triassic megashear between Laurasia and Gondwanaland. *Nature*, **301**, 499–502. [*299*]

Halls, H.C. (1976). A least-squares method to find a remanence direction from converging remagnetization circles. *Geophys. J. R. Astron. Soc.*, **45**, 297–304. [*125*]

Hamano, Y. (1980). An experiment on the post-depositional remanent magnetization in artificial and natural sediments. *Earth Planet. Sci. Lett.*, **51**, 221–232. [*70*]

Hamilton, W.B. (1970). The Uralides and the motion of the Russian and Siberian platforms. *Geol. Soc. Am. Bull.*, **81**, 2553–2576. [*261*]

Hamilton, W.B. (1981). Plate tectonic mechanism of Laramide deformation. *Contrib. Geol.*, **19**, 87–92. [*253*]

Handschumacher, D.W., Sager, W.W., Hilde, T.W.C., and Bracey, D.R. (1988). Pre-Cretaceous evolution of the Pacific plate and extension of the geomagnetic polarity reversal time scale with implications for the origin of the Jurassic "Quiet Zone". *Tectonophysics*, **155**, 365–380. [*207, 211*]

Hanson, R.E., Martin, M.W., Bowring, S.A., and Munyanyiwa, H. (1998). U–Pb zircon age for the Umkondo dolerites, eastern Zimbabwe: 1.1 Ga large igneous province in southern Africa–East Antarctica and posible Rodinia correlations. *Geology*, **26**, 1143–1146. [*321*]

Harbert, W. (1990). Paleomagnetic data from Alaska: Reliability, interpretation and terrane trajectories. *Tectonophysics*, **184**, 111–135. [*306*]

Hargraves, R.B., Hattingh, P.J., and Onstott, T.C. (1994). Palaeomagnetic results from the Timbavati Group in the Kruger National Park, South Africa. *S. Afr. J. Geol.*, **97**, 114–118. [*321*]

Harland, W.B., Armstrong, R.L., Cox, A.V., Craig, L.E., Smith, A.G., and Smith, D.G. (1990). *A Geologic Time Scale 1989*. Cambridge Univ. Press, Cambridge, 263pp. [*150, 209, 252*]

Harland, W.B., Cox, A.V., Llewellyn, P.G., Pickton, C.A.G., Smith, A.G., and Walters, R. (1982). *A Geologic Time Scale*. Cambridge Univ. Press, Cambridge, 131pp. [*150, 209*]

Harper, C.T. (1967). The geological interpretation of potassium–argon ages of metamorphic rocks from the Scottish Caledonides. *Scot. J. Geol.*, **3**, 46–66. [*249*]

Harrison, C.G.A., and Somayajulu, B.L.K. (1966). Behaviour of the earth's magnetic field during a reversal. *Nature*, **212**, 1193–1195. [*173*]

Hartstra, R.L. (1983). TRM, ARM and I_{sr} of two natural magnetites of MD and PSD grain size. *Geophys. J. R. Astron. Soc.*, **73**, 719–737. [*63*]

Haxby, W.F. (1987). *Gravity Field of the World's Oceans*. National Geophysical Data Center,

NOAA, Boulder, CO. [*207*]

Hayes, D.E., and Heirtzler, J.R. (1968). Magnetic anomalies and their relation to the Aleutian island arc. *J. Geophys. Res.*, **73**, 4637–4646. [*206*]

Hayes, D.E., and Pitman, W.C. III (1970). Magnetic lineations in the North Pacific. *Geol. Soc. Amer. Mem.*, **126**, 291–314. [*207*]

Hayes, D.E., and Rabinowitz, P.D. (1975). Mesozoic magnetic lineations and the magnetic quiet zone off northwest Africa. *Earth Planet. Sci. Lett.*, **28**, 105–115. [*211*]

Hays, J.D., and Pitman, W.C. III (1973). Lithospheric plate motion, sea level changes and climatic and ecological consequences. *Nature*, **246**, 18–22. [*210*]

Heider, F., Dunlop, D.J., and Soffel, H.C. (1992). Low-temperature and alternating field demagnetization of saturation remanence and thermoremanence in magnetite grains (0.037μm to 5 mm). *J. Geophys. Res.*, **97**, 9371–9381. [*133*]

Heider, F., Dunlop, D.J., and Sugiura, N. (1987). Magnetic properties of hydrothermally recrystalized magnetic crystals. *Science*, **236**, 1287–1290. [*57*]

Heirtzler, J.R., Le Pichon, X., and Baron, J.G. (1966). Magnetic anomalies over the Reykjanes Ridge. *Deep-Sea Res.*, **13**, 427–443. [*205*]

Heirtzler, J.R., Dickson, G.O., Herron, E.M., Pitman III, W.C., and Le Pichon, X. (1968). Marine magnetic anomalies, geomagnetic reversals and motions of the ocean floors and continents. *J. Geophys. Res.*, **73**, 2119–2136. [*149, 150, 151, 205, 206*]

Helbig, K. (1965). Optimum configuration for the measurement of the magnetic moment of samples of cubical shape with a fluxgate magnetometer. *J. Geomag. Geoelect.*, **17**, 373–380. [*82*]

Heller, F. (1978). Rock magnetic studies of Upper Jurassic limestones from southern Germany. *J. Geophys.*, **44**, 525–543. [*129*]

Heller, F., Lowrie, W., Li, H., and Wang, J. (1988). Magnetostratigraphy of the Permo-Triassic boundary section at Shangsi. *Earth Planet. Sci. Lett.*, **103**, 301–310. [*163*]

Heller, P.L., Anderson, D.L., and Angevine, C.L. (1996). Is the middle Cretaceous pulse of rapid sea-floor spreading real or necessary? *Geology*, **24**, 491–494. [*210*]

Helsley, C.E., and Steiner, M.D. (1969). Evidence for long intervals of normal polarity during the Cretaceous period. *Earth Planet. Sci. Lett.*, **5**, 325–332. [*162, 209*]

Henry, S.G. (1979). Chemical demagnetization; methods, procedures, and applications through vector analysis. *Can. J. Earth Sci.*, **16**, 1832–1841. [*118*]

Herron, E.M., Dewey, J.F., and Pitman, W.C. (1974). Plate tectonic model for the evolution of the Arctic. *Geology*, **2**, 377–380. [*290*]

Hess, H.H. (1960). Evolution of ocean basins. Report to Office of Naval Research on ONR Contract No 1858 (10). [*183*]

Hess, H.H. (1962); History of ocean basins. In *Petrologic Studies: A Volume to Honor A.F. Buddington* (A.E.J. Engel, H. James and B.F. Leonard, Eds.), pp.599–620. Geological Society of America, Boulder, CO. [*149, 183*]

Hilde, T.W.C., Isezaki, N., and Wageman, J.M. (1976). Mesozoic seafloor spreading in the north Pacific. In *The Geophysics of the Pacific Ocean Basin and its Margin* (G.H. Sutton, R. Moberly and M. Manghanani, Eds.), *Am. Geophys. Union Monogr.*, **19**, 205–226. [*207, 225, 226*]

Hildebrand, J.A., and Parker, R.L. (1987). Paleomagnetism of Cretaceous Pacific seamounts revisited. *J. Geophys. Res.*, **92**, 12695–12712. [*215*]

Hillhouse, J.W. (1977). A method for the removal of rotational remanent magnetization acquired during alternating field demagnetization. *Geophys. J. R. Astron. Soc.*, **50**, 29–34. [*115*]

Hillhouse, J.W., and Cox, A. (1976). Brunhes–Matuyama polarity transition. *Earth Planet. Sci. Lett.*, **29**, 51–64. [*166*]

Hillhouse, J.W., and Grommé, C.S. (1984). Northward displacement and accretion of Wrangellia: New paleomagnetic evidence from Alaska. *J. Geophys. Res.*, **89**, 4461–4477. [*304*]

Hirt, A.M., and Lowrie, W. (1988). Paleomagnetism of the Umbrian–Marches orogenic belt. *Tectonophysics*, **146**, 91–103. [*130, 312*]

Hoffman, K.A. (1975). Cation diffusion processes and self reversal of thermoremanent magnetization in the ilmenite–haematite solid solution series. *Geophys. J. R. Astron. Soc.*, **41**, 65–80. [*139, 140*]

Hoffman, K.A. (1979). Behavior of the geodynamo during reversal: A phenomenological model. *Earth Planet. Sci. Lett.*, **44**, 7–17. [*166*]

Hoffman, K.A. (1991a). Long-lived transitional states of the geomagnetic field and the two dynamo families. *Nature*, **354**, 273–277. [*168, 171*]

Hoffman, K.A. (1992a). Dipolar reversal states of the geomagnetic field and core–mantle dynamics. *Nature*, **359**, 789–794. [*138, 171*]

Hoffman, K.A. (1992b). Self-reversal of thermo-remanent magnetization in the ilmenite–hematite system: Order–disorder, symmetry, and spin alignment. *J. Geophys. Res.*, **97**, 10883–10896. [*139, 140*]

Hoffman, K.A. (1996). Transitional paleomagnetic field behavior; preferred paths or patches? *Surv. Geophys.*, **17**, 207–211. [*138*]

Hoffman, K.A. (1999). Temporal aspects of a geomagnetic reversal. *Nature*, in press. [*170*]

Hoffman, K.A., and Day, R. (1978). Separation of multicomponent NRM: A general method. *Earth Planet. Sci. Lett.*, **40**, 433–438. [*123*]

Hoffman, P.F. (1991b). Did the breakout of Laurentia turn Gondwana inside out? *Science*, **252**, 1409–1412. [*315, 316*]

Holcomb, R.T. (1987). Eruptive history and long-term behavior of Kilauea Volcano, Hawaii. In *Volcanism in Hawaii*, (Decker, R.W., T.L. Wright and P.H. Stauffer, Eds.); U.S. Geological Survey Professional Paper *1350*, 261–350. [*170*]

Hospers, J. (1953). Reversals of the main geomagnetic field I, II. *Proc. Kon. Nederl. Akad. Wetensch., B.*, **56**, 467–491. [*143*]

Hospers, J. (1954). Reversals of the main geomagnetic field III. *Proc. Kon. Nederl. Akad. Wetensch., B.*, **57**, 112–121. [*143*]

Hospers, J. (1955). Rock magnetism and polar wandering. *J. Geol.*, **63**, 59–74. [*15, 154*]

Huestis, S.P., and Acton, G.D. (1997). On the construction of geomagnetic time scales from non-prejudicial treatment of magnetic anomaly data from multiple ridges. *Geophys. J. Int.*, **129**, 176–182. [*206*]

Hyodo, H., Dunlop, D.J., and McWilliams, M.O. (1986). Timing and extent of Grenvillian magnetic overprinting near Temagami, Ontario. In *The Grenville Province* (J.M. Moore, A. Davidson, A.J. Baer, Eds.), *Geological Association of Canada, Special Paper*, No. 31, 119–126. [*317*]

Idnurm, M., and Cook, P.J. (1980). Palaeomagnetism of beach ridges in South Australia and the Milankovitch theory of ice ages. *Nature*, **286**, 699–702. [*138*]

Idnurm, M., and Giddings, J.W. (1995). Paleoproterozoic-Neoproterozoic North America–Australia link: New evidence from paleomagnetism. *Geology*, **23**, 149–152 (1995). [*321*]

Irving, E. (1956). Palaeomagnetic and palaeoclimatological aspects of polar wandering. *Geofis. Pura. Appl.*, **33**, 23–41. [*15, 241, 255, 279, 289*]

Irving, E. (1958). Rock magnetism: A new approach to the problems of polar wandering and continental drift. In *Continental Drift – A Symposium* (Ed., S.W. Carey), pp.24–61, Univ. of Tasmania, Hobart. [*287*]

Irving, E. (1959). Palaeomagnetic pole positions: A survey and analysis. *Geophys. J. R. Astron. Soc.*, **2**, 51–79. [*15*]

Irving, E. (1964). *Paleomagnetism and its Application to Geological and Geophysical Problems.* Wiley, New York, 399pp. [*16, 110, 141, 142, 228, 232, 244, 289*]

Irving, E. (1970). The Mid-Atlantic Ridge at 45°N. XVI. Oxidation and magnetic properties of basalt; review and discussion. *Can. J. Earth Sci.*, **7**, 1528–1538. [*193*]

Irving, E. (1977). Drift of the major continental blocks since the Devonian. *Nature*, **270**, 304–309. [*248, 299, 303*]

Irving, E. (1979). Paleopoles and paleolatitudes of North America and speculations about displaced terrains. *Can. J. Earth Sci.*, **16**, 669–694. [*303*]

Irving, E., and Briden, J.C. (1962). Palaeolatitude of evaporite deposits. *Nature*, **196**, 425–428. [*241, 242*]

Irving, E., and Brown, D.A. (1964). Abundance and diversity of the labyrinthodonts as a function of paleolatitude. *Am. J. Sci.*, **262**, 689–708. [*243*]

Irving, E., and Gaskell, T.F. (1962). The palaeogeographic latitude of oil fields. *Geophys. J. R. Astron. Soc.*, **7**, 54–64. [*242, 243*]

Irving, E., and Green, R. (1958). Polar movement relative to Australia. *Geophys. J. R. Astron. Soc.*, **1**, 64–72. [*276*]

Irving, E., and Major, A. (1964). Post-depositional remanent magnetization in a synthetic sediment. *Sedimentology*, **3**, 135–143. [*68, 70*]

Irving, E., and McGlynn, J.C. (1976). Proterozoic magnetostratigraphy and the tectonic evolution of Laurentia. *Philos. Trans. R. Soc. London*, **A280**, 243–265. [*323*]

Irving, E., and Opdyke, N.D. (1965). The palaeomagnetism of the Bloomsburg redbeds and its possible application to the tectonic history of the Appalachians. *Geophys. J. R. Astron. Soc.*, **9**, 153–167. [*117*]

Irving, E., and Parry, L.G. (1963). The magnetism of some Permian rocks from New South Wales. *Geophys. J. R. Astron. Soc.*, **7**, 395–411. [*143, 155, 162*]

Irving, E., and Pullaiah, G. (1976). Reversals of the geomagnetic field, magnetostratigraphy, and relative magnitude of secular variation in the Phanerozoic. *Earth Sci. Rev.*, **12**, 35–64. [*180*]

Irving, E., and Yole, R.W. (1972). Paleomagnetism and the kinematic history of mafic and ultramafic rocks in fold mountain belts. *Publ. Earth Phys. Branch*, Dept. Energy, Mines and Resources, Ottawa, **42**, 87–95. [*303*]

Irving, E., Robertson, W.A., and Stott, P.M. (1963). The significance of the palaeomagnetic results from Mesozoic from eastern Australia. *J. Geophys. Res.*, **68**, 2313–2317. [*276*]

Irving, E., Woodsworth, G.J., Wynne, P.J., and Morrison, A. (1985). Paleomagnetic evidence for displacement from the south of the Coast Plutonic Complex, British Columbia. *Can. J. Earth Sci.*, **22**, 584–598. [*304, 305, 306*]

Irving, E., Robertson, W.A., Stott, P.M., Tarling, D.H., and Ward, M.A. (1961). Treatment of partially stable sedimentary rocks showing planar distribution of directions of magnetization. *J. Geophys. Res.*, **66**, 1927–1933. [*117*]

Isacks, B., Oliver, J., and Sykes, L.R. (1968). Seismology and the new global tectonics. *J. Geophys. Res.*, **73**, 5855–5899. [*185*]

Isezaki, N. (1988). Magnetic anomalies. In *The Ocean basins and Margins, Vol.7B – The Pacific Ocean*, (A.E.M. Nairn and F.G. Stelhi, Eds.) pp.595–624, Plenum, New York. [*207*]

Ishikawa, Y., and Syono, Y. (1963). Order-disorder transformation and reverse thermoremanent magnetism in the $FeTiO_3$-Fe_2O_3 system. *J. Phys. Chem. Solids*, **24**, 517–528. [*139, 140*]

Jackson, K.C. (1990). A paleomagnetic study of Apennine thrusts, Italy: Monte Maiella and Monte Raparo. *Tectonophysics*, **178**, 231–240. [*312*]

Jacobs, J.A. (1994). *Reversals of the Earth's Magnetic Field*. Cambridge Univ. Press, Cambridge, UK, 346pp. [*137*]

Jäger, E., Niggli, E., and Wenk, E. (1967). Rb-Sr Altersbestimmimgen an Glimmern der Zentralalpen. *Beitrage zur Geol. Karte der Schweiz*, Neue Folge, 134, Lieferungen. [*249*]

James, R.W., and Winch, D.E. (1967). The eccentric dipole. *Pure Appl. Geophys.*, **66**, 77–86. [*235*]

Johnson, C.L., and Constable, C.G. (1995).The time-averaged field as recorded by lava flows over the past 5Myr. *Geophys. J. Int.*, **122**, 489–519. [*235*]

Johnson, C.L., and Constable, C.G. (1997). The time-averaged geomagnetic field: Global and regional biases for 0–5 Ma. *Geophys. J. Int.*, **131**, 643–666. [*235*]

Johnson, G.D., Opdyke, N.D., Tandon, S.K., and Nanda, A.C. (1983). The magnetic polarity stratigraphy of the Siwalik Group at Haritalyangar (India) and a new last appearance datum for *Ramapithecus* and *Sivapithecus* in Asia. *Palaeogeog. Palaeoclimatol. Palaeoecol.*, **44**, 223–249. [*156*]

Johnson, H.P., and Atwater, T. (1977). Magnetic study of basalts from the Mid-Atlantic Ridge, lat. 37°N. *Geol. Soc. Am. Bull.*, **88**, 637–647. [*192, 193*]

Johnson, H.P., and Hall, J.M. (1978). A detailed rock magnetic and opaque mineralogical study of the basalts from the Nazca plate. *Geophys. J. R. Astron. Soc.*, **52**, 45–64. [*45*]

Johnson, H.P., and Pariso, J.E. (1993). Variations in oceanic crustal magnetization: Systematic changes in the last 160 million years. *J. Geophys. Res.*, **98**, 435–445. [*193*]

Johnson, H.P., Lowrie, W., and Kent, D.V. (1975). Stability of anhysteretic remanent magnetization in fine and coarse magnetite and maghemite particles. *Geophys. J. R. Astron. Soc.*, **41**, 1–10. [*132*]

Johnson, H.P., Van Patten, D., Tivey, M.A., and Sager, W. (1995). Geomagnetic polarity reversal rate for the Phanerozoic. *Geophys. Res. Lett.*, **22**, 231–234. [*145, 162, 181*]

Jones, D.L., and McElhinny, M.W. (1966). Paleomagnetic correlation of basic intrusions in the Precambrian of southern Africa. *J. Geophys. Res.*, **71**, 543–552. [*321*]

Jones, D.L., and McElhinny, M.W. (1967). Stratigraphic interpretation of paleomagnetic measurements on the Waterberg redbeds of South Africa. *J. Geophys. Res.*, **72**, 4171–4179. [*118*]

Jones, D.L., Robertson, I.D.M., and McFadden, P.L. (1975). A palaeomagnetic study of Precambrian dyke swarms associated with the Great Dyke of Rhodesia. *Trans. Geol. Soc. S. Afr.*, **78**, 57–65. [*125*]

Jones, D.L., Siberling, N.J., and Hillhouse, J. (1977). Wrangellia – A displaced terrain in northwestern North America. *Can. J. Earth Sci.*, **14**, 2565–2577. [*303*]

Jones, G.M. (1977). Thermal interaction of the core and the mantle and long term behaviour of the geomagnetic field. *J. Geophys. Res.*, **82**, 1703–1709. [*179*]

Jordan, T.H. (1975). The continental tectosphere. *Rev. Geophys. Space Phys.*, **13**, 1–12. [*185*]

Jurdy, D.M., and Van der Voo, R. (1974). A method for the separation of true polar wander and continental drift, including results for the last 55 Ma. *J. Geophys. Res.*, **79**, 2945–2953. [*327*]

Jurdy, D.M., and Van der Voo, R. (1975). True polar wander since the Early Cretaceous. *Science*, **187**, 1193–1196. [*327*]

Kearey, P., and Vine, F.J. (1996). *Global Tectonics*. Blackwell Science, Oxford, 333pp. [*197*]

Kelly, P., and Gubbins, D. (1997). The geomagnetic field over the past 5Myr. *Geophys. J. Int.*, **128**, 315–330. [*235*]

Kent, D.V. (1973). Post-depositional remanent magnetization in deep-sea sediment. *Nature*, **246**, 32–34. [*71*]

Kent, D.V. (1988). Further paleomagnetic evidence for oroclinal rotation in the central foldded Appalachians from the Bloomsburg and Mauch Chunk formations. *Tectonics*, **7**, 749–760. [*254*]

Kent, D.V., and Clemmensen, L.B. (1996). Paleomagnetism and cycle stratigraphy of the Triassic Fleming Fjord and Gipsdalen Formations of East Greenland. *Bull. Geol. Soc. Denmark*, **42**, 121–136. [*255*]

Kent, D.V., and Gradstein, F.M. (1986). A Jurassic to recent chronology. In *The Geology of North America, Vol. M*, The Western North Atlantic Region, (P.R. Vogt and B.E. Tucholke, Eds.), pp.45–50, Geological Society of America, Boulder, CO. *[151, 152, 179, 209, 211]*

Kent, D.V., and Olsen, P.E. (1999). Astronomically tuned geomagnetic polarity time scale for the Late Triassic. *J. Geophys. Res.*, in press. *[160, 161, 181]*

Kent, D.V., and Smethurst, M.A. (1998). Shallow bias of paleomagnetic inclinations in the Paleozoic and Precambrian. *Earth Planet. Sci. Lett.*, 160, 391–402. *[237, 238]*

Kent, D.V., and Van der Voo, R. (1990). Palaeozoic palaeogeography from palaeomagnetism of the Atlantic-bordering continents. In . In *Palaeozoic Paleogeography and Biogeography* (W.S. McKerrow and C.R. Scotese, Eds.), *Geol. Soc. London Mem.*, 12, 49–56. *[271, 272]*

Kent, D.V., and Witte, W.K. (1993). Slow apparent polar wander for North America in the Late Triassic and large Colorado Plateau rotation. *Tectonics*, 12, 291–300. *[253, 254]*

Kent, D.V., Olsen, P.E., and Witte, W.K. (1995). Late Triassic–Early Jurassic geomagnetic polarity sequence from drill cores in the Newark rift basin, eastern North America. *J. Geophys. Res.*, 100, 14965–14998. *[157, 159, 160]*

Kent, D.V., Honnerez, B.M., Opdyke, N.D., and Fox, P.J. (1984). Magnetic properties of dredged ocean gabbros and the source of marine magnetic anomalies. *Geophys. J. R. Astron. Soc.*, 55, 513–537. *[216]*

Kent, J.T., Briden, J.C., and Mardia, K.V. (1983). Linear and planar structure in ordered multivariate data as applied to progressive demagnetization of palaeomagnetic remanence. *Geophys. J. R. Astron. Soc.*, 75, 593–622. *[125]*

Kern, J.W. (1961). Effects of moderate stresses on directions thermoremanent magnetization. *J. Geophys. Res.*, 66, 3801–3806. *[74]*

Khramov, A.N. (1955). Study of remanent magnetization and the problem of stratigraphic correlation and subdivision of non-fossiliferous strata. *Akad. Nauk SSSR*, 100, 551–554 (in Russian). *[15, 143, 144, 154]*

Khramov, A.N. (1957). Paleomagnetism as a basis for a new technique of sedimentary rock correlation and subdivision. *Akad. Nauk SSSR*, 112, 849–852 [In Russian]. *[15, 143, 144, 154]*

Khramov, A.N. (1958). *Palaeomagnetism and Stratigraphic Correlation.* Gostoptechizdat, Leningrad, 218pp. [English translation by Lojkine, A.J., published by Dept. of Geophysics, Australian Natl. Univ., Canberra, 1960]. *[15, 143, 144, 154]*

Khramov, A.N. (1987). *Paleomagnetology.* Springer–Verlag, Berlin, 308pp. *[98, 262]*

Khramov, A.N., and Ustritsky, V.I. (1990). Paleopositions of some northern Eurasian tectonic blocks: Paleomagnetic and paleobiological constraints. *Tectonophysics*, 184, 101–109. *[314]*

Khramov, A.N., Petrova, G.N., and Pechersky, D.M. (1981). Paleomagnetism of the Soviet Union. In *Paleoreconstruction of the Continents* (M.W. McElhinny and D.A. Valencio, Eds.), *Am. Geophys. Union Geodynamics Ser.*, 2, 177–194. *[262]*

Khramov, A.N., Goncharov, G.I., Komissarova, R.A., Osipova, E.P., Pogarskaya, I.A., Rodionov, V.P., Slautsitais, I.P., Smirnov, I.S., and Forsh, N.N. (1974). *Paleomagnetism of the Paleozoic.* NEDRA, Moscow [In Russian]. *[163]*

King, R.F. (1955). The remanent magnetism of artificially deposited sediments. *Mon. Not. R. Astron. Soc., Geophys. Suppl.*, 7, 115–134. *[70]*

Kirschvink, J.L. (1978). The Precambrian–Cambrian boundary problem: Paleomagnetic directions from the Amadeus Basin, central Australia. *Earth Planet. Sci. Lett.*, 40, 91–100. *[100, 111, 112]*

Kirschvink, J.L. (1980). The least squares line and plane and the analysis of palaeomagnetic data. *Geophys. J. R. Astron. Soc.*, 62, 699–718. *[124, 157]*

Kirschvink, J.L., Ripperdan, R.L., and Evans, D.A. (1997). Evidence for a large-scale reorganization of Early Cambrian continental masses by inertial interchange true polar wander. *Science*, 277, 541–545. *[329]*

Klitgord, K.D., Huestis, S.P., Mudie, J.D., and Parker, R.L. (1975). An analysis of near-bottom magnetic anomalies: Sea floor spreading and the magnetized layer. *Geophys. J. R. Astron. Soc.*, **43**, 387–424. [*150*]

Klootwijk, C.T. (1979). A review of palaeomagnetic data from the Indo–Pakistani fragment of Gondwanaland. In *Geodynamics of Pakistan* (A. Farah and K.A. De Jong, Eds.), pp.41–80. Geolpgical Survey of Pakistan, Quetta. [*221*]

Klootwijk, C.T., and Peirce, J.W. (1979). India's and Australia's pole path since the late Mesozoic and the India–Asia collision. *Nature*, **282**, 605–607. [*221*]

Klootwijk, C.T., and Radhakrishnamurty, C. (1981). Phanerozoic paleomagnetism of the Indian plate and the India–Asia collision. In *Paleoreconstruction of the Continents* (M.W. McElhinny and D.A. Valencio, Eds.), *Am. Geophys. Union Geodynamics Ser.*, **2**, 93–105. [*280*]

Klootwijk, C.T., Conaghan, P.J., and Powell, C. McA. (1985). The Himalayan arc: Large-scale continental subduction, oroclinal bending and back-arc spreading. *Earth Planet. Sci. Lett.*, **75**, 167–183. [*314*]

Kobayashi, K. (1959). Chemical remanent magnetization of ferromagnetic minerals and its application in rock magnetism. *J. Geomag. Geoelect.*, **10**, 99–117. [*116*]

Kono, M. (1972). Mathematical models of the Earth's magnetic field. *Phys. Earth Planet. Int.*, **5**, 140–150. [*25, 175*]

Kono, M. (1980a). Paleomagnetism of DSDP Leg 55 basalts and implications for the tectonics of the Pacific plate. *Init. Rept. Deep Sea Drill. Proj.*, **55**, 737–752. [*216, 224*]

Kono, M. (1980b). Statistics of paleomagnetic inclination data. *J. Geophys. Res.*, **85**, 3878–3882. [*97, 221*]

Kono, M. (1997). Distributions of paleomagnetic directions and poles. *Phys. Earth Planet. Int.*, **103**, 313–327. [*99*]

Kono, M., and Tanaka, H. (1995). Intensity of the geomagnetic field in geological time: A statistical study. In *The Earth's Central Part: Its Structure and Dynamics* (Ed., T. Yukutake), pp.75–94, Terrapub, Tokyo. [*25, 26, 211, 239*]

Kono, M., and Uchimura, H. (1994). Inverse problem of paleomagnetic reconstruction: Formulation. *J. Geomag. Geoelect.*, **40**, 311–328. [*288*]

Kosterov, A.A., and Perrin, M. (1996). Paleomagnetism of the Lesotho basalt, southern Africa. *Earth Planet. Sci. Lett.*, **139**, 63–78. [*165*]

Kosterov, A.A., Perrin, M., Glen, J.M., and Coe, R.S. (1998). Paleointensity of the Earth's magnetic field in Early Cretaceous time: The Parana basalt, Brazil. *J. Geophys. Res.*, **103**, 9739–9753. [*239*]

Kosterov, A.A., Prévot, M., Perrin, M., and Shashkanov, V.A. (1997). Paleointensity of the Earth's magnetic field in Jurassic: New results from a Thellier study of the Lesotho basalt, southern Africa. *J. Geophys. Res.*, **102**, 24859–24872. [*25, 134, 211*]

Kosygin, Y.A., and Parfenov, L.M. (1981). Tectonics of the Soviet far east. In *The Ocean Basins and Margins* (A.E.M. Nairn, M. Churkin, and F.G. Stelhi, Eds.), *Vol. 5*, pp.377–412, Plenum Press, New York. [*295*]

Kristjansson, L. (1985). Some statistical properties of paleomagnetic directions in Icelandic lava flows, *Geophys. J. R. Astron. Soc.*, **80**, 57–71. [*173*]

LaBrecque, J.L., Kent, D.V., and Cande, S.C. (1977). Revised magnetic polarity time scale for Late Cretaceous and Cenozoic time. *Geology*, **5**, 330–335. [*150, 206, 209*]

Laj, C., Nordemann, D., and Pomeau, Y. (1979). Correlation function analysis of geomagnetic field reversals. *J. Geophys. Res.*, **84**, 4511–4514. [*175*]

Laj, C., Guitton, S., Kissel, C., and Mazaud, A. (1988). Complex behavior of the geomagnetic field during three successive polarity reversals 11–12 m.y. B.P., *J. Geophys. Res.*, **93**, 11655–11666. [*166*]

Laj, C., Mazaud, A., Weeks, R., Fuller, M., and Herrero-Bervera, E. (1991). Geomagnetic reversal paths. *Nature*, **351**, 447. [*138, 167*]

Laj, C., Rais, A., Surmont, J., Gillot, P.Y., Guillou, H., Kissel, C., and Zanella, E. (1997). Changes of the geomagnetic field vector obtained from lava sequences on the island of Vulcano (Aeolian Islands, Sicily). *Phys. Earth Planet. Int.*, **99**, 161–177. [*134*]

Langel, R.A. (1987). The main field. In *Geomagnetism Vol. 1* (Ed., J.A. Jacobs), pp.249–512. Academic Press, London. [*12*]

Langereis, C.G., Dekkers, M.J., de Lange, G.J., Paterne, M., and van Santvoort, P.J.M. (1997). Magnetostratigraphy and astronomical calibration of the last 1.1 Myr from an eastern Mediterranean piston core and dating of short events in the Brunhes. *Geophys. J. Int.*, **129**, 75–94. [*174*]

Larmor, J. (1919). How could a rotating body such as the sun become a magnet? *Rept. Brit. Assoc. Adv. Sci. 1919*, 159–160. [*9*]

Larson, R.L. (1991). Latest pulse of the Earth: Evidence for a mid-Cretaceous superplume. *Geology*, **19**, 547–50. [*210*]

Larson, R.L., and Chase, C.G. (1972). Late Mesozoic evolution of the Western Pacific Ocean. *Geol. Soc. Am. Bull.*, **83**, 3627–3644. [*207, 213, 225, 226*]

Larson, R.L., and Hilde, T.W.C. (1975). A revised time scale of magnetic reversals for the Early Cretaceous and Late Jurassic. *J. Geophys. Res.*, **80**, 2586–2594. [*207, 211*]

Larson, R.L., and Pitman, W.C. III. (1972). World-wide correlation of Mesozoic magnetic anomalies, and its implications. *Geol. Soc. Am. Bull.*, **83**, 3645–3662. [*207, 210*]

Larson, R.L., Smith, S.M., and Chase, C.G. (1972). Magnetic lineations of Early Cretaceous age in the western equatorial Pacific Ocean. *Earth Planet. Sci. Lett.*, **15**, 315–319. [*207*]

Le Pichon, X. (1968). Sea-floor spreading and continental drift. *J. Geophys. Res.*, **73**, 3661–3697. [*183, 184, 187, 281, 328*]

Le Pichon, X., and Heirtzler, J.R. (1968). Magnetic anomalies in the Indian Ocean and seafloor spreading. *J. Geophys. Res.*, **73**, 2101–2117. [*205*]

Le Pichon, X., Sibuet, J.C., and Francheteau, J. (1977). The fit of the continents around the North Atlantic. *Tectonophysics*, **38**, 169–209. [*290*]

Leaton, B.R., and Malin, S.R.C. (1967). Recent changes in the magnetic dipole moment of the earth. *Nature*, **213**, 1110. [*12*]

Lewis, T., and Fisher, N.I. (1982). Graphical methods for investigating the fit of a Fisher distribution to spherical data, *Geophys. J.R. Astron. Soc.*, **69**, 1–13. [*96*]

Li, Y., McWilliams, M.O., Cox, A., Sharps, R., Li, Y.A., Gao, Z., Zhang, Z., and Zhai, Y. (1988). Late Permian paleomagnetic pole from dikes of the Tarim craton, China. *Geology*, **16**, 275–278. [*268*]

Li, Z.X. (1999). New palaeomagnetic results from the "cap dolomite" of the Neoproterozoic Walsh Tillite, northwestern Australia. *Precambrian Res.* (in press). [*319*]

Li, Z.X., Zhang, L., and Powell, C.McA. (1995). South China in Rodinia: Part of the missing link between Australia–East Antarctica and Laurentia? *Geology*, **23**, 407–410. [*316, 317*]

Li, Z.X., Powell, C.McA., Embleton, B.J.J., and Schmidt, P.W. (1991). New palaeomagnetic results from the Amadeus Basin and their implications for stratigraphy and tectonics. *Bur. Min. Resour. Geol. Geophys. Austr. Bull.*, **236**, 349–368. [*297*]

Li, Z.X., Powell, C.McA., Thrupp, G.A., and Schmidt, P.W. (1990). Australian Palaeozoic palaeomagnetism and tectonics – II. A revised apparent polar wander path and palaeogeography. *J. Struct. Geol.*, **12**, 567–575. [*228, 271, 272, 277*]

Lin, J.L., Fuller, M., and Zhang, W. (1985). Preliminary Phanerozoic polar wander paths for the North and South China blocks. *Nature*, **313**, 444–449. [*265*]

Lindsley, D.H. (1976). The crystal chemistry and structure of oxide minerals as exemplified by Fe–Ti oxides, and experimental studies of oxide minerals. In *Oxide Minerals*, (Ed., D. Rumble), pp.L1–84, Minerological Society of America, Washington, DC. [*38*]

Lindsley, D.H. (1991). Experimental studies of oxide minerals. In *Oxide Minerals, Petrologic and Magnetic Significance*, 2nd edn. (Ed., D.H. Lindsley), pp.69–106, Min. Soc. Amer., Washington DC. [*38*]

Livermore, R.A., Vine, F.J., and Smith, A.G. (1983). Plate motions and the geomagnetic field. I. Quaternary and late Tertiary. *Geophys. J. R. Astron. Soc.*, **73**, 153–171. [*326*]

Livermore, R.A., Vine, F.J., and Smith, A.G. (1984). Plate motions and the geomagnetic field. II. Jurassic to Tertiary. *Geophys. J. R. Astron. Soc.*, **79**, 939–961. [*236, 326*]

Lock, J., and McElhinny, M.W. (1991). The global paleomagnetic database: Design, installation and use with ORACLE. *Surv. Geophys.*, **12**, 317–491. [*16, 162, 231*]

Loper, D.E. (1984). The dynamical structure of D" and deep plumes in a non-Newtonian mantle. *Phys. Earth Planet. Int.*, **34**, 56–57. [*222*]

Loper, D.E., and Stacey, F.D. (1983). The dynamical and thermal structure of deep-mantle plumes. *Phys. Earth Planet. Int.*, **33**, 304–317. [*222*]

Lottes, A.L., and Rowley, D.B. (1990). Early and Late Permian reconstructions of Pangaea. In *Palaeozoic Palaeogeography and Biogeography* (W.A. McKerrow and C.R. Scotese, Eds.), *Geol. Soc. London Mem.*, **12**, 383–395. [*270, 295, 296, 298*]

Love, J.J. (1998). Paleomagnetic volcanic data and geomagnetic regularity of reversals and excursions. *J. Geophys. Res.*, **103**, 12435–12452. [*170, 171*]

Love, J.J., and Mazaud, A. (1997). A database for the Matuyama–Brunhes magnetic reversal. *Phys. Earth Planet. Int.*, **103**, 207–245. [*170*]

Lowrie, W. (1974). Oceanic basalt magnetic properties and the Vine and Matthews hypothesis. *J. Geophys.*, **40**, 513–536. [*216*]

Lowrie, W. (1986). Paleomagnetism and the Adriatic promontory: A reappraisal. *Tectonics*, **5**, 797–807. [*312*]

Lowrie, W. (1990). Identification of ferromagnetic minerals in a rock by coercivity and unblocking temperature properties. *Geophys. Res. Lett.*, **17**, 159–162. [*130, 131*]

Lowrie, W., and Alvarez, W. (1977). Upper Cretaceous-Paleocene magnetic stratigraphy at Gubbio, Italy. III. Upper Cretaceous magnetic stratigraphy. *Geol. Soc. Am. Bull.*, **88**, 374–377. [*159*]

Lowrie, W., and Alvarez, W. (1981). One hundred million years of geomagnetic polarity history. *Geology*, **9**, 392–397. [*205*]

Lowrie, W., and Fuller, M.D. (1971). On the alternating field demagnetization characteristics of multidomain thermoremanent magnetization in magnetite. *J. Geophys. Res.*, **76**, 6339–6349. [*131*]

Lowrie, W., and Heller, F. (1982). Magnetic properties of marine limestones. *Rev. Geophys. Space Phys.*, **20**, 171–192. [*129, 158*]

Lowrie, W., Hirt, A.M., and Kligfield, R. (1986). Effects of tectonic deformation on the remanent magnetization of rocks. *Tectonics*, **5**, 713–722. [*108*]

Lowrie, W., Alvarez, W., Napoleone, G., Perch-Nielson, K., Primoli-Silva, I., and Toumarkine, M. (1982). Paleogene magnetic stratigraphy in Umbrian pelagic carbonate rocks: The Contessa sections, Gubbio. *Geol. Soc. Am. Bull.*, **93**, 414–432. [*155*]

Lund, S.P., Acton, G., Clement, B., Hastedt, M., Okada, M., Williams, T., and the ODP Leg 172 Scientific Party (1998). Geomagnetic field excursions occurred often during the last million years. *EOS, Trans. Am. Geophys. Union*, **79**, 178–179. [*173, 174*]

Luterbacher, H.P., and Premoli Silva, I. (1964). Biostratigraphia del limite Cretacio–Terziario nelli Appennino Centrale. *Riv. Ital. Palaeontol. Stud.*, **70**, 67–77. [*159*]

Lutz, T.M. (1985). The magnetic reversal record is not periodic. *Nature*, **317**, 404–407. [*180*]

Lutz, T.M., and Watson, G.S. (1988). Effects of long-term variation on the frequency spectrum of the geomagnetic reversal record. *Nature*, **334**, 240–242. [*180*]

Luyendyk, B.P., Mudie, J.D., and Harrison, C.G.A. (1968). Lineations of magnetic anomalies in the northeast Pacific observed near the ocean floor. *J. Geophys. Res.*, **73**, 5951–5957. [*190*]

Macdonald, K.C. (1977). Near-bottom magnetic anomalies, asymmetrical spreading, oblique spreading, and tectonics of the Mid-Atlantic Ridge near lat 37°N. *Geol. Soc. Am. Bull.*, **88**, 541–555. [*190, 193*]

MacDonald, W.D. (1980). Net tectonic rotation, apparent tectonic rotation, and the structural tilt correction in paleomagnetic studies. *J. Geophys. Res.*, **85**, 3659–3569. [*101*]

MacFadden, B.J., Swisher, C.C. III, Opdyke, N.D., and Woodburn, M.O. (1990). Paleomagnetism, geochronology, and possible tectonic rotation of the middle Miocene Barstow Formation, Mojave Desert, southern California. *Geol. Soc. Am. Bull.*, **102**, 478–493. [*157*]

MacFadden, B.J., Whitelaw, M.J., McFadden, P., and Rich, T.H.V. (1987). Magnetic polarity stratigraphy of the Pleistocene section at Portland (Victoria), Australia. *Quaternary Res.*, **28**, 364–373. [*138*]

Mackereth, F.J.H. (1971). On the variation in direction of the horizontal component of remanent magnetisation in lake sediments. *Earth Planet. Sci. Lett.*, **12**, 332–338. [*27*]

Mac Niocaill, C., van der Pluijm, B.A., and Van der Voo, R. (1997). Ordovician paleogeography and the evolution of the Iapetus Ocean. *Geology*, **25**, 159–162. [*309*]

Magnus, G., and Opdyke, N.D. (1991). A paleomagnetic investigation of the Minturn Formation, Colorado: A study in establishing the timing of remanence acquisition. *Tectonophysics*, **187**, 181–189. [*163*]

Malod, J.A. (1989). Iberides et plaque Iberique. *Bull. Soc. Géol. France*, Ser. 8, **5**, 927–934. [*311*]

Mammerickx, J., and Sharman, G.F. (1988). Tectonic evolution of the North Pacific during the Cretaceous quiet period. *J. Geophys. Res.*, **93**, 3009–3024. [*220*]

Mankinen, E.A., and Dalrymple, G.B. (1979). Revised geomagnetic polarity time scale for the interval 0–5 m.y.b.p. *J. Geophys. Res.*, **84**, 615–626. [*156*]

Mankinen, E.A., Prévot, M., Grommé, C.S., and Coe, R. (1985). The Steens Mountain (Oregon) geomagnetic polarity transition. 1. Directional history, duration of episodes, and rock magnetism, *J. Geophys. Res.*, **90**, 10393–10416. [*173*]

Marcano, M.C., Van der Voo, R., and Mac Niocaill, C. (1999). True polar wander during the Permo-Triassic. *J. Geodynamics*, in press. [*326, 329*]

March, A. (1932). Mathematische Theorie der Regelung nach der Korngestalt bei Affiner Deformation. *Z. Kristallogr.*, **81**, 285–???. [*108*]

Mardia, K.V. (1972). *Statistics of Directional Data*. Academic Press, London, 375pp. [*93*]

Marzocchi, W. (1997). Missing reversals in the geomagnetic polarity timescale: Their influence on the analysis and in constraining the process that generates geomagnetic reversals. *J. Geophys. Res.*, **102**, 5157–5171, [*191*]

Marzocchi, W., and Mulargia, F. (1990). Statistical analysis of the geomagnetic reversal sequences. *Phys. Earth Planet. Int.*, **61**, 149–164. [*180*]

Mascle, J., and Phillips, J. (1972). Smooth zones in the south Atlantic. *Nature*, **240**, 80–84. [*211*]

Matthews, D.H., and Williams, C.A. (1968). Linear magnetic anomalies in the Bay of Biscay: A qualitative interpretation. *Earth Planet. Sci. Lett.*, **4**, 315–320. [*206*]

Matuyama, M. (1929). On the direction of magnetisation of basalt in Japan, Tyosen and Manchuria. *Proc. Imp. Acad. Jap.*, **5**, 203–205. [*15, 139, 143*]

Maxwell, A.E., Von Herzen, R.P., Hsu, K.J., Andrews, J.E., Saito, T., Percival, S.F., Milow, E.D., and Boyce, R.E. (1970), Deep sea drilling in the South Atlantic. *Science*, **168**, 1047–1059. [*206*]

May, S.R., and Butler, R.F. (1986). North American Jurassic apparent polar wander, implications for plate motions, paleogeography and Cordilleran tectonics. *J. Geophys. Res.*, **91**, 11519–11544. [*256*]

Mazaud A., and Laj, C. (1991). The 15 m.y. geomagnetic reversal periodicity: A quantitative test. *Earth Planet. Sci. Lett.*, **107**, 689–696. *[180]*

Mazaud A., Laj, C., de Seze, L., and Verosub, K.L. (1983). 15-Myr periodicity in the reversal frequency of geomagnetic reversals since 100 Ma. *Nature*, **304**, 328–330. *[180]*

McCabe, C., and Elmore, R.D. (1989). The occurrence and origin of Late Paleozoic remagnetization in the sedimentary rocks of North America. *Rev. Geophys.*, **27**, 471–494. *[67]*

McCabe, C., Van der Voo, R., Peacor, D.R., Scotese, C.R., and Freeman, R. (1983). Diagenetic magnetite carries ancient yet secondary remanence in some Paleozoic carbonates. *Geology*, **11**, 221–223. *[105]*

McClelland Brown, E. (1983). Palaeomagnetic studies of fold development and propagation in the Pembrokeshire Old Red Sandstone. *Tectonophysics*, **98**, 131–149. *[100, 106]*

McClelland, E. (1996). Theory of CRM acquired by grain growth, and its implications for TRM discrimination and palaeointensity determination in igneous rocks. *Geophys. J. Int.*, **126**, 271–280. *[65, 116, 117]*

McClelland, E., and Goss, C. (1993). Self-reversal of chemical remanent magnetization on the transformation of maghemite to haematite. *Geophys. J. Int.*, **112**, 517–532. *[139, 140]*

McDonald, K.L., and Gunst, R.H. (1968). Recent trends in the earth's magnetic field. *J. Geophys. Res.*, **73**, 2057–2067. *[12, 13]*

McDougall, I., and Tarling, D.H. (1963). Dating of polarity zones in the Hawaiian Islands. *Nature*, **200**, 54–56. *[144]*

McDougall, I., and Tarling, D.H. (1964). Dating geomagnetic polarity zones. *Nature*, **202**, 171–172. *[144]*

McDougall, I., Watkins, N.D., and Kristjansson, L. (1976a). Geochronology and paleomagnetism of a Miocene–Pliocene lava sequence at Bessatadaá, eastern Iceland. *Am. J. Sci.*, **276**, 1078–1095. *[149]*

McDougall, I., Watkins, N.D., Walker, G.P.L., and Kristjansson, L. (1976b). Potassium–argon and paleomagnetic analysis of Icelandic lava flows: Limits on the age of anomaly 5. *J. Geophys. Res.*, **81**, 1505–1512. *[149]*

McDougall, I., Saemundsson, K., Johannesson, H., Watkins, N.D., and Kristjansson, L. (1977). Extension of the geomagnetic polarity time scale to 6.5 my: K–Ar dating, geological and paleomagnetic study of a 3,500 m lava succession in western Iceland. *Geol. Soc. Am. Bull.*, **88**, 1–15. *[149]*

McElhinny, M.W. (1964). Statistical significance of the fold test in palaeomagnetism. *Geophys. J. R. Astron. Soc.*, **8**, 338–340. *[102]*

McElhinny, M.W. (1966). The palaeomagnetism of the Umkondo lavas, eastern Southern Rhodesia. *Geophys. J. R. Astron. Soc.*, **10**, 375–381. *[321]*

McElhinny, M.W. (1968). Northward drift of India – Examination of recent palaeomagnetic results. *Nature*, **217**, 342–344. *[280]*

McElhinny, M.W. (1971). Geomagnetic reversals during the Phanerozoic. *Science*, **172**, 157–159. *[180]*

McElhinny, M.W. (1973a). *Palaeomagnetism and Plate Tectonics.* Cambridge Univ. Press, Cambridge, UK, 358pp. *[5, 16, 91, 129, 141, 142, 228, 232, 233, 265, 278, 284, 288, 289, 330]*

McElhinny, M.W. (1973b). Mantle plumes, palaeomagnetism and polar wandering. *Nature*, **241**, 523–524. *[327, 328, 329]*

McElhinny, M.W. (1978). Limits to Earth expansion. *Bull. Aust. Soc. Explor. Geophys.*, **9**, 149–152. *[330]*

McElhinny, M.W., and Brock, A. (1975). A new palaeomagnetic result from East Africa and estimates of the Mesozoic palaeoradius. *Earth Planet. Sci. Lett.*, **27**, 321–328. *[331]*

McElhinny, M.W., and Embleton, B.J.J. (1974). Australian palaeomagnetism and the Phanerozoic plate tectonics of eastern Gondwanaland. *Tectonophysics*, **22**, 411–429. [*276*]

McElhinny, M.W., and Embleton, B.J.J. (1976). Precambrian and early Palaeozoic palaeomagnetism in Australia. *Philos. Trans. R. Soc. London*, **A280**, 417–432. [*323*]

McElhinny, M.W., and Lock, J. (1990a). IAGA global palaeomagnetic database. *Geophys. J. Int.*, **101**, 763–766. [*16, 231*]

McElhinny, M.W., and Lock, J. (1990b). Global palaeomagnetic database project. *Phys. Earth Planet. Int.*, **63**, 1–6. [*16, 231*]

McElhinny, M.W., and Lock, J. (1993). Global paleomagnetic database supplement number one, update to 1992. *Surv. Geophys.*, **14**, 303–329. [*162, 231*]

McElhinny, M.W., and Lock, J. (1996). IAGA paleomagnetic databases with Access. *Surv. Geophys.*, **17**, 575–591. [*16, 165, 231*]

McElhinny, M.W., and McFadden, P.L. (1997). Palaeosecular variation over the past 5 Myr based on a new generalized database. *Geophys. J. Int.*, **131**, 240–252. [*29, 95, 96, 99*]

McElhinny, M.W., and McWilliams, M.O. (1977). Precambrian geodynamics – a palaeomagnetic view. *Tectonophysics*, **40**, 137–159. [*323*]

McElhinny, M.W., and Opdyke, N.D. (1964). The paleomagnetism of the Precambrian dolerites of eastern Southern Rhodesia: An example of geologic correlation by rock magnetism. *J. Geophys. Res.*, **69**, 2465–2493. [*321*]

McElhinny, M.W., and Opdyke, N.D. (1968). Paleomagnetism of some Carboniferous glacial varves from central Africa. *J. Geophys. Res.*, **73**, 689–696. [*272*]

McElhinny, M.W., and Senanayake, W.E. (1982). Variations in the geomagnetic dipole 1: The past 50 000 years. *J. Geomag. Geoelect.*, **34**, 39–51. [*24, 239*]

McElhinny, M.W., McFadden, P.L., and Merrill, R.T. (1996). The time-averaged paleomagnetic field 0–5 Ma. *J. Geophys. Res.*, **101**, 25007–25027. [*233, 234, 235, 236*]

McElhinny, M.W., Opdyke, N.D., and Pisarevsky, S.A. (1998). A global database for magnetostratigraphy. *EOS, Trans. Am. Geophys. Union*, **79**, 167. [*157*]

McElhinny, M.W., Taylor, S.R., and Stevenson, D.J. (1978). Limits to the expansion of the Earth, Moon, Mars and Mercury and to changes in the gravitational constant. *Nature*, **271**, 316–321. [*331, 332*]

McElhinny, M.W., Briden, J.C., Jones, D.L., and Brock, A. (1968). Geological and geophysical implications of paleomagnetic results from Africa. *Rev. Geophys.*, **6**, 201–328. [*270*]

McElhinny, M.W., Embleton, B.J.J., Daly, L., and Pozzi, J.P. (1976). Paleomagnetic evidence for the location of Madagascar in Gondwanaland. *Geology*, **4**, 455–457. [*273*]

McElhinny, M.W., Embleton, B.J.J., Ma, X., and Zhang, Z. (1981). Fragmentation of Asia in the Permian. *Nature*, **293**, 212–216. [*265, 295*]

McFadden, P.L. (1980a). The best estimate of Fisher's precision parameter κ. *Geophys. J. R. Astron. Soc.*, **60**, 397–407. [*90*]

McFadden, P.L. (1980b). Determination of the angle in a Fisher distribution which will be exceeded with a given probability. *Geophys. J. R. Astron. Soc.*, **60**, 391–396. [*90*]

McFadden, P.L. (1981). A theoretical investigation of the effect of individual grain anisotropy in alternating field demagnetization. *Geophys. J. R. Astron. Soc.*, **67**, 35–51. [*115*]

McFadden, P.L. (1982). Rejection of palaeomagnetic observations. *Earth Planet. Sci. Lett.*, **61**, 392–395. [*92*]

McFadden, P.L. (1984a). Statistical tools for the analysis of geomagnetic reversal sequences. *J. Geophys. Res.*, **89**, 3363–3372. [*177, 179, 180*]

McFadden, P.L. (1984b). 15-Myr periodicity in the frequency of geomagnetic reversals since 100 Myr. *Nature*, **311**, 396. [*180*]

McFadden, P.L. (1987). Comment on "Aperiodicity of magnetic reversals?" *Nature*, **330**, 27. [*180*]

McFadden, P.L. (1990). A new fold test for palaeomagnetic studies. *Geophys. J. Int.*, **103**, 163–169. [*104*]

McFadden, P.L. (1998). The fold test as an analytical tool. *Geophys. J. Int.*, **135**, 329–338. [*102, 104, 105, 107*]

McFadden, P.L., and Jones, D.L. (1981). The fold test in palaeomagnetism. *Geophys. J. R. Astron. Soc.*, **67**, 53–58. [*102, 103, 104, 105*]

McFadden, P.L., and Lowes, F.J. (1981). The discrimination of mean directions drawn from Fisher distributions. *Geophys. J. R. Astron. Soc.*, **67** 19–33. [*93, 94*]

McFadden, P.L., and McElhinny, M.W. (1982). Variations in the geomagnetic dipole 2: Statistical analysis of VDMs for the past 5 million years. *J. Geomag. Geoelectr.*, **34**, 163–189. [*25, 175, 239*]

McFadden, P.L., and McElhinny, M.W. (1988). The combined analysis of remagnetization circles and direct observations in palaeomagnetism. *Earth Planet. Sci. Lett.*, **87**, 161–172. [*125, 126*]

McFadden, P.L., and McElhinny, M.W. (1990). Classification of the reversals test in palaeomagnetism. *Geophys. J. Int.*, **103**, 725–729. [*113, 114, 157*]

McFadden, P.L., and McElhinny, M.W. (1995). Combining groups of paleomagnetic directions or poles. *Geophys. Res. Lett.*, **22**, 2191–2194. [*247, 254, 331*]

McFadden, P.L., and Merrill, R.T. (1984). Lower mantle convection and geomagnetism. *J. Geophys. Res.*, **89**, 3354–3362. [*138, 178, 179, 180, 191*]

McFadden, P.L., and Merrill, R.T. (1993). Inhibition and geomagnetic field reversals. *J. Geophys. Res.*, **98**. 6189–6199. [*173, 178, 179*]

McFadden, P.L., and Merrill, R.T. (1995). History of Earth's magnetic field and possible connections to core–mantle boundary processes. *J. Geophys. Res.*, **100**, 317–316. [*162, 166, 169, 179, 180*]

McFadden, P.L., and Merrill, R.T. (1997). Asymmetry in the reversal rate before and after the Cretaceous Normal Polarity Superchron, *Earth Planet. Sci. Lett.*, **149**, 43–47. [*138, 179*]

McFadden, P.L., and Reid, A.B. (1982). Analysis of palaeomagnetic inclination data. *Geophys. J. R. Astron. Soc.*, **69**, 307–319. [*97, 221*]

McFadden, P.L., and Schmidt, P.W. (1986). The accumulation of palaeomagnetic results from multicomponent analysis. *Geophys. J. R. Astron. Soc.*, **86**, 965–979. [*125*]

McFadden, P.L., Brock, A., and Partridge, T.C. (1979). Palaeomagnetism and the age of the Makapansgat hominid site, *Earth Planet. Sci. Lett.*, **44**, 373–382. [*138*]

McFadden, P.L., Merrill, R.T., and Barton, C.E. (1993). Do virtual geomagnetic poles follow preferred paths during geomagnetic reversals? *Nature*, **361**, 342–344. [*169*]

McFadden, P.L., Merrill, R.T., and McElhinny, M.W. (1988a). Dipole/ Quadrupole family modelling of paleosecular variation. *J. Geophys. Res.*, **93**, 11583–11588. [*28*]

McFadden, P.L., Ma, X.H., McElhinny, M.W., and Zhang, Z.K. (1988b). Permo-Triassic magnetostratigraphy in China: Northern Tarim. *Earth Planet. Sci. Lett.*, **87**, 152–160. [*126, 163, 268*]

McFadden, P.L., Merrill, R.T., Lowrie, W., and Kent, D.V. (1987). The relative stabilities of the reverse and normal polarity states of the earth's magnetic field. *Earth Planet. Sci. Lett.*, **82**, 373–383. [*180*]

McIntosh, W.C, Hargraves, R.B., and West, C.L. (1985). Paleomagnetism and oxide mineralogy of upper Triassic to lower Jurassic red beds and basalts in the Newark Basin. *Geol. Soc. Am. Bull.*, **96**, 463–480. [*159*]

McKenzie, D.P. (1972). Plate tectonics. In *The Nature of the Solid Earth* (Ed., E. Robertson), pp.323–360, McGraw-Hill, New York. [*326*]

McKenzie, D.P., and Bickle, M.J. (1988). The volume and composition of melt generated by extension of the lithosphere. *J. Petrol.*, **29**, 625–679. [*185*]

McKenzie, D.P., and Morgan, W.J. (1969). Evolution of triple junctions. *Nature*, **224**, 125–133. [*187, 225, 281*]

McKenzie, D.P., and Parker, R.L. (1967). The North Pacific: An example of tectonics on a sphere. *Nature*, **216**, 1276–1280. [*183, 281*]

McKenzie, D.P., and Sclater, J.G. (1971). The evolution of the Indian Ocean since the Late Cretaceous. *Geophys. J. R. Astron. Soc.*, **24**, 437–528. [*206, 210*]

McMenamin, M.A.S., and McMenamin, D.L.S. (1990). *The Emergence of the Animals – The Cambrian Breakthrough*. Columbia Univ. Press, New York, 217pp. [*315*]

McNutt, M. (1986). Nonuniform magnetization of seamounts: A least squares approach. *J. Geophys. Res.*, **91**, 3686–3700. [*215*]

McNutt, M.K., Caress, D.W., Reynolds, J., Jordahl, K.A., and Duncan, R.A. (1997). Failure of plume theory to explain midplate volcanism in the southern Austral islands. *Nature*, **389**, 479–482. [*224*]

McWilliams, M.O., and Dunlop, D.J. (1978). Grenville paleomagnetism and tectonics. *Can. J. Earth Sci.*, **15**, 687–695. [*317, 318*]

Meert, J.G., and Van der Voo, R. (1997). The assembly of Gondwana 800–550 Ma. *J. Geodynamics*, **23**, 223–235. [*295, 322*]

Meert, J.G., Van der Voo, R., Powell, C.McA., Li, Z.X., McElhinny, M.W., Chen, Z., and Symons, D.T.A. (1993). A plate-tectonic speed limit? *Nature*, **363**, 216–217. [*287*]

Menard, H.W. (1969). Elevation and subsidence of oceanic crust. *Earth Planet. Sci. Lett.*, **6**, 275–284. [*210*]

Menning, M., and Jin, Y.G. (1998). Comment on 'Permo-Triassic magnetostratigraphy in China: The type section near Taiyuan, Shanxi Province, North China' by B.J.J. Embleton, M.W. McElhinny, X. Ma, Z. Zhang and Z.X. Li. *Geophys. J. Int.*, **133**, 213–216. [*163*]

Menning, M., Katzung, G., and Lutzner, H. (1988). Magnetostratigraphic investigation in the Rotliegendes (300–252 Ma) of central Europe. 2. *Geol. Wiss. Berlin*, **16**, 1045–1063. [*163*]

Mercanton, P.L. (1926). Inversion de l'inclinaison magnétique terrestre aux âges géologiques. *Terr. Magn. Atmosph. Elec.*, **31**, 187–190. [*14, 15, 143*]

Merrill, R.T. (1985). Correlating magnetic field polarity changes with geologic phenomena. *Geology*, **13**, 487–490. [*141*]

Merrill, R.T., and McElhinny, M.W. (1977). Anomalies in the time-averaged paleomagnetic field and their implications for the lower mantle. *Rev. Geophys. Space Phys.*, **15**, 309–323. [*233*]

Merrill, R.T., and McElhinny, M.W. (1983). *The Earth's Magnetic Field: Its History, Origin and Planetary Perspective*. Academic Press, London, 401pp. [*6, 23, 39, 60, 101, 142, 167, 178, 233, 235*]

Merrill, R.T., and McFadden, P.L. (1994). Geomagnetic field stability: Reversal events and excursions. *Earth Planet. Sci. Lett.* **121**, 57–69. [*140, 174, 180, 181*]

Merrill, R.T., and McFadden, P.L. (1999). Geomagnetic polarity transitions. *Rev. Geophys.*, **37**, 201–226. [*164, 166, 168, 170, 172, 173*]

Merrill, R.T., McElhinny, M.W., and McFadden, P.L. (1996). *The Magnetic Field of the Earth: Paleomagnetism, the Core, and the Deep Mantle*. Academic Press, San Diego, 531pp. [*1, 3, 4, 8, 9, 12, 18, 20, 22, 24, 25, 27, 28, 138, 146, 177, 179, 193, 235, 239*]

Merrill, R.T., McElhinny, M.W., and Stevenson, D.J. (1979). Evidence for long-term asymmetries in the earth's magnetic field and possible implications for dynamo theories. *Phys. Earth Planet. Int.*, **20**, 75–82. [*138, 180*]

Miller, J.D., and Opdyke, N.D. (1985). The magnetostratigraphy of the Red Sandstone Creek section, Vail, Colorado. *Geophys. Res. Lett.*, **12**, 133–136. [*163*]

Miller, K.C., and Hargraves, R.B. (1994). Paleomagnetism of some Indian kimberlites and lamproites. *Precambrian Res.*, **69**, 259–267. [*319*]

Minster, J.B., and Jordan, T.H. (1978). Present day plate motions. *J. Geophys. Res.*, **83**, 5331–5354. [*187, 223*]

Minster, J.B., Jordan, T.H., Molnar, P., and Haines, E. (1974). Numerical modelling of instantaneous plate motions. *Geophys. J. R. Astron. Soc.*, **36**, 541–576. [*223*]

Molina Garza, R.S., Acton, G.D., and Geissman, J.W. (1998). Carboniferous through Jurassic paleomagnetic data and their bearing on rotation of the Colorado plateau. *J. Geophys. Res.*, **103**, 24179–24188. [*253*]

Molina Garza, R.S., Van der Voo, R., and Geissman, J. (1989). Paleomagnetism of the Dewey Lake Formation, northwest Texas: End of the Kiaman superchron in North America. *J. Geophys. Res.*, **94**, 17881–17888. [*163*]

Molnar, P., Atwater, T., Mammerickx, J., and Smith, S. (1975). Magnetic anomalies, bathymetry and the tectonic evolution of the South Pacific since the Late Cretaceous. *Geophys. J. R. Astron. Soc.*, **40**, 383–420. [*207*]

Molostovsky, E.A. (1992). Paleomagnetic stratigraphy of the Permian system. *Int. Geol. Rev.*, **34**, 1001–1007. [*163*]

Monger, J.W.H., and Irving, E. (1980). Northward displacement of north–central British Columbia. *Nature*, **285**, 289–294. [*304*]

Moon, T.S., and Merrill, R.T. (1984). The magnetic moments of non-uniformly magnetized grains. *Phys. Earth Planet. Inter.*, **34**, 186–194. [*57*]

Moon, T.S., and Merrill, R.T. (1985). Nucleation theory and domain states in multidomain magnetic material. *Phys. Earth Planet. Inter.*, **37**, 214–222. [*57*]

Moores, E.M. (1991). Southwest U.S.–East Antarctica (SWEAT) connection: A hypothesis. *Geology*, **19**, 598–601. [*315*]

Morel, P., and Irving, E. (1978). Tentative paleocontinental maps for the Early Phanerozoic and Proterozoic. *J. Geol.*, **86**, 535–561. [*271*]

Morel, P., and Irving, E. (1981). Paleomagnetism and the evolution of Pangea. *J. Geophys. Res.*, **86**, 1858–1872. [*299, 300*]

Morgan, W.J. (1968). Rises, trenches, great faults, and crustal blocks. *J. Geophys. Res.*, **73**, 1959–1982. [*183, 186, 187*]

Morgan, W.J. (1971). Convection plumes in the lower mantle. *Nature*, **230**, 42–43. [*222*]

Morgan, W.J. (1972). Deep mantle convection plumes and plate motions. *Geol. Soc. Amer. Mem.*, **132**, 7–22. [*222, 223*]

Morgan, W.J. (1981). Hotspot tracks and the opening of the Atlantic and Indian Oceans. In *The Sea*, Vol. 7 (Ed., E. Emiliani.), pp.443–475, Wiley–Interscience, New York. [*223*]

Morley, L.W., and Larochelle, A. (1964). Paleomagnetism as a means of dating geological events. In *Geochronology in Canada* (Ed., F.F. Osborne), *Royal Society of Canada Special Publication*, No. 8, pp.39–51, Univ. Toronto Press, Toronto. [*188*]

Moskowitz, B.M., Halgedahl, S.L., and Lawson, C.A. (1988). Magnetic domains on unpolished and polished surfaces of titanium-rich titanomagnetite. *J. Geophys. Res.*, **93**, 3372–3386. [*50*]

Moskowitz, B.M., Frankel, R.B., Walton, S.A., Dickson, D.P.E., Wong, K.K.W., Douglas, T., and Mann, S. (1997). Determination of the preexponential frequency factor for superparamagnetic maghemite particles in magnetoferritin. *J. Geophys. Res.*, **102**, 22671–22680. [*55*]

Mound, J.E., Mitrovica, J.X., Evans, D.A.D., and Kirschvink, J.L. (1999). A sea-level test for inertial interchange true polar wander events. *Geophys. J. Int.*, **136**, F5–F10. [*329*]

Müller, R.D., Roest, W.R., Royer, J-Y., Gahagan, L.M., and Sclater, J.G. (1997). Digital isochrons of the world's ocean floor. *J. Geophys. Res.*, **102**, 3211–3214. [*207*]

Muttoni, G., Kent, D.V., and Channell, J.E.T. (1996). Evolution of Pangea: Paleomagnetic constraints from the southern Alps, Italy. *Earth Planet. Sci. Lett.*, **140**, 97–112. [*301*]

Nagata, T. (1952). Reverse thermal-remanent magnetism. *Nature*, **169**, 704. [*140*]

Nagata, T. (1969). Lengths of geomagnetic polarity intervals. *J. Geomag. Geoelect.*, **21**, 701–704. [*176*]

Nagata, T., Uyeda, S., and Ozima, M. (1957). Magnetic interaction between ferro-magnetic minerals contained in rocks. *Philos. Mag., Suppl. Adv. Phys.*, **6**, 264–287. [*140*]

Nagy, E.A., and Valet, J.P. (1993). New advances in paleomagnetic studies of sediment cores using U-channels. *Geophys. Res. Lett.*, **20**, 671–674. [*81*]

Naidu, P.S. (1971). Statistical structure of geomagnetic field reversals. *J. Geophys. Res.*, **76**, 2649–2662. [*176*]

Naidu, P.S. (1974). Are geomagnetic reversals independent? *J. Geomag. Geoelect.*, **26**, 101–104. [*176*]

Needham, J. (1962). *Science and Civilisation in China, Vol. 4. Physics and Physical Technology, Part 1. Physics.* Cambridge Univ. Press, Cambridge, UK. [*1, 2*]

Néel, L. (1947). Propriétés d'un ferromagnétique cubique en grains fins. *C. R. Acad. Sci. Paris*, **224**, 1488–1490. [*58*]

Néel, L. (1949). Théorie du trainage magnétique des ferromagnétiques grains fins avec applications aux terres cuites. *Ann. Géophys.*, **5**, 99–136. [*51, 62, 249*]

Néel, L. (1955). Some theoretical aspects of rock magnetism. *Adv. Phys.*, **4**, 191–243. [*51, 139, 249*]

Negi, J.G., and Tiwari, R.K. (1983). Matching long term periodicities of geomagnetic reversals and galactic motions of the solar system. *Geophys. Res. Lett.*, **10**, 713–716. [*180*]

Newell, A.J., and Merrill, R.T. (1999). Single-domain critical sizes for coercivity and remanence. *J. Geophys. Res.*, **104**, 617–628. [*58, 59*]

Newell, A.J., Dunlop, D.J., and Williams, W. (1993). A two-dimensional micromagnetic model of magnetization and fields in magnetite. *J. Geophys. Res.*, **98**, 9533–9549. [*57*]

Niitsuma, N. (1971). Detailed study of the sediments recording the Matuyama–Brunhes geomagnetic reversal. *Tohoku Univ. Sci. Rep 2nd Ser. (Geology)*, **43**, 1–39. [*173*]

Ninkovich, D., Opdyke, N.D., Heezen, B.C., and Foster, J.H. (1966). Paleomagnetic stratigraphy, rates of deposition and tephrachronology in North Pacific deep-sea sediments. *Earth Planet. Sci. Lett.*, **1**, 476–492. [*147*]

Norton, I.O., and Sclater, J.G. (1979). A model for the evolution of the Indian Ocean and the breakup of Gondwanaland. *J. Geophys. Res.*, **84**, 6803–6830. [*206*]

Nyblade, A.A., Lei, Y., Shive, P.N., and Tesha, A. (1993). Paleomagnetism of Permian sedimentary rocks from Tanzania and the Permian paleogeography of Gondwana. *Earth Planet. Sci. Lett.*, **118**, 181–194. [*272*]

Ogg, J.C. (1995) Magnetic polarity time scale of the Phanerozoic. In *Global Earth Physics: A Handbook of Physical Constants* (Ed., T.J. Ahrens), pp.240–270, American Geophysical Union, Washington, DC. [*162*]

Ohno, M., and Hamano, Y. (1992). Geomagnetic poles over the past 10,000 years. *Geophys. Res. Lett.*, **19**, 1715–1718. [*23*]

Okada, M., and Niitsuma, N. (1989). Detailed paleomagnetic records during the Brunhes–Matuyama geomagnetic reversal and a direct determination of depth lag for magnetization in marine sediments, *Phys. Earth Planet. Int.*, **56**, 133–150. [*166*]

Olivet, J.L., Bonnin, J., Beuzart, P., and Auzende, J.M. (1984). Cinématique de l'Atlantique nord et central. *Rapports Scientifiques.* CNEXO, Paris. [*255*]

Olsen, P.E., and Kent, D.V. (1996). Milankovitch climate forcing in the tropics of Pangea during the Late Triassic. *Palaeogeogr. Palaeoclimatol. Palaeoecol.*, **122**, 1–26. [*160*]

Olsen, P.E., Schlische, R.W., and Fedosh, M.S. (1996b). 580 ky Duration of the Early Jurassic flood basalt event in eastern North America using Milankovitch cyclostratigraphy. In *The Continental Jurassic* (Ed., M. Morales), pp.11–12, Museum of Northern Arizona. [*159*]

Olsen, P.E., Kent, D.V., Cornet, B., Witte, W.K., and Schlische, R.W. (1996a). High-resolution stratigraphy of the Newark rift basin (early Mesozoic, eastern North America). *Geol. Soc. Am. Bull.*, **108**, 40–77. [*160*]

Onstott, T.C. (1980). Application of the Bingham distribution function in paleomagnetic studies. *J. Geophys. Res.*, **85**, 1500–1510. [*99*]

Opdyke, N.D. (1961). The palaeoclimatological significance of desert sandstone. In *Descriptive Palaeoclimatology* (Ed., A.E.M. Nairn), pp.54–60, Wiley–Intercsience, New York. [*243*]

Opdyke, N.D. (1962). Palaeoclimatology and continental drift. In *Continental Drift* (Ed., S.K. Runcorn), pp.41–65, Academic Press, London. [*241*]

Opdyke, N.D. (1964). The paleomagnetism of the Permian redbeds of southwest Tanganyika. *J. Geophys. Res.*, **69**, 2477–2487. [*272*]

Opdyke, N.D. (1972). Paleomagnetism of deep-sea cores. *Rev. Geophys. Space Phys.*, **10**, 213–249. [*173*]

Opdyke, N.D. (1995). Magnetostratlgraphy of Permo-Carboniferous time. In *Geochronology Time Scales and Global Stratigraphic Correlation*, SEPM Special Publication No. 54, pp.41–47. [*163*]

Opdyke, N.D., and Channell, J.E.T. (1996). *Magnetic Stratigraphy*. Academic Press, San Diego, Calif., 346pp. [*80, 138, 151, 154, 155, 157, 162, 180, 182, 209, 211, 212*]

Opdyke, N.D., and Glass, B.P. (1969). The paleomagnetism of sediment cores from the Indian Ocean. *Deep-Sea Res.*, **16**, 249–261. [*147*]

Opdyke, N.D., and Henry, K.W. (1969). A test of the dipole hypothesis. *Earth Planet. Sci. Lett.*, **6**, 139–151. [*71, 232*]

Opdyke, N.D., and Runcorn, S.K. (1956). New evidence for reversal of the geomagnetic field near the Pliocene–Pleistocene boundary. *Science*, **123**, 1126–1127. [*143*]

Opdyke, N.D., and Runcorn, S.K. (1960). Wind direction in the western United States in the late Paleozoic. *Geol. Soc. Am. Bull.*, **71**, 959–972. [*243*]

Opdyke, N.D., Kent, D.V., and Lowrie, W. (1973). Details of magnetic polarity transitions recorded in a high deposition rate deep-sea core. *Earth Planet. Sci. Lett.*, **20**, 315–324. [*165*]

Opdyke, N.D., Roberts, J., Claoué-Long, J., Irving, E., and Jones, P.J. (1999). The base of the Kiaman, its definition and global stratigraphic significance. *Geol. Soc. Am. Bull.*, in press. [*155, 162, 163*]

O'Reilly, W. (1984). *Rock and Mineral Magnetism*. Blackie, Chapman & Hall, New York, 220pp. [*38, 54*]

O'Reilly, W., and Banerjee, S.K. (1966). Oxidation of titanomagnetites and self-reversal. *Nature*, **211**, 26–28. [*139*]

Özdemir, Ö., and Dunlop, D.J. (1985). An experimental study of chemical remanent magnetizations of synthetic monodomain titanomaghemites with initial thermoremanent magnetizations. *J. Geophys. Res.*, **90**, 11513–11523. [*193*]

Özdemir, Ö., and Dunlop, D.J. (1998). Single-domain-like behavior in a 3–mm natural single crystal of magnetite. *J. Geophys. Res.*, **103**, 2549–2562. [*57*]

Packer, D.R., and Stone, D.B. (1972). An Alaska Jurassic palaeomagnetic pole and the Alaskan orocline. *Nature Phys. Sci.*, **237**, 25–26. [*303*]

Panuska, B.C., and Stone, D.B. (1985). Latitudinal motion of the Wrangellia and Alexander terranes and the Southern Alaska superterrane. In *Tectono-stratigraphic Terranes of the Circum-Pacific Region 1* (Ed., D.G. Howell), pp.109–120. Circum-Pacific Council for Energy and Mineral Resources, Houston. [*304*]

Park, J.K. (1970). Acid leaching of red beds and the relative stability of the red and black magnetic components. *Can. J. Earth Sci.*, **7**, 1086–1092. [*118*]

Park, J.K. (1981). Analysis of the multicomponent magnetization of the Little Dal Group,

Mackenzie Mountains, Northwest Territories, Canada. *J. Geophys. Res.*, **86**, 5134–5146. [*319*]

Park, J.K. (1984). Paleomagnetism of the Mudcracked formation of the Precambrian Little Dal Group, Mackenzie Mountains, Northwest Territories, Canada. *Can. J. Earth Sci.*, **21**, 371–375. [*319*]

Park, J.K. (1997). Paleomagnetic evidence for low-latitude glaciation during deposition of the Neoproterozoic Rapitan Group, Mackenzie Mountains, N.W.T., Canada. *Can. J. Earth Sci.*, **34**, 34–49. [*244, 319*]

Park, J.K., and Aitken, J.D. (1986a). Paleomagnetism of the Late Proterozoic Tzezotene Formation of northwest Canada. *J. Geophys. Res.*, **91**, 4955–4970. [*319*]

Park, J.K., and Aitken, J.D. (1986b). Paleomagnetism of the Katherine Group in the Mackenzie Mountains: Implications for post-Grenville (Hadrynian) apparent polar wander. *Can. J. Earth Sci.*, **23**, 308–323. [*319*]

Park, J.K., and Jefferson, C.W. (1991). Magnetic and tectonic history of the Late Proterozoic Upper Little Dal and Coates Lake Groups of northwestern Canada. *Precambrian Res.*, **52**, 1–35. [*319*]

Parker, E.N. (1955). Hydromagnetic dynamo models. *Astrophys. J.*, **122**, 293–314. [*10, 12*]

Parker, R.L. (1977). Understanding inverse theory. *Annu. Rev. Earth Planet. Sci.*, **5**, 35–64. [*196*]

Parker, R.L. (1988). A statistical theory of seamount magnetism. *J. Geophys. Res.*, **93**, 3105–3115. [*215*]

Parker, R.L. (1991). A theory of ideal bodies for seamount magnetism. *J. Geophys. Res.*, **96**, 16,101–16,112. [*216*]

Parker, R.L. (1994). *Geophysical Inverse Theory*. Princeton Univ. Press, Princeton, NJ, 386pp. [*196*]

Parker, R.L. (1997). Coherence of signals from magnetometers on parallel paths. *J. Geophys. Res.*, **102**, 5111–5117. [*178, 190, 191*]

Parker, R.L., Shure, L, and Hildebrand, J.A. (1987). The application of inverse theory to seamount magnetism. *Rev. Geophys.*, **25**, 17–40. [*215*]

Parrish, J.T., Ziegler, A.M., and Scotese, C.R. (1982). Rainfall patterns and the distribution of coals and evaporites in the Mesozoic and Cenozoic. *Palaeogeogr. Palaeoclimatol. Palaeoecol.*, **40**, 67–101. [*241*]

Parry, L.G. (1965). Magnetic properties of dispersed magnetite powders. *Philos. Mag.*, **11**, 303–312. [*63*]

Parry, L.G. (1982). Magnetization of immobilized particle dispersions with two distinct particle sizes. *Phys. Earth Planet. Int.*, **28**, 230–241. [*135*]

Parsons, B., and Sclater, J.G. (1977). An analysis of the variation of ocean floor bathymetry and heat flow with age. *J. Geophys. Res.*, **82**, 803–828. [*210*]

Pavlov, V.E., and Petrov, P.Yu. (1996). Paleomagnetic investigation of the Riphean sediments of the Turukhan region. *Fizika Zemli*, No.3, 70–81 [in Russian]. [*263*]

Peirce, J.W. (1976). Assessing the reliability of DSDP paleolatitudes. *J. Geophys. Res.*, **81**, 4173–4187. [*221*]

Peirce, J.W. (1978). The northward motion of India since the Late Cretaceous. *Geophys. J. R. Astron. Soc.*, **52**, 277–311. [*221*]

Perigo, R., Van der Voo, R., Auvray, B., and Bonhommet, N. (1983). Palaeomagnetism of late Precambrian–Cambrian volcanics and intrusives from the Armorican Massif, France. *Geophys. J. R. Astron. Soc.*, **75**, 236–260. [*308*]

Perrin, M., and Shcherbakov, V. (1997). Paleointensity of the Earth's magnetic field for the past 400 Ma: Evidence for a dipole structure during the Mesozoic low. *J. Geomag. Geoelect.*, **49**, 601–614. [*25, 211, 239, 240*]

Perrin, M., Prévot, M., and Mankinen, E. (1991). Low intensity of the geomagnetic field in Early

Jurassic time. *J. Geophys. Res.*, **96**, 14197–14210. [*25, 211*]

Perroud, H. (1983). Palaeomagnetism of Palaeozoic rocks from the Cabo de Peñas, Asturia, Spain. *Geophys. J. R. Astron. Soc.*, **75**, 201–215. [*122*]

Pesonen, L.K., and Neuvonen, K.J. (1981). Palaeomagnetism of the Baltic Shield – implications for Precambrian tectonics. In *Precambrian Plate Tectonics* (Ed., A. Kröner), pp.623–648, Elsevier, Amsterdam. [*323*]

Petronotis, K.E., and Gordon, R.G. (1989). Age dependence of skewness of magnetic anomalies above seafloor formed at the Pacific-Kula spreading center. *Geophys. Res. Lett.*, **16**, 315–318. [*202*]

Petronotis, K.E., Gordon, R.G., and Acton, G.D. (1992). Determining palaeomagnetic poles and anomalous skewness from marine magnetic anomaly skewness data from a single plate. *Geophys. J. Int.*, **109**, 209–224. [*201, 203, 213, 219*]

Petronotis, K.E., Gordon, R.G., and Acton, G.D. (1994). A 57 Ma Pacific plate palaeomagnetic pole determined from a skewness analysis of crossings of marine magnetic anomaly 25r. *Geophys. J. Int.*, **118**, 529–554. [*220*]

Phillips, J.D. (1977). Time variation and asymmetry in the statistics of geomagnetic reversal sequences. *J. Geophys. Res.*, **82**, 835–843. [*176*]

Phillips, J.D., and Cox, A. (1976). Spectral analysis of geomagnetic reversal time scales. *Geophys. J. R. Astron. Soc.*, **45**, 19–33. [*176*]

Phillips, J.D., Blakely, R.J., and Cox, A. (1975). Independence of geomagnetic polarity intervals. *Geophys. J. R. Astron. Soc.*, **43**, 19–33. [*176*]

Pindell, J., and Dewey, J.F. (1982). Permo-Triassic reconstruction of western Pangea and the evolution of the Gulf of Mexico/Caribbean region. *Tectonics*, **1**, 179–211. [*270*]

Piper, J.D.A. (1976). Palaeomagnetic evidence for a Proterozoic supercontinent. *Philos. Trans. R. Soc. London*, **A280**, 469–490. [*325*]

Piper, J.D.A. (1980). Palaeomagnetic study of the Swedish Rapakivi suite: Proterozoic tectonics of the Baltic Shield. *Earth Planet. Sci. Lett.*, **46**, 443–461. [*323*]

Piper, J.D.A. (1982). The Precambrian palaeomagnetic record: The case for the Proterozoic supercontinent. *Earth Planet. Sci. Lett.*, **59**, 61–89. [*325*]

Piper, J.D.A. (1987). *Palaeomagnetism and the Continental Crust*. Open Univ. Press, Milton Keynes, 434pp. [*325*]

Piper, J.D.A., and Grant, S. (1989). A palaeomagnetic test of the axial dipole assumption and implications for continental distribution throughout geological time. *Phys. Earth. Planet. Int.*, **55**, 37–53. [*237*]

Piper, J.D.A., Briden, J.C., and Lomax, K. (1973). Precambrian Africa and South America as a single continent. *Nature*, **245**, 244–248. [*323*]

Pisarevsky, S.A., and Bylund, G. (1998). Neoproterozoic palaeomagnetic directions in rocks from a key section of the Protogine Zone, southern Sweden. *Geophys. J. Int.*, **133**, 185–200. [*320*]

Pitman, W.C. III and Hayes, D.E. (1968). Sea floor spreading in the Gulf of Alaska. *J. Geophys. Res.*, **73**, 6571–6580. [*206*]

Pitman, W.C. III and Heirtzler, J.R. (1966). Magnetic anomalies over the Pacific–Antarctic Ridge. *Science*, **154**, 1164–1171. [*149, 204*]

Pitman, W.C. III and Talwani, M. (1972). Sea-floor spreading in the North Atlantic. *Geol. Soc. Am. Bull.*, **83**, 619–646. [*206*]

Pitman, W.C. III, Herron, E.M., and Heirtzler, J.R. (1968). Magnetic anomalies in the Pacific and sea floor spreading. *J. Geophys. Res.*, **73**, 2069–2085. [*205*]

Pitman, W.C. III, Talwani, M., and Heirtzler, J.R. (1971). Age of the North Atlantic Ocean from magnetic anomalies. *Earth Planet. Sci. Lett.*, **11**, 195–200. [*206*]

Plouff, D. (1976). Gravity and magnetic fields of polygonal prisms and application to magnetic terrain corrections. *Geophysics*, **41**, 727–741. [*215*]

Poehls, K.A., Luyendyk, B.P., and Heirtzler, J.R. (1973). Magnetic smooth zones in the world's oceans. *J. Geophys. Res.*, **78**, 6985–6997. [*211*]

Powell, C.McA. (1995). Are Neoproterozoic glacial deposits preserved on the margins of Laurentia related to the fragmentation of two supercontinents?: Comment and reply. *Geology*, **23**, 1053–1054. [*321*]

Powell, C.McA., Johnson, B.D., and Veevers, J.J. (1980). A revised fit of east and west Gondwanaland. *Tectonophysics*, **63**, 13–29. [*221*]

Powell, C.McA., Dalziel, I.W.D., Li, Z.X., and McElhinny, M.W. (1995). Did Pannotia, the latest Proterozoic southern supercontinent, really exist? *EOS, Trans. Am. Geophys. Union*, **76**, 172, [Fall Meeting]. [*321*]

Powell, C.McA., Li, Z.X., Thrupp, G.A., and Schmidt, P.W. (1990). Australian Palaeozoic palaeomagnetism and tectonics, I. Tectonostratigraphic terrane constraints from the Tasman Fold Belt. *J. Struct. Geol.*, **12**, 553–565. [*276, 277*]

Powell, C.McA., Li, Z.X., McElhinny, M.W., Meert, J.G., and Park, J.K. (1993). Paleomagnetic constraints on timing of the Neoproterozoic breakup of Rodinia and the Cambrian formation of Gondwana. *Geology*, **21**, 889–892. [*295, 317, 318, 319*]

Premoli Silva, I. (1977). Upper Cretaceous–Paleocene magnetic stratigraphy at Gubbio, Italy. II. Biostratigraphy. *Geol. Soc. Am. Bull.*, **88**. 371–374. [*159*]

Prévot, M., and Camps, P. (1993). Absence of preferred longitudinal sectors for poles from volcanic records of geomagnetic reversals. *Nature*, **366**, 53–57. [*168, 169, 170*]

Prévot, M., and Perrin, M. (1992). Intensity of the Earth's magnetic field since Precambrian time from Thellier-type paleointensity data and inferences on the thermal history of the core. *Geophys. J. Int.*, **108**, 613–620. [*25, 211*]

Prévot, M., Lecaille, A., and Hekinian, R. (1979). Magnetism of the Mid-Atlantic Ridge crest near 37°N from FAMOUS and DSDP results: A review. In *Deep Drilling Results in the Atlantic Ocean: Ocean Crust* (M. Talwani, C.G.A. Harrison and D.E. Hayes, Eds.), pp.210–219, Am. Geophys. Union, Washington, DC. [*192*]

Prévot, M., Derder, M.E., McWilliams, M.O., and Thompson, J. (1990). Intensity of the Earth's magnetic field: Evidence for a Mesozoic dipole low. *Earth Planet. Sci. Lett.*, **97**, 129–139. [*25, 211*]

Prévot, M., Mankinen, E., Coe, R.S., and Grommé, C.S. (1985). The Steens Mountain (Oregon) geomagnetic polarity transition; 2, Field intensity variations and discussion of reversal models. *J. Geophys. Res.*, **90**, 10417–10448. [*165, 173*]

Pullaiah, G.E., Irving, E., Buchan, K.L., and Dunlop, D.J. (1975). Magnetization changes caused by burial and uplift. *Earth Planet. Sci. Lett.*, **28**, 133–143. [*36, 72, 73*]

Quidelleur, X., and Courtillot, V. (1996). On low-degree spherical harmonic models of paleosecular variation. *Phys. Earth Planet. Int.*, **95**, 55–77. [*233, 235*]

Quidelleur, X., and Valet, J.-P. (1994). Paleomagnetic records of excursions and reversals: Possible biases caused by magnetization artefacts. *Phys. Earth. Planet. Int.*, **82**, 27–48. [*170*]

Quidelleur, X., Valet, J.-P., Courtillot,V., and Hulot, G. (1994). Long-term geometry of the geomagnetic field for the last 5 million years: An updated secular variation database from volcanic sequences. *Geophys. Res. Lett.*, **21**, 1639–1624. [*233*]

Quidelleur, X., Valet, J.-P., Le Goff, M., and Bouldoire, X. (1995). Field dependence on magnetization of laboratory-redeposited deep-sea sediments: First results. *Earth. Planet. Sci. Lett.*, **133**, 311–325. [*170*]

Rabinowitz, P.D., Coffin, M.F., and Falvey, D. (1983). The separation of Madagascar and Africa. *Science*, **220**, 67–69. [*206*]

Rahman, A.A., Duncan, A.D., and Parry, L.G. (1973). Magnetization of multidomain magnetite particles. *Riv. Ital. Geofis.*, **22**, 259–266. *[63]*

Rampino, M.R., and Caldeira, K. (1993). Major episodes of geologic change: Correlations, time structure and possible causes. *Earth Planet. Sci. Lett.*, **114**, 215–227. *[180]*

Ramsay, J.G. (1967). *Folding and Fracturing of Rocks*. McGraw-Hill, New York, 568pp. *[108]*

Rapalini, A.E., and Astini, R.A. (1998). Paleomagnetic confirmation of the Laurentian origin of the Argentine Precordillera. *Earth Planet. Sci. Lett.*, **155**, 1–14. *[322]*

Raup, D.M. (1985). Magnetic reversals and mass extinctions. *Nature*, **314**, 341–343. *[180]*

Raymond, C.A., and LaBrecque, J.L. (1987). Magnetization of the oceanic crust: Thermoremanent magnetization or chemical remanent magnetization? *J. Geophys. Res.*, **92**, 8077–8088. *[203]*

Rea, D.K., and Dixon, J.M. (1983). Late Cretaceous and Paleogene tectonic evolution of the north Pacific Ocean. *Earth Planet. Sci. Lett.*, **65**, 145–166. *[220]*

Regan, J., and Anderson, D.L. (1984). Anisotropy models of the upper mantle. *Phys. Earth Planet. Int.*, **35**, 227–263. *[185]*

Richards, M.A., Ricard, Y., Lithgow-Bertelloni, C., Spada, G., and Sabadini, R. (1997). An explanation for the Earth's long-term rotational stability. *Science*, **275**, 372–375. *[326]*

Roberts, P.H, (1971). Dynamo theory. In *Mathematical Problems in the Geophysical Sciences*. (Ed., W.H. Reid), pp.129–206, American Mathematical Society, Providence, RI. *[12]*

Roberts, P.H. (1992). Geomagnetism, Origin. *Encyclopedia of Earth System Science Vol. 2*, Academic Press, San Diego, CA, pp.295–309. *[12]*

Roberts, P.H., and Stix, M. (1972). α–effect dynamos by the Bullard–Gellman formalism. *Astron. Astrophys.*, **18**, 453–466. *[12]*

Robins, B.W. (1972). Remanent magnetization in spinel iron oxides. PhD. thesis, Univ. of New South Wales, Sydney. *[63]*

Robinson, P.L. (1973). Palaeoclimatology and continental drift. In *Implications of continental drift to the earth sciences* (D.H. Tarling and S.K. Runcorn, Eds.), pp.451–476, Academic Press, New York. *[244]*

Roche, A. (1951). Sur les inversions de l'aimantation remanente des roches volcaniques dans les monts d'Auvergne. *C. R. Acad. Sci. Paris*, **233**, 1132–1134. *[143]*

Roche, A. (1956). Sur la date de la dernière inversion du champ magnétique terrestre. *C. R. Acad. Sci. Paris*, **243**, 812–814. *[143]*

Roest, W.R., Arkani-Hamed, J., and Verhoef, J. (1992). The seafloor spreading rate dependence of the anomalous skewness of marine magnetic anomalies. *Geophys. J. Int.*, **109**, 653–669. *[202]*

Roggenthen, W.M., and Napoleone, G. (1977). Upper Cretaceous–Paleocene magnetic stratigraphy at Gubbio, Italy. IV. Upper Maastrichtian–Paleocene magnetic stratigraphy. *Geol. Soc. Am. Bull.*, **88**. 378–382. *[159]*

Rona, P.A., Brakl, J., and Heirtzler, J.R. (1970). Magnetic anomalies in the northeast Atlantic between the Canary and Cape Verde Islands. *J. Geophys. Res.*, **75**, 7412–7420. *[206]*

Roots, W.D. (1976). Magnetic smooth zones and slope anomalies: A mechanism to explain both. *Earth Planet. Sci. Lett.*, **31**, 113–118. *[211]*

Roquet, J. (1954). Sur les rémanences des oxydes de fer et leur intérêt en géomagnetisme. *Ann. Géophys.*, **10**, 226–247 and 282–325. *[63]*

Rowley, D.B., and Lottes, A.L. (1988). Plate-kinematic reconstructions of the North Atlantic and Arctic: Late Jurassic to present. *Tectonophysics*, **155**, 73–120. *[290]*

Roy, J.L., and Lapointe, R.L. (1976). The palaeomagnetism of Huronian red beds and Nipissing diabase: Post-Huronian igneous events and apparent polar wander path for the interval -2300 to -1300 for Laurentia. *Can. J. Earth Sci.*, **13**, 749–773. *[323]*

Roy, J.L., and Park, J.K. (1974). The magnetization of certain red beds: Vector analysis of chemical and thermal results. *Can. J. Earth Sci.*, **11**, 437–471. *[118, 119]*

Runcorn, S.K. (1956). Paleomagnetic comparisons between Europe and North America. *Proc. Geol.*

Assoc. Canada, **8**, 77–85. [*15, 255, 289*]

Runcorn, S.K. (1961). Climatic change through geological time in the light of palaeomagnetic evidence for polar wandering and continental drift. *Q. J. R. Meteorolog. Soc.*, **87**, 282–313. [*241*]

Rutten, M.G. (1959). Paleomagnetic reconnaissance of mid-Italian volcanoes. *Geol. Mijnbouw*, **21**, 373–374. [*143*]

Sager, W.W. (1987). Late Eocene and Maastrichtian paleomagnetic poles for the Pacific plate: Implications for the validity of seamount paleomagnetic data. *Tectonophysics*, **144**, 301–314. [*216, 219*]

Sager, W.W., and Pringle, M.S. (1987). Paleomagnetic constraints on the origin and evolution of the Musicians and South Hawaiian seamounts, central Pacific Ocean. In *Seamounts, Islands, and Atolls*, (B. Keating, P. Fryer, R. Batiza and G. Boehlert, Eds.), *Geophysics Monograph Series, No. 43*, 133–162 (AGU, Washington, DC). [*220*]

Sager, W.W., and Pringle, M.S. (1988). Mid-Cretaceous to Early Tertiary apparent polar wander path of the Pacific plate. *J. Geophys. Res.*, **93**, 11753–11771. [*215, 216, 218, 219*]

Sager, W.W., Weiss, C.J., Tivey, M.A., and Johnson, H.P. (1998). Geomagnetic polarity reversal model of deep-tow profiles from the Pacific Jurassic Quiet Zone. *J. Geophys. Res.*, **103**, 5269–5286. [*190, 207, 211, 212*]

Sandwell, D.T., and Smith, W.H.F. (1997). Marine gravity anomaly from Geosat and ERS 1 satellite altimetry. *J. Geophys. Res.*, **102**, 10039–10054. [*207*]

Savostin, L.A., Sibuet, J.C., Zonenshain, L.P., Le Pichon, X., and Roulet, M.J. (1986). Kinematic evolution of the Thethys belt from the Atlantic Ocean to the Pamirs since the Triassic. *Tectonophysics*, **123**, 1–35. [*290*]

Schlich, R. (1982). The Indian Ocean: Aseismic ridges, spreading centers, and oceanic basins. In *The Ocean Basins and Margins* (A.E.M. Nairn and F.G. Stelhi, Eds.), *Vol.6 – The Indian Ocean*, pp.51–147, Plenum, New York. [*206*]

Schmidt, P.W. (1977). The non-uniqueness of the Australian Mesozoic palaeomagnetic pole position. *Geophys. J. R. Astron. Soc.*, **47**, 285–300. [*278*]

Schmidt, P.W. (1982). Linearity spectrum analysis of multi-component magnetizations and its application to some igneous rocks from south-eastern Australia. *Geophys. J. R. Astron. Soc.*, **70**, 647–665. [*124*]

Schmidt, P.W. (1985). Bias in converging great circle methods. *Earth Planet. Sci. Lett.*, **72**, 427–432. [*125*]

Schmidt, P.W., and Clark, D.A. (1980). The response of palaeomagnetic data to Earth expansion. *Geophys. J. R. Astron. Soc.*, **61**, 95–100. [*332*]

Schmidt, P.W., and Williams, G.E. (1995). The Neoproterozoic climatic paradox: Equatorial palaeolatitude for Marinoan glaciation near sea level in South Australia. *Earth Planet. Sci. Lett.*, **134**, 107–124. [*244*]

Schmidt, P.W., Williams, G.E., and Embleton, B.J.J. (1991). Low palaeolatitude of Late Proterozoic glaciation: Early timing of remanence in haematite of the Elatina Formation, South Australia. *Earth Planet. Sci. Lett.*, **105**, 355–367. [*244*]

Schmidt, P.W., Powell, C.McA., Li, Z.X., and Thrupp, G.A. (1990). Relaibility of Palaeozoic palaeomagnetic poles and the APWP of Gondwanaland. *Tectonophysics*, **184**, 87–100. [*271, 272, 277*]

Schmidt, V.A. (1976). The variation of the blocking temperature in models of thermoremanence. *Earth Planet. Sci. Lett.*, **29**, 146–154. [*132*]

Schneider, D.A., and Kent, D.V. (1990). The time-averaged paleomagnetic field, *Rev. Geophys.*, **28**, 71–96. [*71*]

Schouten, H. (1971). A fundamental analysis of magnetic anomalies over ocean ridges. *Marine*

Geophys. Res., **1**, 111–144. [*199, 200*]

Schouten, H., and Candy, S.C. (1976). Palaeomagnetic poles from marine magnetic anomalies. *Geophys. J. R. Astron. Soc.*, **44**, 567–575. [*200, 201, 212, 213, 214*]

Schouten, H., and McCamy, K. (1972). Filtering marine magnetic anomalies. *J. Geophys. Res.*, **77**, 7089–7099. [*199, 200, 213*]

Sclater, J.G., and Fisher, R.L. (1974). The evolution of the east–central Indian Ocean, with emphasis on the tectonic setting of the Ninetyeast Ridge. *Geol. Soc. Am. Bull.*, **85**, 683–702. [*206*]

Sclater, J.G., Anderson, R.N., and Bell, M.L. (1971). The elevation of ridges and the evolution of the central eastern Pacific. *J. Geophys. Res.*, **76**, 7888–7915. [*210*]

Sclater, J.G., Hellinger, S., and Tapscott, C. (1977). The paleobathymetry of the Atlantic Ocean from the Jurassic to the present. *J. Geol.*, **85**, 509–552. [*290*]

Scotese, C.R. (1997). Paleogeographic Atlas, PALEOMAP Project Progress Report No. 90–0497. Department of Geology, University of Texas at Arlington, Arlington, 45pp. [*301*]

Scotese, C.R., and Barrett, S.F. (1990). Gondwana's movement over the South Pole during the Palaeozoic: Evidence from lithological indicators of climate. In *Palaeozoic Palaeogeography and Biogeography* (W.S. McKerrow and C.R. Scotese, Eds.), *Geol. Soc. London Mem.*, **12**, 75–85. [*241*]

Scotese, C.R., and McKerrow, W.S. (1990). Revised world maps and introduction. In *Palaeozoic Palaeogeography and Biogeography* (C.R. Scotese, W.S. McKerrow, Eds.), *Geol. Soc. London Mem.*, **12**, 1–21. [*301*]

Sengör, A.M.C., Yilmaz, Y., and Sungurlu, O. (1984). Tectonics of the Mediterranean Cimmerides: Nature and evolution of the western termination of Palaeo-Tethys. *Geol. Soc. London Spec. Publ.*, **17**, 77–112. [*301*]

Shanley, R.J., and Mahtab, M.A. (1976). Delineation and analysis of clusters in orientation data. *Math. Geol.*, **8**, 9–23. [*104*]

Shaw, J. (1974). A new method for determining the magnitude of the palaeomagnetic field, *Geophys. J. R. Astron. Soc.*, **39**, 133–141. [*171*]

Sheldon, R.P. (1964). Paleolatitudinal and paleogeographic distribution of phosphate. *U.S. Geological Survey Professional Paper No. 501–C*, pp.C106–C113. [*245*]

Shipunov, S.V. (1997). Synfolding magnetization: Detection, testing and geological applications. *Geophys. J. Int.*, **130**, 405–410. [*106*]

Shipunov, S.V., Muraviev, A.A., and Bazhenov, M.L. (1998). A new conglomerate test in palaeomagnetism. *Geophys. J. R. Astron. Soc.*, **133**, 721–725. [*108*]

Shive, P.N. (1985). Alignment of magnetic grains in fluids. *Earth Planet. Sci. Lett.*, **72**, 117–124. [*70*]

Simpson, R.W. (1975). Relations between a criterion for polar wander and some conditions for absolute plate motion. *J. Geophys. Res.*, **80**, 4823–4824. [*326*]

Singer, B.S., and Pringle, M.S. (1996). Age and duration of the Matuyama–Brunhes geomagnetic polarity reversal from $^{40}Ar/^{39}Ar$ incremental heating analyses of lavas, *Earth Planet. Sci. Lett.*, **139**, 47–61. [*173*]

Smethurst, M.A., Khramov, A.N., and Torsvik, T.H. (1998). The Neoproterozoic and Palaeozoic palaeomagnetic data for the Siberian platform: From Rodinia to Pangea. *Earth Sci. Rev.*, **43**, 1–24. [*262, 263, 320*]

Smith, A.G., and Hallam, A. (1970). The fit of the southern continents. *Nature*, **225**, 1328–1333. [*298*]

Smith, A.G., and Livermore, R.A. (1991). Pangea in Permian to Jurassic time. *Tectonophysics*, **187**, 135–179. [*298, 300, 301*]

Smith, A.G., Hurley, A.M., and Briden, J.C. (1981). *Phanerozoic Palaeo-continental World Maps*.

Cambridge Univ. Press, Cambridge, UK, 102pp. [*300*]

Smith, G.M., and Banerjee, S.K. (1986). Magnetic structure of the upper kilometer of the marine crust at Deep Sea Drilling Project Hole 504B, eastern Pacific Ocean. *J. Geophys. Res.*, **91**, 10337–10354. [*192*]

Smith, G.M., and Merrill, R.T. (1980). The origin of rotational remanent magnetization. *Geophys. J. R. Astron. Soc.*, **61**, 329–336. [*115*]

Smith, P.J. (1968). Pre-Gilbertian conceptions of terrestrial magnetism. *Tectonophysics*, **6**, 499–510. [*1*]

Smith, P.J. (1970a). Petrus Peregrinus Epistola: The beginning of experimental studies of magnetism in Europe. *Atlas* (*News Suppl. to Earth Sci. Rev.*), **6**, A11–A17. [*1, 2*]

Smith, P.J. (1970b). Do magnetic polarity-oxidation state correlations imply self-reversal? *Commun. Earth Sci. Geophys.*, **1**, 74–85. [*141*]

Smith, P.J., and Needham, J. (1967). Magnetic declination in medieval China. *Nature*, **214**, 1213–1214. [*2*]

Smith, W.H.F., and Sandwell, D.T. (1994). Bathymetry prediction from dense satellite altimetry and sparse shipboard bathymetry. *J. Geophys. Res.*, **99**, 21803–21824. [*207*]

Soffel, H.C., and Förster, H.G. (1981). Apparent polar wander path of central Iran and its geotectonic implications. In *Global Reconstruction and the Geomagnetic Field during the Palaeozoic* (M.W. McElhinny, A.N. Khramov, M. Ozima, D.A. Valencio, Eds.), *Adv. Earth Planet. Sci.*, **10**, pp.117–135, Reidel, Dordrecht. [*312*]

Soffel, H.C., Davoudzadeh, M., Rolf, C., and Schmidt, S. (1996). New paleomagnetic data from central Iran and a Triassic reconstruction. *Geol. Rundsch.*, **85**, 293–302. [*312*]

Solomon, S.C., Sleep, N.H., and Jurdy, D.M. (1977). Mechanical models for absolute plate motions. *J. Geophys. Res.*, **82**, 203–212. [*326*]

Soroka, W., and Beske-Diehl, S. (1984). Variation of magnetic directions within pillow basalts. *Earth Planet Sci. Lett.*, **69**, 215–223. [*193*]

Srivastava, S.P. (1978). Evolution of the Labrador Sea and its bearing on the early evolution of the North Atlantic. *Geophys. J. R. Astron. Soc.*, **52**, 313–357. [*206*]

Srivastava, S.P., and Tapscott, C. (1986). Plate kinematics of the North Atlantic. In *The Geology of North America*, Vol. M, *The Western Atlantic Region* (B.E. Tucholke and P.R. Vogt, Eds.), pp.379–404, Geolical Society of America, Boulder, CO. [*290*]

Srivastava, S.P., Roest, W.R., Kovacs, L.C., Oakey, G., Levesque, S., Verhoef, J., and Macnab, R. (1990). Motion of Iberia since the Late Jurassic: Results from detailed aeromagnetic measurements in the Newfoundland Basin. *Tectonophysics*, **184**, 229–260. [*311*]

Stacey, F.D. (1958). Thermoremanent magnetization (TRM) of multidomain grains in igneous rocks. *Philos. Mag.*, **3**, 1391–1401. [*62*]

Stacey, F.D. (1962). A generalized theory of thermoremanence, covering the transition from single-domain to multidomain magnetic grains. *Philos. Mag.*, **7**, 1887–1990. [*56*]

Stacey, F.D. (1963). The physical theory of rock magnetism. *Adv. Phys.*, **12**, 45–133. [*54, 65, 69*]

Stacey, F.D., and Banerjee, S.K. (1974). *The Physical Principles of Rock Magnetism*. Elsevier, Amsterdam, 195pp. [*38*]

Stacey, F.D., and Loper, D.E. (1983). The thermal boundary layer interpretation of D" and its role as a plume source. *Phys. Earth. Planet. Int.*, **33**, 45–55. [*222*]

Stein, C.A., and Stein, S. (1992). A model for the global variation in oceanic depth and heat flow with lithospheric origin. *Nature*, **359**, 123–129. [*210*]

Steinberger, B., and O'Connell, R.J. (1997). Changes in the Earth's rotation axis owing to advection of mantle density heterogeneities. *Nature*, **387**, 169–173. [*326*]

Steinberger, B., and O'Connell, R.J. (1998). Advection of plumes in mantle flow: Implications for hotspot motion, mantle viscosity and plume distribution. *Geophys. J. Int.*, **132**, 412–434. [*224*]

Steiner, M.B. (1986). Rotation of the Colorado Plateau. *Tectonics*, **5**, 649–660. [253]

Steiner, M.B. (1988). Paleomagnetism of the Late Pennsylvanian and Permian: A test of the rotation of the Colorado Plateau. *J. Geophys. Res.*, **93**, 2201–2215. [163, 253]

Steiner, M.B., and Ogg, J.G. (1988). Early and Middle Jurassic magnetic polarity time scale. In *2nd International Symposium on Jurassic Stratigraphy, Vol.2* (R.B. Rocha and A.F. Soares, Eds.), pp.1097–1111. Cent. de Estrarigrafia e Paleobiol. Da Univ. Nova de Lisboa, Lisbon. [211]

Steiner, M.B., Ogg, J., Zhang, A., and Sun, S. (1989). The Late Permian/Early Triassic magnetic polarity timescale and plate motions of South China. *J. Geophys. Res.*, **84**, 7343–7363. [163]

Stephens, M.A. (1969). Multi-sample tests for the Fisher distribution for directions. *Biometrika*, **56**, 169–181. [93]

Stephenson, A. (1980). Rotational remanent magnetization and the torque exerted on a rotating rock in an alternating magnetic field. *Geophys. J. R. Astron. Soc.*, **62**, 113–132. [115]

Stephenson, A. (1993). Three-axis static alternating field demagnetization of rocks and the identification of natural remanent magnetization, gyroremanent magnetization, and anisotropy. *J. Geophys. Res.*, **98**, 373–381. [115]

Stigler, S. (1987). Aperiodicity of magnetic reversals? *Nature*, **330**, 26–27. [180]

Stothers, R.B. (1986). Periodicity of the Earth's magnetic reversals. *Nature*, **332**, 444–446. [180]

Stott, P.M., and Stacey, F.D. (1960). Magnetostriction and paleomagnetism of igneous rocks. *J. Geophys. Res.*, **65**, 2419–2424. [74]

Strangway, D.W., Larson, E.E., and Goldstein, M. (1968). A possible cause of high magnetic stability in volcanic rocks. *J. Geophys. Res.*, **73**, 3787–3795. [60]

Suk, D., Peacor, D.R., and Van der Voo, R. (1990). Replacement of pyrite framboids by magnetite in limestones and implications for palaeomagnetism. *Nature*, **345**, 611–613. [67]

Sutter, J.F. (1988). Innovative approaches to the dating of igneous events in the early Mesozoic basins of the eastern United States. *U.S. Geol. Surv. Bull.*, **1776**, 194–200. [159]

Symons, D.T.A., and Litalien, C.R. (1984). Paleomagnetism of the Lower Jurassic Copper Mountain intrusions and the geodynamics of Terrane I, British Columbia. *Geophys. Res. Lett.*, **10**, 1065–1068. [304]

Talwani, M. (1965). Computation with the help of a digital computer of magnetic anomalies caused by bodies of arbitrary shape. *Geophysics*, **30**, 797–817. [215]

Tanaka, H., and Idnurm, M. (1994). Palaeomagnetism of Proterozoic mafic intrusions and host rocks of the Mount Isa Inlier, Australia: Revisited. *Precambrian Res.*, **69**, 241–258. [319]

Tanaka, H., Kono, M., and Uchimura, H. (1995). Some global features of palaeointensity in geological time. *Geophys. J. Int.*, **120**, 97–102. [25, 171, 172, 240]

Tarduno, J.A., and Cottrell, R.D. (1997). Paleomagnetic evidence for motion of the Hawaiian hotspot during formation of the Emperor seamounts. *Earth Planet. Sci. Lett.*, **153**, 171–180. [223, 224]

Tarduno, J.A., and Gee, J. (1995). Large-scale motion between Pacific and Atlantic hotspots. *Nature*, **378**, 477–480. [223]

Tarduno, J.A., Lowrie, W., Sliter, W.V., Bralower, T.J., and Heller, F. (1992). Reversed polarity characteristic magnetizations in the Albian Contessa section, Umbrian Apenines, Italy: Implications for the existence of a mid-Cretaceous mixed polarity interval. *J. Geophys. Res.*, **97**, 241–271. [181]

Tauxe, L. (1998). *Paleomagnetic Principles and Practice*. Kluwer Acad. Publ., Dordrecht. [100]

Tauxe, L., and Watson, G.S. (1994). The fold test: An eigen analysis approach. *Earth Planet. Sci. Lett.*, **122**, 331–341. [105, 106]

Tauxe, L., Kylstra, N., and Constable, C. (1991). Bootstrap statistics for paleomagnetic data. *J. Geophys. Res.*, **96**, 11723–11740. [100]

Tauxe, L., Mullender, T.A.T., and Pick, T. (1996). Potbellies, wasp-waists, and superparamagnetism in magnetic hysteresis. *J. Geophys. Res.*, **101**, 571–583. [*135*]

Tauxe, L., LaBrecque, J.L., Dodson, R., and Fuller, M. (1983a). U-channels – a new technique for paleomagnetic analysis of hydraulic piston cores, *EOS, Trans. Am. Geophys. Union*, **64**, 219. [*81*]

Tauxe, L., Tucker, P., Peterson, N.P., and LaBrecque, J.L. (1983b). The magnetostratigraphy of Leg 73 sediments. *Palaeogeogr., Palaeoclimatol., Palaeoecol.*, **42**, 65–90. [*159*]

Thellier, E. (1937). Aimantation des terres ciutes; application à la recherche de l'intensité du champ magnétique terrestre dans le passé. *C. R. Acad. Sci. Paris*, **204**, 184–186. [*22*]

Thellier, E. (1938). Sur l'aimantation des terres cuites et ses applications géophysiques. *Ann. Inst. Phys. Globe Univ. Paris*, **16**, 157–302. [*61*]

Thellier, E. (1966). Le champ magnétique terrestre fossile. *Nucleus*, **7**, 1–35. [*22*]

Thellier, E., and Thellier, O. (1959a). Sur l'intensité du champ magnétique terrestre dans le passé historique et géologique. *Ann. Geophys.*, **15**, 285–376. [*24, 171*]

Thellier, E., and Thellier, O. (1959b). The intensity of the geomagnetic field in the historical and geological past. *Akad. Nauk. SSR. Izv. Geophys. Ser.*, 1296–1331. [*24, 171*]

Thomas, W.A., and Astini, R.A. (1996). The Argentine Precordillera: A traveler from the Ouachita embayment of North American Laurentia. *Science*, **283**, 752–757. [*322*]

Thompson, R., and Clark, R.M. (1981). Fitting polar wander paths. *Phys. Earth Planet. Int.*, **27**, 1–7. [*248*]

Tikku, A.A., and Cande, S.C. (1999). The oldest magnetic anomalies in the Australian–Antarctic Basin: Are they isochrons? *J. Geophys. Res.*, **104**, 661–677. [*206, 298*]

Tivey, M.A., and Johnson, H.P. (1995). Alvin magnetic survey of zero-age crust: CoAxial Segment eruption, Juan de Fuca Ridge 1993. *Geophys. Res. Lett.*, **22**, 171–174. [*194*]

Torcq, F., Besse, J., Vaslet, D., Marcoux, J., Ricou, L.E., Halawani, M., and Basahel, M. (1997). Paleomagnetic results from Saudi Arabia and the Permo-Triassic Pangea configuration. *Earth Planet. Sci. Lett.*, **148**, 553–567. [*301*]

Torsvik, T.H., and Trench, A. (1991). The Ordovician history of the Iapetus Ocean in Britain: New palaeomagnetic constraints. *J. Geol. Soc. London*, **148**, 423–425. [*259*]

Torsvik, T.H., Meert, J.G., and Smethurst, M.A. (1998). Polar wander and the Cambrian. *Science*, **279**, 9a. [*329*]

Torsvik, T.H., Smethurst, M.A., Briden, J.C., and Sturt, B.A. (1990). A review of Palaeozoic palaeomagnetic data from Europe and their palaeogeographical implications. In *Palaeozoic Palaeogeography and Biogeography* (W.A. McKerrow and C.R. Scotese, Eds.), *Geol. Soc. London Mem.*, **12**, 25–41. [*259*]

Torsvik, T.H., Trench, A., Svensson, I., and Walderhaug, H. (1993). Palaeogeographic significance of mid-Silurian palaeomagnetic results from southern Britain – major revision of the apparent polar wander path for eastern Avalonia. *Geophys. J. Int.*, **113**, 651–668. [*306*]

Torsvik, T.H., Smethurst, M.A., Van der Voo, R., Trench, A., Abrahamsen, N., and Halvorsen, E. (1992). Baltica. A synopsis of Vendian–Permian palaeomagnetic data and their palaeotectonic implications. *Earth Sci.Revs.*, **33**, 133–152. [*259, 261*]

Torsvik, T.H., Smethurst, M.A., Meert, J.G., Van der Voo, R., McKerrow, W.S., Brasier, M.D., Sturt, B.A., and Walderhaug, H.J. (1996). Continental break-up and collision in the Neoproterozoic and Palaeozoic – A tale of Baltica and Laurentia. *Earth Sci. Rev.*, **40**, 229–258. [*309, 310, 322*]

Trench, A., and Torsvik, T.H. (1991). A revised Palaeozoic apparent polar wander path for southern Britain (Eastern Avalonia). *Geophys. J. Int.*, **104**, 227–233. [*259, 306*]

Tric, E., Laj, C., Jehanno, C., Valet, J-P., Kissel, C., Mazaud, A., and Iaccarino, S. (1991). High-resolution record of the upper Olduvai transition from Po Valley (Italy) sediments; support for dipolar transition geometry? *Phys. Earth Planet. Int.*, **65**, 319–336. [*167*]

Tucker, P. (1980). A grain mobility model of post-depositional realignment. *Geophys. J. R. Astron. Soc.*, **63**, 149–163. [*70*]

Tucker, P., and O'Reilly, W. (1980). Reversed thermoremanent magnetization in synthetic titanomagnetites as a consequence of high temperature oxidation. *J. Geomag. Geoelect.*, **32**, 341–355. [*139*]

Tucker, R.D., and McKerrow, W.S. (1995). Early Paleozoic chronology: A review in light of new U–Pb zircon ages from Newfoundland and Britain. *Can. J. Earth Sci.*, **32**, 368–379. [*252*]

Turnbull, G. (1959). Some palaeomagnetic measuremnts in Antarctica. *Arctic*, **12**, 151–157. [*279*]

Turner, G.M., and Thompson, R. (1981). Lake sediment record of the geomagnetic secular variation in Britain during Holocene times. *Geophys. J. R. Astr. Soc.*, **65**, 703–725. [*27*]

Turner, G.M., and Thompson, R. (1982). Detransformation of the British geomagnetic secular variation record for Holocene times. *Geophys. J. R. Astr. Soc.*, **70**, 789–792. [*27*]

Ullrich, L., and Van der Voo, R. (1981). Minimum continental velcocities with respect to the pole since the Archean. *Tectonophysics*, **74**, 17–27. [*286, 287*]

Uyeda, S. (1958). Thermoremanent magnetism as a medium of paleomagnetism with special references to reverse thermoremanent magnetism. *Jap. J. Geophys.*, **2**, 1–123. [*139, 140*]

Vacquier, V. (1963). A machine method for computing the magnitude and direction of magnetization of a uniformly magnetized body from its shape and a magnetic survey. In *Proceedings of the Benedum Earth Magnetism Symposium, 1962* (Ed., T. Nagata), pp.123–137, Univ. of Pittsburgh Press, Pittsburg. [*214, 215*]

Valet, J.–P., and Laj, C. (1981). Paleomagnetic record of two successive Miocene geomagnetic reversals in western Crete. *Earth Planet. Sci. Lett.*, **54**, 53–63. [*165*]

Valet, J.–P., Laj, C., and Tucholka, P. (1986). High-resolution sedimentary record of a geomagnetic reversal. *Nature*, **322**, 27–32. [*165*]

Valet, J.–P., Tauxe, L., and Clement, B. (1989). Equatorial and mid-latitude records of the last geomagnetic reversal from the Atlantic Ocean. *Earth Planet. Sci.*, **94**, 371–384. [*165*]

Valet, J.–P., Tucholka, P., Courtillot, V., and Meynadier, L. (1992). Palaeomagnetic constraints on the geometry of the geomagnetic field during reversals. *Nature*, **356**, 400–407. [*168*]

Van Andel, S.I., and Hospers, J. (1968). Palaeomagnetism and the hypothesis of an expanding Earth: A new calculation method and its results. *Tectonophysics*, **5**, 273–285. [*331*]

van Andel, T.H. (1974). Cenozoic migration of the Pacific plate, northward shift of the axis of deposition, and paleobathymetry of the central equatorial Pacific. *Geology*, **1**, 507–510. [*217*]

Vandamme, D., Courtillot, V., Besse, J., and Montigny, R. (1991). Paleomagnetism and age of the Deccan traps (India): Results of a Nagpur-Bombay traverse and review of earlier work. *Rev. Geophys.*, **29**, 159–190. [*280*]

VandenBerg, J. (1983). Reappraisal of paleomagnetic data from Gargano (south Italy). *Tectonophysics*, **98**, 29–41. [*312*]

VandenBerg, J., Klootwijk, C.T., and Wonders, A.A.H. (1978). The late Mesozoic and Cenozoic movements of the Italian Peninsula: Further paleomagnetic results from the Umbrian sequence. *Geol. Soc. Am. Bull.*, **89**, 133–150. [*312*]

van der Pluijm, B.A. (1987). Grain-scale deformation and the fold test – evaluation of syn-folding remagnetization. *Geophys. Res. Lett.*, **14**, 155–157. [*106*]

Van der Voo, R. (1981). Paleomagnetism of North America – A brief review. In *Paleoreconstructions of the Continents* (M.W. McElhinny and D.A. Valencio, Eds.), *Am. Geophys. Union Geodynam. Ser.*, **2**, 159–176. [*255*]

Van der Voo, R. (1982). Pre-Mesozoic paleomagnetism and plate tectonics. *Annu. Rev. Earth Planet. Sci.*, **10**, 191–220. [*257, 307, 308*]

Van der Voo, R. (1988). Palezoic paleogeography of North America, Gondwana and intervening displaced terranes: Comparisons of paleomagnetism with paleoclimatology and biogeographical patterns. *Geol. Soc. Am. Bull.*, **100**, 311–324. [*257, 308*]

Van der Voo, R. (1990a). The reliability of paleomagnetic data. *Tectonophysics*, **184**, 1–9. [*157, 229, 230, 231, 251*]

Van der Voo, R. (1990b). Phanerozoic paleomagnetic poles from Europe and North America and comparisons with continental reconstructions. *Rev. Geophys.*, **28**, 167–206. [*252, 253, 255, 259, 289, 290*]

Van der Voo, R. (1993). *Paleomagnetism of the Atlantic, Tethys and Iapetus Oceans.* Cambridge Univ. Press, Cambridge, UK, 411pp. [*109, 228, 248, 251, 252, 259, 261, 265, 271, 280, 289, 300, 301, 306, 307, 308, 311, 312, 313*]

Van der Voo, R. (1994). True polar wander during the middle Paleozoic? *Earth Planet. Sci. Lett.*, **122**, 239–243. [*329*]

Van der Voo, R., and French, R.B. (1974). Apparent polar wandering for the Atlantic bordering continents: Late Carboniferous to Eocene. *Earth Sci. Rev.*, **10**, 99–119. [*299, 300*]

Van der Voo, R., Jones, M., Grommé, C.S., Eberlein, G.D., and Churkin, M. (1980). Paleozoic paleomagnetism and northward drift of the Alexander terrane. *J. Geophys. Res.*, **85**, 5331–5343. [*304*]

Van Fossen, M.C., and Kent, D.V. (1992). Paleomagnetism of 122 Ma plutons in New England and the Mid-Cretaceous paleomagnetic field in North America: True polar wander or large-scale differential mantle motion? *J. Geophys. Res.*, **97**, 19651–19661. [*256*]

van Hoof, A.A.M., and Langereis, C.G. (1992a). The upper Kaena sedimentary geomagnetic reversal record from Sicily. *J. Geophys. Res.*, **97**, 6941–6957. [*165*]

van Hoof, A.A.M., and Langereis, C.G. (1992b). The upper and lower Thvera sedimentary geomagnetic reversal record from southern Sicily. *Earth Planet. Sci. Lett.*, **114**, 59–75. [*165*]

van Zijl, J.S.V., Graham, K.W.T., and Hales, A.L. (1962a). The paleomagnetism of the Stormberg lavas of South Africa I. Evidence for a genuine reversal of the earths's magnetic field in Triassic–Jurassic times. *Geophys. J. R. Astron. Soc.*, **7**, 23–39. [*165*]

van Zijl, J.S.V., Graham, K.W.T., and Hales, A.L. (1962b). The paleo-magnetism of the Stormberg lavas of South Africa II. The behaviour of the magnetic field during a reversal. *Geophys. J. R. Astron. Soc.*, **7**, 169–182. [*165*]

Veevers, J.J., Stagg, H.M.J., Wilcox, J.B., and Davies, H.L. (1990). Pattern of slow seafloor spreading (<4 mm/year) from breakup (96 Ma) to A20 (44.5 Ma) off the southern margin of Australia. *BMR J. Aust. Geol. Geophys.*, **11**, 499–507. [*206*]

Verhoogen, J. (1956). Ionic ordering and self-reversal of impure magnetites. *J. Geophys. Res.*, **61**, 201–209. [*139*]

Verosub, K.L., and Banerjee, S.K. (1977). Geomagnetic excursions and their paleomagnetic record. *Rev. Geophys. Space Phys.*, **15**, 145–155. [*174*]

Verosub, K.L., and Moores, E.M. (1981). Tectonic rotations in extensional regimes and their paleomagnetic consequences for ocean basalts. *J. Geophys. Res.*, **86**, 6335–6349. [*203*]

Vilas, J.F.A. (1981). Paleomagnetism of South American rocks and the dynamic processes related to the fragmentation of western Gondwana. In *Paleoreconstructions of the Continents* (M.W. McElhinny and D.A. Valencio, Eds.), *Am. Geophys. Union Geodynam. Ser.*, **2**, 106–114. [*273*]

Vine, F.J. (1966). Spreading of the ocean floor: New evidence. *Science*, **154**, 1405–1415. [*149, 197, 198, 199*]

Vine, F.J., and Matthews, D.H. (1963). Magnetic anomalies over oceanic ridges. *Nature*, **199**, 947–949. [*15, 149, 188, 189*]

Vine, F.J., and Wilson J.T. (1965). Magnetic anomalies over a young ocean ridge off Vancouver Island. *Science*, **150**, 485–489. [*149*]

Vogt, P.R., and Johnson, G.L. (1971). Cretaceous sea-floor spreading in the western North Atlantic.

Nature, **234**, 22–25. [*206*]

Vogt, P.R., Anderson, C.N., and Bracey, D.R. (1971). Mesozoic magnetic anomalies, sea-floor spreading, and geomagnetic reversals in the southwestern North Atlantic. *J. Geophys. Res.*, **76**, 4796–4823. [*206*]

Vogt, P.R., Anderson, C.N., Bracey, D.R., and Schneider, E.D. (1970). North Atlantic smooth zones. *J. Geophys. Res.*, **75**, 3955–3966. [*206*]

Wallace, W.K., Hanks, C.L., and Rogers, J.F. (1989). The southern Kahiltna terrane: Implications for the tectonic evolution of southwestern Alaska. *Geol. Soc. Am. Bull.*, **101**, 1389–1407. [*304*]

Ward, M.A. (1963). On detecting changes in the Earth's radius. *Geophys. J. R. Astron. Soc.*, **8**, 217–225. [*330, 331, 332*]

Watkins, N.D. (1969). Non-dipole behaviour during an Upper Miocene geomagnetic polarity transition in Oregon. *Geophys. J. R. Astron. Soc.*, **17**, 121–149. [*165*]

Watkins, N.D., and Haggerty, S.E. (1967). Primary oxidation variation and petrogenesis in a single lava. *Contr. Miner. Petrol.*, **15**, 251–271. [*43*]

Watson, G.S. (1956a). Analysis of dispersion on a sphere. *Mon. Not. R. Astron. Soc. Geophys. Suppl.*, **7**, 153–159. [*88, 92, 93*]

Watson, G.S. (1956b). A test for randomness of directions. *Mon. Not. R. Astron. Soc. Geophys. Suppl.*, **7**, 160–161. [*88, 94*]

Watson, G.S. (1983). Large sample theory of the Langevin distributions. *J. Stat. Planning Inference*, **8**, 245–256. [*114*]

Watson, G.S., and Enkin, R.J. (1993). The fold test in paleomagnetism as a parameter estimation problem. *Geophys. Res. Lett.*, **20**, 2135–2137. [*105*]

Watson, G.S., and Irving, E. (1957). Statistical methods in rock magnetism. *Mon. Not. R. Astron. Soc. Geophys. Suppl.*, **7**, 289–300. [*88, 94*]

Weeks, R., Laj, C., Edignoux, L., Fuller, M., Roberts, A., Manganne, R., Blanchard, E., and Goree, W.S. (1993). Improvements in long-core measurement techniques: Applications in palaeomagnetism and palaeoceanography. *Geophys. J. Int.*, **114**, 651–662. [*84*]

Wei, W. (1995). Revised age calibration points for the geomagnetic polarity time scale. *Geophys. Res. Lett.*, **22**, 957–960. [*151, 153*]

Weil, A.B., Van der Voo, R., MacNiocaill, C., and Meert, J.G. (1998). The Proterozoic supercontinent Rodinia: Paleomagnetically derived reconstructions for 1100 to 800 Ma. *Earth Planet. Sci. Lett.*, **154**, 13–24. [*320*]

Weissel, J.K., and Hayes, D.E. (1972). Magnetic anomalies in the Southeast Indian Ocean. In *Antarctic Oceanologiy, II. The Australian–New Zealand Sector* (Ed., D.E. Hayes), American Geophysical Union, Washington, DC, *Antarctic Res. Ser.*, **19**, 165–196. [*202, 206*]

Weissel, J.K., Hayes, D.E., and Herron, E.M. (1977). Plate tectonic synthesis: The displacements between Australia, New Zealand and Antarctica since the Late Cretaceous. *Mar. Geol.*, **25**, 231–277. [*207*]

Wensink, H. (1973). Newer paleomagnetic results on the Deccan Traps, India. *Tectonophysics*, **17**, 41–59. [*280*]

Wensink, H. (1979). The implications of some paleomagnetic data from Iran for its structural history. *Geol. Mijnbouw*, **58**, 175–185. [*312*]

Wensink, H. (1982). Tectonic inferences of paleomagnetic data from some Mesozoic formations in central Iran. *J. Geophys.*, **51**, 12–23. [*312*]

Wensink, H. (1983). Paleomagnetism of some red beds of Early Devonian age from central Iran. *Earth Planet. Sci. Lett.*, **63**, 325–334. [*312*]

Wessel, P., and Kroenke, L.W. (1998). The geometric relationship between hotspots and seamounts: Implications for Pacific hotspots. *Earth Planet. Sci. Lett.*, **158**, 1–18. [*224*]

Westphal, M. (1993). Did a large departure from the geocentric axial dipole field occur during the Eocene? Evidence from the magnetic polar wander path of Eurasia. *Earth Planet. Sci. Lett.*, **117**, 15–28. [*314*]

Whitelaw, M.J. (1991a). Magnetic polarity stratigraphy of the Fisherman's Cliff and Bone Gulch vertebrate fossil faunas from the Murray Basin, New South Wales, Australia. *Earth Planet. Sci. Lett.*, **104**, 417–423. [*138*]

Whitelaw, M.J. (1991b). Magnetic polarity stratigraphy of Plio and Pleistocene fossil vertebrate localities in southeastern Australia. *Geol. Soc. Am. Bull.*, **103**, 1493–1503. [*138*]

Whitelaw, M.J. (1992). Magnetic polarity stratigraphy of the Pliocene sections and inferences for the ages of vertebrate fossil sites near Bacchus Marsh, Victoria, Australia. Australian 3. *Earth Sci.*, **39**, 521–528. [*138*]

Wiens, D.A., DeMets, C., Gordon, R.G., Stein, S., Argus, D., Engeln, J.F., Lundgren, P., Quible, D., Stein, C., Weinstein, S., and Woods, D.F. (1985). A diffuse plate boundary model for Indian Ocean tectonics. *Geophys. Res. Lett.*, **12**, 429–432. [*184, 328*]

Williams, C.A., and McKenzie, D.P. (1971). The evolution of the north-east Atlantic. *Nature*, **232**, 167–173. [*206*]

Wilson, J.T. (1963). Evidence from islands on the spreading of the ocean floor. *Nature*, **197**, 536–538. [*222*]

Wilson, R.L. (1962a). The palaeomagnetic history of a doubly-baked rock. *Geophys. J. R. Astron. Soc.*, **6**, 397–399. [*142*]

Wilson, R.L. (1962b). The palaeomagnetism of baked contact rocks and reversals of the earth's magnetic field. *Geophys. J. R. Astron., Soc.*, **7**, 194–202. [*141, 142*]

Wilson, R.L. (1962c). An instrument for measuring vector magnetization at high temperatures. *Geophys. J. R. Astron. Soc.*, **7**, 125–130. [*117*]

Wilson, R.L. (1970). Permanent aspects of the earth's non-dipole magnetic field over Upper Tertiary times. *Geophys. J. R. Astron. Soc.*, **19**, 417–437. [*233, 234, 235*]

Wilson, R.L. (1971). Dipole offset – The time-averaged palaeomagnetic field over the past 25 million years. *Geophys. J. R. Astron. Soc.*, **22**, 491–504. [*233, 234*]

Wilson, R.L. (1972). Palaeomagnetic differences between normal and reversed field sources and the problem of far-sided and right-handed pole positions. *Geophys. J. R. Astron. Soc.*, **28**, 295–304. [*234*]

Wilson, R.L., and Ade-Hall, J.M. (1970). Palaemagnetic indications of a permanent aspect of the non-dipole field. In *Palaeogeophysics* (Ed., S.K. Runcorn), pp.307–312, Academic Press, New York. [*232*]

Wilson, R.L., and Haggerty, S.E. (1966). Reversals of the earth's magnetic field. *Endeavour*, **25**, 104–109. [*42*]

Wilson, R.L., and Lomax, A. (1972). Magnetic remanence related to slow rotation of ferromagnetic material in alternating magnetic fields. *Geophys. J. R. Astron. Soc.*, **30**, 295–303. [*115*]

Wilson, R.L., and McElhinny, M.W. (1974). Investigation of the large scale palaeomagnetic field over the past 25 million years; eastward shift of the Icelandic spreading ridge. *Geophys. J. R. Astron. Soc.*, **39**, 571–586. [*233*]

Wilson, R.L., and Watkins, N.D. (1967). Correlation of petrology and natural magnetic polarity in Columbia Plateau basalts. *Geophys. J. R. Astron. Soc.*, **12**, 405–424. [*43, 141*]

Witte, W.K., and Kent, D.V. (1989). A middle Carnian to early Norian (~225 Ma) paleopole from sediments of the Newark Basin Pennsylvania. *Geol. Soc. Am. Bull.*, **101**, 1118–1126. [*159*]

Witte, W.K., and Kent, D.V. (1990). The paleomagnetism of red beds and basalts of the Hettangian Extrusive Zone, Newark Basin, New Jersey. *J. Geophys. Res.*, **95**, 17533–17545. [*159*]

Witte, W.K., Kent, D.V., and Olsen, P.E. (1991). Magnetostratigraphy and paleomagnetic poles from Late Triassic–earliest Jurassic strata of the Newark Basin. *Geol. Soc. Am. Bull.*, **103**,

1648–1662. [*159*]

Wittpenn, N.A., Harrison, C.G.A., and Handschumacher, D.W. (1989). Crustal magnetization in the South Atlantic from inversion of magnetic anomalies. *J. Geophys. Res.*, **94**, 15463–15480. [*193*]

Witzke, B.J. (1990). Palaeoclimatic constraints for Palaeozoic palaeolatitudes of Laurentia and Euramerica. In *Palaeozoic Palaeogeography and Biogeography* (W.S. McKerrow and C.R. Scotese, Eds.), *Geol. Soc. London Mem.*, **12**, 57–73. [*241*]

Woods, M.T., and Davies, G.F. (1982). Late Cenozoic genesis of the Kula plate. *Earth Planet. Sci. Lett.*, **58**, 161–166. [*225, 226*]

Worm, H-U. (1986). Herstellung und magnetische Eigenschaften kleiner Titanomagnetit-Ausscheidungen in Silikaten. Doctoral thesis. University of Bayreuth, Beyreuth, Germany. [*63*]

Worm, H-U. (1997). A link between geomagnetic reversals and events and glaciations, *Earth Planet. Sci. Lett.*, **147**, 55–67. [*166*]

Worm, H-U., and Markert, H. (1987). Magnetic hysteresis properties of fine particle titanomagnetites precipitated in a silicate matrix. *Phys. Earth Planet. Int.*, **46**, 84–92. [*57*]

Yang, Z.Y., Cheng, Y.Q., and Wang, H.Z. (1986). *The Geology of China*. Clarendon, Oxford. [*295*]

York, D. (1978). A formula describing both magnetic and isotopic blocking temperatures. *Earth Planet. Sci. Lett.*, **39**, 89–93. [*249*]

Zhao, X., and Coe, R.S. (1987). Palaeomagnetic constraints on the collision and rotation of North and South China. *Nature*, **327**, 141–144. [*265, 294*]

Zhao, X., Coe, R.S., Gilder, S.A., and Frost, G.M. (1996). Palaeomagnetic constraints on the palaeogeography of China: Implications for Gondwanaland. *Austr. J. Earth Sci.*, **43**, 643–672. [*265, 294, 295*]

Zhao, X., Coe, R.S., Zhou, Y., Wu, H., and Wang, J. (1990). New paleomagnetic results from northern China: Collision and suturing with Siberia and Kazakhstan. *Tectonophysics*, **181**, 43–81. [*265*]

Zhu, R., Ding, Z., Wu, H., Huang, B., and Jiang, L. (1993). Details of magnetic polarity transition recorded in Chinese loess, *J. Geomag. Geoelect.*, **45**, 289–299. [*173*]

Zijderveld, J.D.A. (1967). A.C. demagnetization of rocks. In *Methods in Palaeomagnetism* (D.W. Collinson, K.M. Creer and S.K. Runcorn, Eds.), pp.256–286, Elsevier, New York. [*119*]

Zonenshain, L.P., Kuzmin, M.I., and Natapov, L.M. (1990). *Geology of the USSR: A Plate Tectonic Synthesis*. Geodynamics Series No. 21. American Geophysical Union, Washington, DC. [*261, 263*]

Zonenshain, L.P., Verhoef, J., Macnab, R., and Meyers, H. (1991). Magnetic imprints of continental accretion in the U.S.S.R. *EOS, Trans. Am. Geophys. Union*, **72**, 305, 310. [*262*]

Index

International Geophysics Series

EDITED BY

RENATA DMOWSKA
Division of Engineering and Applied Science
Harvard University
Cambridge, Massachusetts

JAMES R. HOLTON
Department of Atmospheric Sciences
University of Washington
Seattle, Washington

H. THOMAS ROSSBY
Graduate School of Oceanography
University of Rhode Island
Narragansett, Rhode Island

* Out of Print

Printed and bound by CPI Group (UK) Ltd, Croydon, CR0 4YY

03/10/2024

01040419-0004